ADVANCES IN SPATIO-TEMPORAL ANALYSIS

International Society for Photogrammetry and Remote Sensing (ISPRS) Book Series

Book Series Editor

Paul Aplin
School of Geography
The University of Nottingham
Nottingham, UK

Advances in Spatio-Temporal Analysis

Editors

Xinming Tang
Key Laboratory of Geo-informatics of SBSM, China

Yaolin Liu
Wuhan University, China

Jixian Zhang
Chinese Academy of Surveying and Mapping, China

Wolfgang Kainz
Department of Geography and Regional Research, University of Vienna, Austria

Taylor & Francis
Taylor & Francis Group

LONDON / LEIDEN / NEW YORK / PHILADELPHIA / SINGAPORE

Taylor & Francis is an imprint of the Taylor & Francis Group, an informa business

© 2008 Taylor & Francis Group, London, UK

Typeset by Charon Tec Ltd (A Macmillan Company), Chennai, India
Printed and bound in Great Britain by Antony Rowe Ltd (CPI-group), Chippenham, Wiltshire

Published by: Taylor & Francis/Balkema
                P.O. Box 447, 2300 AK Leiden, The Netherlands
                e-mail: Pub.NL@tandf.co.uk
                www.balkema.nl, www.taylorandfrancis.co.uk, www.crcpress.com

British Library Cataloguing in Publication Data
A catalogue record for this book is available from the British Library

Library of Congress Cataloging-in-Publication Data
Advances in spatio-temporal analysis / edited by Xinming Tang ... [et al.].
    p. cm. – (International Society for Photogrammetry and Remote Sensing book series)
  Includes bibliographical references and index.
  ISBN 978-0-415-40630-7 (alk. paper)
  1. Geographic information systems. 2. Spatial analysis (Statistics)
I. Tang, Xinming.

G70.212.A42 2008
910.1'5195–dc22   2007024275

ISBN 978-0-415-40630-7 (Hbk)
ISBN 978-0-203-93755-6 (Ebook)

Advances in Spatio-Temporal Analysis – Tang et al. (eds)
© 2008 Taylor & Francis Group, London, ISBN 978-0-415-40630-7

Contents

Preface   vii

Contributors   ix

Advances in spatio-temporal analysis: An introduction   1
Xinming Tang, Yaolin Liu, Jixian Zhang & Wolfgang Kainz

Spatial modelling

An integrated model for topology and qualitative distance in discrete space   9
Jianhua He & Yaolin Liu

Boolean operations on triangulated solids and their applications in 3D geological modelling   21
Honggang Qu, Mao Pan, Bin Wang, Yong Wang & Zhangang Wang

Towards uncertainty-based geographic information science – theories of
modelling uncertainties in spatial analyses   29
Wenzhong Shi

The expression of spherical entities and generation of voronoi diagrams based on
truncated icosahedron DGG   41
Xiaochong Tong, Jin Ben, Yongsheng Zhang & Li Meng

A fuzzy spatial region model based on flou set   51
Qiangyuan Yu, Dayou Liu & Jianzhong Chen

Spatio-temporal modelling

A spatial database extension of analytical fields representing dynamic and
continuous phenomena   59
Bin Chen, Fengru Huang & Yu Fang

Design and implementation of a unified spatio-temporal data model   67
Peiquan Jin, Lihua Yue & Yuchang Gong

Spatio-temporal indexing mechanism based on snapshot-increment   77
Lin Li, Zhangcai Yin & Haihong Zhu

Temporal topological relationships of convex spaces in space syntax theory   85
Hani Rezayan, Farid Karimipour, Mahmoud R. Delavar & Andrew U. Frank

Spatio-temporal data model: An approach based on version-difference   101
Huibing Wang, Xinming Tang, Bing Lei & Liang Zhai

An algebra for moving objects
Shengsheng Wang, Dayou Liu, Jie Liu & Xinying Wang   111

Spatio-temporal analysis

Interfacing GIS with a process-based agro-ecosystem model – Case study
North China Plain
Georg Bareth & Zhenrong Yu   123

An activity-based spatio-temporal data model for epidemic transmission analysis   135
Tao Cheng, Zhilin Li & Jianhua Gong

Visual exploration of time series of remote sensing data   147
Ulanbek D. Turdukulov, Menno J. Kraak & Valentyn Tolpekin

Time series analysis of land cover change – The National Carbon Accounting
System trial in Fujian province, China   155
Xiaoliang Wu, Peter Caccetta, Suzanne Furby, Jeremy Wallace & Min Zhu

3D modelling of groundwater based on volume visualization technology   163
Zhiyong Xu, Xiaofang Wu, Guorui Zhu & Huiwu Yan

Spatial reasoning and data mining

Spatial data mining and knowledge discovery   171
Deren Li & Shuliang Wang

Genetic neural network based data mining and its application in police case analysis 193
Hanli Liu, Lin Li & Haihong Zhu

Qualitative spatial reasoning about Internal Cardinal Direction relations   203
Yu Liu, Yi Zhang, Xiaoming Wang & Xin Jin

Data mining technology in predicting the demand for cultivated land   217
Zuohua Miao & Yaolin Liu

Statistical data mining for NFGIS database content refinement   227
Liang Zhai, Xinming Tang, Lan Wu & Lin Li

Author index   239

Advances in Spatio-Temporal Analysis – Tang et al. (eds)
© 2008 Taylor & Francis Group, London, ISBN 978-0-415-40630-7

Preface

Almost everything in the world changes over time. In order to represent and operate upon real world geographical phenomena, researchers in the geospatial area have always been trying to find the most suitable way to incorporate time into geographical information systems. This is known as Temporal Geographical Information Systems (TGIS). TGIS, which are capable of handling temporal as well as spatial information, will greatly expand current GIS applications and allow new information to be obtained.

The past two decades have witnessed a significant advancement, as well as a growth in popularity, in TGIS. Various specialized technical meetings have been held on the subject. Chief among these was the International Symposium on Spatial-temporal Modelling, Spatial Reasoning, Analysis, Data Mining and Data Fusion(STM'05) which took place from 27 to 29 August 2005 in Beijing, China. STM'05 was a joint workshop of ISPRS WGII/1,2,7 and WG VII/6, providing an interdisciplinary forum for international scientists and researchers to present their latest research results and share experiences in TGIS, especially in spatio-temporal analysis. STM'05 attracted about 120 papers from more than 20 countries, and 21 of these papers were carefully selected, updated and peer-reviewed to form this book – Advances in Spatio-temporal Analysis. Spatio-temporal analysis is here considered to embody spatial modelling, spatio-temporal modelling, spatio-temporal analysis, and spatial reasoning and data mining.

This book contributes to the field of spatio-temporal analysis, presenting innovative ideas and examples that reflect the current progress and achievements. It will be a useful reference for advanced GIS students, as well as professionals, engaged in TGIS. I trust readers will find the book of benefit in understanding the developments in the emerging field of spatio-temporal analysis.

Vienna, January 2007

Wolfgang Kainz
President, ISPRS TC II

Contributors

Georg Bareth, University of Cologne, Cologne, Germany, E-mail: g.bareth@uni-koeln.de

Jin Ben, Institute of Surveying and Mapping, Information Engineering University, Zhengzhou, 450052

Peter Caccetta, CSIRO Mathematical and Information Sciences, Private Bag 5, Wembley, WA 6913, Australia, E-mail: Peter.Caccetta@csiro.au

Bin Chen, Institute of Remote Sensing and Geographic Information System, Peking University, Beijing, China, E-mail: gischen@pku.edu.cn

Jianzhong Chen, Department of Computing, Imperial College London South Kensington, London SW7 2BZ, UK, E-mail: jz.chen@imperial.ac.uk

Tao Cheng, ①Department of Land Surveying and GeoInformatics, The Hong Kong Polytechnic University, Hong Kong, ②School of Geography and Planning, Sun Yat-sen University, Guang Zhou, China, E-mail: lstc@polyu.edu.hk

Mahmoud R. Delavar, Dept. of Surveying and Geomatic Eng., Eng. Faculty, University of Tehran, Tehran, Iran, E-mail: mdelavar @ut.ac.ir

Yu Fang, Institute of Remote Sensing and Geographic Information System, Peking University, Beijing, China

Andrew U. Frank, Dept. of Geo-Information E-127, Technische University Wien, Gusshausstr. 27-29, A-1040 Vienna Austria, E-mail: frank@geoinfo.tuwien.ac.at

Suzanne Furby, CSIRO Mathematical and Information Sciences, Private Bag 5, Wembley, WA 6913, Australia, Email: Suzanne.Furby@csiro.au

JianHua Gong, State Key Laboratory of Remote Sensing Science, Institute of Remote Sensing Applications, Chinese Academy of Sciences, Datun Road, Chaoyang District, Beijing, China, E-mail: jhgong@irsa.ac.cn

Yuchang Gong, Dept. of Computer Science and Technology, University of Science and Technology of China, Jinzhai Road 96#, Hefei, 230027, China, E-mail: ycgong@ustc.edu.cn

Jianhua He, School of Resource and Environment Science, Wuhan University, 129 Luoyu Road, Wuhan, 430079, China, E-mail: hjianh@126.com

Fengru Huang, Institute of Remote Sensing and Geographic Information System, Peking University, Beijing, China

Peiquan Jin, Dept. of Computer Science and Technology, University of Science and Technology of China, Jinzhai Road 96#, Hefei, 230027, China, E-mail: jpq@ustc.edu.cn

Xin Jin, Institute of Remote Sensing and Geographic Information Systems, Peking University, Beijing, 100871, China, E-mail: terrible2@sina.com

Wolfgang Kainz, Department of Geography and Regional Research, University of Vienna, Austria, Email: wolfgang.kainz@univie.ac.at

Farid Karimipour, Dept. of Surveying and Geomatic Eng., Eng. Faculty, University of Tehran, Tehran, Iran, E-mail: karimipr @ut.ac.ir

Menno J. Kraak, ITC, GIP dept, PO Box 6, 7500 AA Enschede, the Netherlands, Email: kraak@itc.nl

Bing Lei, Key Laboratory of Geo-informatics of State Bureau of Surveying and Mapping, Chinese Academy of Surveying and Mapping, 16 Beitaiping Road, Haidian District, Beijing, 100039, China, E-mail: leibing@casm.ac.cn

Deren Li, National Laboratory for Information Engineering in Surveying, Mapping and Remote Sensing, Wuhan University, Wuhan, 430079, China, E-mail: drli@whu.edu.cn

Lin Li, School of Resource and Environment Science, Wuhan University, Wuhan China 430079, E-mail: lilin@telecarto.com

Zhilin Li, Department of Land surveying and GeoInformatics, The Hong Kong Polytechnic University, Hong Kong, E-mail: lszlli@polyu.edu.hk

Dayou Liu, College of Computer Science and Technology, Key Laboratory of Symbolic Computation and Knowledge Engineering of Ministry of Education, Jilin University, Changchun, 130012, China, E-mail: dyliu@jlu.edu.cn

Hanli Liu, School of Resource and Environment Science, Wuhan University, 129 Luoyu Road, Wuhan, P.R.China, 430079, E-mail: liuhl000@sohu.com

Jie Liu, College of Computer Science and Technology, Key Laboratory of Symbolic Computation and Knowledge Engineering of Ministry of Education, Jilin University, Changchun, 130012, China

Yaolin Liu, School of Resource and Environment Science, Wuhan University, 129 Luoyu Road, Wuhan, 430079, China, E-mail: Yaolin610@163.com

Yu Liu, Institute of Remote Sensing and Geographic Information Systems, Peking University, Beijing, 100871, China, E-mail: liuyu@urban.pku.edu.cn

Li Meng, Institute of Surveying and Mapping, Information Engineering University, Zhengzhou, 450052, China

Zuohua Miao, School of Resource and Environment Science, Wuhan University, China, Wuhan, E-mail: whmzh@hotmail.com

Mao Pan, School of Earth and Space Science, Peking University, Beijing, 100871

Honggang Qu, School of Earth and Space Science, Peking University, Beijing, 100871, E-mail: xqugang@pku.edu.cn

Hani Rezayan, Dept. of Surveying and Geomatic Eng., Eng. Faculty, University of Tehran, Tehran, Iran, E-mail: rezayan@ut.ac.ir

Wenzhong Shi, Advanced Research Centre for Spatial Information Technology, Department of Land Surveying and Geo-informatics, The Hong Kong Polytechnic University, Email:lswzshi@polyu.edu.hk

Xinming Tang, Key Laboratory of Geo-informatics of State Bureau of Surveying and Mapping, Chinese Academy of Surveying and Mapping, 16 Beitaiping Road, Haidian District, Beijing, 100039, China, E-mail: tang@casm.ac.cn

Valentyn Tolpekin, ITC, EOS dept, PO Box 6, 7500 AA, Enschede, The Netherlands, E-mail: tolpekin@itc.nl

Xiaochong Tong, Institute of Surveying and Mapping, Information Engineering University, Zhengzhou, 450052, China, E-mail: txchr@yahoo.com.cn

Ulanbek D.Turdukulov, ITC, GIP dept, PO Box 6, 7500 AA Enschede The Netherlands, Email: turdukulov@itc.nl

Jeremy Wallace, CSIRO Mathematical and Information Sciences, Private Bag 5, Wembley, WA 6913, Australia, Email: Jeremy.Wallace@csiro.au

Bin Wang, School of Earth and Space Science, Peking University, Beijing

Huibing Wang, Key Laboratory of Geo-informatics of State Bureau of Surveying and Mapping, Chinese Academy of Surveying and Mapping, 16 Beitaiping Road, Haidian District, Beijing, 100039, China, E-mail: wanghb@casm.ac.cn

Shengsheng Wang, College of Computer Science and Technology, Key Laboratory of Symbolic Computation and Knowledge Engineering of Ministry of Education, Jilin University, Changchun, 130012, China, E-mail: wss@ jlu.edu.cn

Shuliang Wang, ①International School of Software, Wuhan University, Wuhan, 430072, China, ②School of Economics and Management, Tsinghua University, Beijing, 100084, China, E-mail: slwang2005@whu.edu.cn

Xiaoming Wang, Institute of Remote Sensing and Geographic Information Systems, Peking University, Beijing, 100871, China, E-mail: wangxiaoming@pku.edu.cn

Xinying Wang, College of Computer Science and Technology, Key Laboratory of Symbolic Computation and Knowledge Engineering of Ministry of Education, Jilin University, Changchun, China

Yong Wang, School of Earth and Space Science, Peking University, Beijing

Zhangang Wang, School of Earth and Space Science, Peking University, Beijing

Lan Wu, State Bureau of Surveying and Mapping, 9 Sanlihe Road, Haidian District Beijing, 100830, China, E-mail: wulan@sbsm.gov.cn

Xiaofang Wu, College of Information, South China Agriculture University, Guangzhou, China

Xiaoliang Wu, CSIRO Mathematical and Information Sciences, Private Bag 5, Wembley, WA 6913, Australia, Email: Xiaoliang.Wu@csiro.au

Zhiyong Xu, Key Laboratory of GIS, Ministry of Education, School of Resources and Environment Science, Wuhan University, Wuhan, China, E-mail: XuSirZY@163.com

Huiwu Yan, Key Laboratory of GIS, Ministry of Education, School of Resources and Environment Science, Wuhan University, Wuhan, China

Zhangcai Yin, ①School of Resource and Environment Science, Wuhan University, Wuhan China, 430079, ②School of Resources and Environmental Engineering, Wuhan University Technology, Wuhan, China, 430070, E-mail: yinzhangcai@163.com

Qiangyuan Yu, ①College of Computer Science and Technology, Jilin University, ②Key Laboratory of Symbolic Computation and Knowledge Engineering of Ministry of Education, Jilin University, 130012, ChangChun, China, E-mail: qiangyuan@jlu.edu.cn

Zhenrong Yu, China Agricultural University, Beijing, China

Lihua Yue, Dept. of Computer Science and Technology, University of Science and Technology of China, Jinzhai Road 96#, Hefei, 230027, China, E-mail: llyue@ustc.edu.cn

Liang Zhai, Key Laboratory of Geo-informatics of State Bureau of Surveying and Mapping, Chinese Academy of Surveying and Mapping, 16 Beitaiping Road, Haidian District, Beijing, 100039, China, E-mail: zhailiang@126.com

Jixian Zhang, Chinese Academy of Surveying and Mapping, 16 Beitaiping Road, Haidian District, Beijing, 100039, China, E-mail: zhangjx@casm.ac.cn

Yi Zhang, Institute of Remote Sensing and Geographic Information Systems, Peking University, Beijing 100871, China, E-mail: zy@pku.edu.cn

Yongsheng Zhang, Institute of Surveying and Mapping, Information Engineering University, Zhengzhou, China

Guorui Zhu, Key Laboratory of GIS, Ministry of Education, School of Resources and Environment Science, Wuhan University, Wuhan, China

Haihong Zhu, School of Resource and Environment Science, Wuhan University, Wuhan, China, 430079

Min Zhu, CSIRO Mathematical and Information Sciences, Private Bag 5, Wembley, WA 6913, Australia, Email: Min.Zhu@csiro.au

Advances in Spatio-Temporal Analysis – Tang et al. (eds)
© *2008 Taylor & Francis Group, London, ISBN 978-0-415-40630-7*

Advances in spatio-temporal analysis: An introduction

Xinming Tang
Key Laboratory of Geo-informatics of State Bureau of Surveying and Mapping,
Chinese Academy of Surveying and Mapping, Beijing, China

Yaolin Liu
School of Resource and Environment Science, Wuhan University, Wuhan, China

Jixian Zhang
Chinese Academy of Surveying and Mapping, Beijing, China

Wolfgang Kainz
Department of Geography and Regional Research, University of Vienna, Austria

ABSTRACT: This paper serves as a brief introduction to this book, which includes 21 peer-reviewed papers selected from the joint workshop of ISPRS WGII/1, 2, 7 and WG VII/6 on spatial-temporal modelling, spatial reasoning, analysis, data mining and data fusion (STM'05), held from 27 to 29 August 2005 at Peking University, China. The background to this workshop is presented and the developments in spatial modelling, spatio-temporal modelling, spatio-temporal analysis, and spatial reasoning and data mining are briefly reviewed and the contents of the book are introduced.

1 BACKGROUND

The advances in geographical information systems have increased rapidly since the subject first appeared in the 1960s. This subject has now developed into a geographical information science (Goodchild et al. 1999), dealing with spatial data acquisition, management, processing and applications. Many specialists have summarized the development stages of GIS. From our point of view, there are three main breakthroughs. The first landmark was the advent of commercial GIS software, of which ARC/INFO was a leader in the early 1980s. At that time, the emphasis of GIS was on spatial data structures and general spatial data analysis, with topological structures and spatial relations being the most popular topics. The second landmark was the application of large database software. Since at that time the volume of spatial data was not too great, small database management systems could handle it. However, as spatial data is at least 2-dimensional and often 3-dimensional, the volumes of data increased at tremendous speed. Because a general database cannot manage such large volumes of data, large database software began to play a role in the management of spatial data. Professional spatial database management is being replaced by relational databases. Nowadays, many commercial database suppliers provide spatial data engines to manage the spatial data. The third landmark was the emergence of car navigation systems and the recent Google Earth. If we say that car navigation systems are still oriented to the car-owner and transportation services, Google Earth enables spatial data to serve the public in general. It means that GIS can not only help governmental and professional applications but can also serve the ordinary people and affect their lives. GIS is coming to everybody's life.

With widespread applications, people expect more and more from GIS. They wish spatial data to be presented not only as a static map, but also as a dynamic one. They have come to believe that a 2-dimensional map is not as intuitive as a 3-dimensional display. They wish the satellite images to be clearer; until they can count the walkers on the streets! They also think that from very large spatial databases they can dig out something useful. This enhancement of requirements deeply affects the present studies in GIS.

Nowadays, GIS is developing towards being dynamic, 3-dimensional and networked, which demands that GIS specialists study and set up more practical and more efficient systems.

There have been many attempts to raise awareness of the importance of time within the GIS community, and to develop models that can be used to represent dynamics (Goodchild 2006). Time is of great importance in all aspects of our world. Space and time together form a four-dimensional space in which other properties are organized (Frank 2001). As discussed in the preface, many researchers have been seeking appropriate ways to have the time component as an important part in GIS. The inclusion of time in GIS allows a variety of applications, such as natural resource management, urban and regional management, transportation, environmental monitoring, cadastre management, etc. There is a real necessity for future GIS to incorporate the temporal data element to allow users to track the evolution or changes of dynamic phenomena and analyse the past, present, and future states of geographic features. Against such a background, STM'05 was held from 27 to 29 August 2005 in Beijing, China.

This opening paper serves as an introduction to the papers included in this book. The introduction comprises four main sections. Spatial modelling, spatio-temporal modelling, spatio-temporal analysis, and spatial reasoning and data mining form an organic whole of broad spatio-temporal analysis. Each section includes a short review and an introduction to each paper. Spatial modelling is the basis for spatio-temporal modelling and analysis, while spatial reasoning and data mining are application extensions for spatio-temporal data. In other words, spatial modelling provides basic methods and techniques for spatio-temporal modelling and analysis, such as topology, uncertainty and statistical analysis, etc; and with the increasing amount of spatio-temporal data that has been collected over the years, spatial reasoning and data mining have already been in development for several years and offer a valuable approach for finding potential relationships or patterns among spatial or spatio-temporal datasets. In this book, spatial modelling is presented first, followed by a section on spatio-temporal modelling. The next section links these two issues, focusing on spatio-temporal analysis. The last section collects some representative research on spatial reasoning and data mining.

2 SPATIAL MODELLING

The essence of spatio-temporal modelling is the extension of spatial modelling with time. A better understanding of "spatial" facilitates this extension. On the way from "spatial" to "spatio-temporal", some problems of spatial modelling, such as spatial uncertainty and 3-dimensional modelling, still need to be addressed. In this section, some representative research activities in spatial modelling are selected, including spatial relations, spatial uncertainty and spherical models, since these lead on to spatio-temporal relations, temporal uncertainty, and spatio-temporal spherical models, etc. Although these topics have been widely discussed elsewhere, the selected papers contain some novelty. Spatio-temporal modelling needs to deal with these topics and the following papers may give some beneficial insights.

Topological relations are always a fundamental issue in GIS. The topological relationship of disjointedness is ubiquitous in the physical world, while the traditional topological relationship models do not give much consideration to this. Point set topology (Egenhofer 1991) is exhaustive for a topological relation set and rational for topological computation; nevertheless, it is not good at spatial reasoning (Tang et al. 2005). Although both RCC5 and RCC8 are competent for spatial reasoning, they are not fit for discrete space. He and Liu present a paper on combining qualitative distance with topological relations. They work out a topological model for both representing and reasoning spatial entities in discrete space. It is actually an extension of the RCC5 model with the relationship of pathhood, which integrates the qualitative distance into topological relation representation. They firstly fuzzify the discrete crisp region with k-neighbourhood using a Voronoi diagram. Then they use the tri-tuple basic relations to represent the topological relations, whose k-neighbourhood can imply the qualitative distance. Finally, a semantic model is reached for the spatial topological relation and qualitative distance relation representation.

Spatial analysis is being extended from 2-dimensional modelling to 3-dimensional modelling (Egenhofer 2005), especially in Digital Earth and geological applications. Qu et al. present a novel method for Boolean operations on triangulated solids. Their method concerns co-planar triangles (compare with Möller (1997), dealing with non-planar triangles). Combined with geological knowledge, this method can reconstruct complex geological bodies and human engineering activities. Three examples–intrusions, ore bodies and tunnels, and bifurcated geological bodies–are given in this paper and show that the method has the ability to build complex geological models.

One of the most important issues in spatial analysis is spatial data quality. Spatial data, unlike non-spatial data that can usually be collected in a relatively reliable way, is always accompanied by uncertainty. This uncertainty includes accuracy, precision, etc., which can be affected by different factors, such as observation error, errors from survey instruments, imprecision in defining a spatial entity, difficulty in classification, errors in process and discontinuous observation data, etc. All these greatly affect the correctness and reliability of spatial and temporal analysis. Actually, whether spatial analysis is successful or not largely depends on the integrity and reliability of spatial data. Due to incompleteness and impreciseness, spatial analysis is not very persuasive. Shi analyses systematically the uncertainty modelling for spatial data. The paper covers modelling uncertainties in integrating multiple sources of data, modelling uncertainty in overlay analysis, modelling uncertainty in line simplification, uncertainty-based spatial data mining, uncertainty-based spatial queries, the theory and methods for controlling the quality of spatial data, modelling uncertainties in topological relations and the dissemination of uncertainty information.

It is well known that the shape of the Earth can best be approximated by an ellipsoid. With the increase of small-scale representations, more requirements are directly oriented towards spherical subdivisions. Tong et al. discuss a model that expresses spherical entities. They propose a sphere-based subdivision model that uses the inverse Snyder polyhedron equal area projection on the surfaces of truncated icosahedrons, and then obtains global multi-resolution overlays by hierarchical subdivision on the initial unfolded projection plane according to a hexagonal grid. The paper puts forward an essential clue to the managing of the three leaf nodes and tiles coding on the basis of a hexagonal grid, and establishes the hexagonal grid expression modes of different spherical entities. Then the paper puts forward an algorithm for generating Voronoi diagrams based on spherical hexagonal overlays and verifies the exactness and efficiency of the algorithm through experiments.

As we know, uncertainty has been discussed by probability theory, fuzzy set theory, evidence theory, rough set, vague set theory, and so on. Yu et al. provide a fuzzy spatial region model based on Flou sets. The model is different from the original fuzzy model that uses membership values; it uses the relative relations between point sets and represents a fuzzy region as a Flou set. The model can obtain the properties and relations of fuzzy regions by using the operational properties of the Flou set. The fuzzy region model based on a Flou set is valuable in fields such as GIS, geography and spatial databases.

3 SPATIO-TEMPORAL MODELLING

As we know, spatial information is becoming more involved in conventional and important projects related to spatial distribution. However, many GIS still remain at the stage of spatial data inventory, only being used for storing spatial data. Even this function is still regarded suspiciously in many application departments. Is your model able to monitor and store all the geographical information that I need?

There is a growing body of work showing that, in many application domains, a treatment of the dynamic aspects of geographic phenomena is essential for useful explanatory and predictive models (Worboys 2005). From the earliest days, GIS were not designed or pre-conceptualized as dynamic modelling tools (Reitsma & Albrecht 2005). Dynamic aspects of spatial data (Egenhofer 1999) – in other words, spatio-temporal issues – were not systematically emphasized until Langran first proposed temporal GIS (Langran 1989). Since then, numerous spatio-temporal data models have been proposed, dealing with data storage and management. These models include the snapshot model, space time composite, spatio-temporal object model, event-based/state-based model, object-oriented model, version-difference model, etc.

Perhaps the simplest model is the snapshot model (Langran 1992) and current spatio-temporal information systems typically represent change implicitly using a series of static snapshots (Worboys & Duckham 2006). In the snapshot model, every layer is a collection of temporally homogeneous units of one theme. It shows the states of a geographic distribution at different times without explicit temporal relations among layers. Time intervals between any two layers may vary and there is no implication for whether changes occur within the time lag of any two layers. The advantage is that we do not have to change the existing database structure. However, data redundancy and data inconsistency are obvious. In the space time composite (STC) model (Langran & Chrisman 1988), the real world is a collection of spatially homogeneous units in a 2D space that changes over time from one unit to another. Each STC has its unique period of change and can be obtained from temporal overlays of snapshot layers. STC can model favourable properties of an object – such as situation – but it requires reconstruction of thematic and temporal attribute tables whenever operations involve any changes in spatial objects (Yuan 1997). In spatio-temporal object modelling (Worboys 1992, Worboys 1994), the real world is considered as a set of spatio-temporal atoms that are constructed from the integration of a temporal dimension orthogonal to 2D space. Each of these spatio-temporal atoms is the largest homogeneous unit that can store specific properties in space and time. Thus, this model can store changes in both temporal and spatial dimensions. Other models, such as the event-based model (Peuquet 1994, Pequet & Duan 1995, Peuquet 2002, Chen & Jiang 2000), version-difference model and tupu model (Chen et al. 2000), etc. are also well documented by many researchers.

In this section, some spatio-temporal models are extensions of existing models; others are novel in their representation of different kinds of spatio-temporal information. These models have a common characteristic – they are application-oriented. In addition, two related issues are discussed: temporal relationships and spatio-temporal indexing.

Chen et al. introduce a method of representing spatial phenomena as dynamic and continuous fields and implement a spatial SQL extension for field information querying. The so-called analytical field is adopted as a representation framework for dynamic and continuous phenomena. In the spatial SQL extension, an abstract data type, named the Analytical_Dynamic_Continuous Field (ADCF), is defined and continuous functions and operators are defined on it to support spatio-temporal querying. It is shown that the extended SQL functions help us to use analytical functions to represent dynamic and continuous field-like phenomena. Using this method, we can make integrative transactions on vector data and field data and have a unified way of representing and handling spatial information based on a field model, and closely integrating extended ADCF with other spatial data types present in spatial databases. The SQL extension of ADCF provides support for dealing with spatial information transaction problems by question-oriented and non-procedural ways.

Aiming to provide general support for spatio-temporal applications in DBMS, Jin et al. designed a unified spatio-temporal data model. They have studied several issues concerning the unified spatio-temporal data model, including the data structure, querying operations and implementation issues. The unified spatio-temporal data model is based on an object-relational data model, in which spatio-temporal objects are represented as spatio-temporal relations and spatio-temporal tuples. Queries on spatio-temporal relations are implemented through an extended relational algebra. The spatio-temporal changes are represented by spatio-temporal data types, and queries on spatio-temporal data and changes are supported by the operations defined on those data types. The unified spatio-temporal data model has been partly implemented on Informix using a real example with the China Historical Geographic Information System (CHGIS).

Spatio-temporal indexing is one of the key techniques applied to spatio-temporal database engines. Li et al. put forward the snapshot-increment spatio-temporal indexing mechanism, focusing on efficiency in both temporal queries and spatial queries by regarding time and space as dimensions of equal importance. This mechanism can perform quick searches based on time points or intervals and be extended into other application areas.

Rezayan et al. discuss in detail temporal topological relationships for convex spaces based on space syntax theory. They study time lifting for a topological characterization of convex spaces in the real world that is described by space syntax theory. This theory illustrates human settlements and societies as a strongly connected space-time relational system between convex spaces. Such a system is represented by a connectivity graph with some morphologic analyses for deriving the graph's properties that illustrate how space and time are overcome by the relational systems and convex spaces. Because there is more dynamics among activities at the local scale, their specific problem is defined as computational modelling of integrated static and dynamic analyses for an activity based scenario at the local scale and they study how effectively these activities overcome space and time. The model is also implemented and executed to analyse a simple simulated urban public transportation system using a functional programming language known as Haskell.

Wang et al. propose a version-difference data model to manage spatial data. This model is very useful for managing national fundamental spatial databases. China has already set up 1:1M, 1:250K and 1:50K topographic databases. Some of these databases have two versions. How to integrate the different versions into one database is an important application-oriented task. This model regards the up-to-date data as the default "version", and all the previous data "differences" from the current data. Data at any time can be reconstructed by version and difference. In this way, they built a so-called version-difference model. The model has been developed especially to manage the fundamental spatial data in China.

Moving object databases are becoming more and more popular due to the increasing number of application domains that deal with moving entities. Wang et al. propose a new spatial algebra for a road network of moving objects, employing interval algebra theory. Their work improves the application of theoretical results of interval algebra and proposes an executable qualitative representation and reasoning method for road networks and moving objects.

4 SPATIO-TEMPORAL ANALYSIS

Spatial and spatio-temporal analyses are important since GIS is widely adopted in spatial information management (Oosterom & Lemmen 2001) and spatio-temporal analysis is the core of TGIS. Analysis integrating spatial and temporal aspects will lead to a more fundamental understanding of the dynamic processes involved and thereby help to develop actual solutions. Thus, spatio-temporal analysis, which can be applied in water management, traffic congestion analysis, railroad/bus-transportation plans, crime analysis, geomorphology and many other fields, acts as a linkage between the spatial database and the spatial decision process. Because spatio-temporal analysis covers many aspects, this book covers some representative ones.

Analysis of dynamic process is a good example of spatio-temporal analysis. Bareth and Yu model the changes in the emission of N_2O, the volatilization of NH_3, and the leaching of NO_3 from a winter wheat/summer maize area in the North China Plain. Although the paper discusses the problem at a regional level, their approach linking agro-ecosystem models to analysis is valuable for process-oriented modelling.

The representation of spatio-temporal behaviour, however, presents unique challenges to geographic information science (Bennett & Tang 2006). Cheng et al. propose an activity-based spatio-temporal data model. This model conceptualizes the spatial and temporal interaction of travel and activity behaviours using the concept of mobility. The activities are modelled as a sequence of "staying at" or "travelling between" activity locations, while the transmission of epidemic disease is modelled as spreading through the common activities involved by two or more persons. This model is process oriented and particularly useful for handling spatio-temporal transmission of serious diseases like SARS.

In spatio-temporal analysis, image-based analysis is relevant due to the abundance of time-series remote sensing data. Turdukulov et al. describe an approach for such data. They decompose each spatial region into a number of 2-dimensional Gaussian functions and compare the attributes of each function in successive frames to find the continuous paths of the spatial object through time. These paths describe the characteristics of the object in each time step.

Land Use and Land Cover Change (LUCC) are central themes of global environmental change research and the International Journal of Geographical Information Science (IJGIS) dedicated a special issue to this theme (Verburg & Veldkamp 2005). Wu et al. summarize their time series analysis method to form multi-temporal classifications of presence/absence of forest cover. The method, which is called the joint model approach, is composed of image rectification, image calibration (top-of-atmosphere reflectance calibration, bi-directional reflectance distribution function (BRDF) calibration, terrain illumination correction), forest classification and spatio-temporal models. They demonstrate that their method is not only applicable in Australia, but also in China.

At present, groundwater is being heavily extracted, which causes a series of grave environmental problems such as land subsidence, seawater inbreak and deterioration of water quality. It is extremely important to study the groundwater resources and make best use of it. Xu et al. introduce volume visualization technology to the field of groundwater resources. The spatio-temporal distribution and dynamic change characteristics of the hydrogeologic layer and its inner physical and chemical attributes are represented by a hybrid volume-rendering technique. This research provides a scientific foundation for decision support in the extraction of groundwater.

5   SPATIAL REASONING AND DATA MINING

Spatial and temporal analysis has opened many avenues for research in areas concerning spatio-temporal data. The above section presents several applications of spatial and temporal analysis. With continuous increase and accumulation, the large amounts of spatio-temporal data available have far exceeded the human capability to interpret and use them. Thus, many challenging problems are awaiting solutions. Under these circumstances, growing attention has been paid to spatial reasoning and spatial data mining (Li et al. 2006). Spatial reasoning and data mining are analyses that probe the internal relations within a spatial dataset or database, which is a typical application of spatio-temporal modelling.

Li and Wang present the principles of spatial data mining and knowledge discovery (SDMKD), propose three new techniques and show their applicability and examples. The first is SDMKD-based image classification that integrates spatial inductive learning from a GIS database and Bayesian classification; the second is a cloud model that integrates randomness and fuzziness; the last is a data field that radiates the energy of observed data for the universe of discourse. SDMKD-based image classification integrates spatial inductive learning from a GIS database and Bayesian classification. In the experimental results of remote sensing image classification, the overall accuracy is increased by more than 11 per cent and the accuracy of some classes, such as garden and forest, is increased further by about 30 per cent. For intelligent integration of GIS and remote sensing, it encourages the discovery of knowledge from spatial databases and its application in image interpretation for spatial data updating. The cloud model integrates the fuzziness and randomness in a unified way by means of the algorithms of forward and backward cloud generators in the contexts of three numerical characteristics, {Ex, En, He}. It takes advantage of human natural language, and is able to search for qualitative concepts described by natural language to generalize a given set of quantitative data with the same feature category. Moreover, the cloud model can act as an uncertainty transition between a qualitative concept and its quantitative data. With this method, it is easy to build the mapping relationship inseparably and interdependently between a qualitative concept and quantitative data, and the knowledge with its hierarchy finally discovered can satisfy different demands from different levels of users. The experimental results on Baota landslide monitoring show that cloud model-based spatial data mining can reduce the task complexity, improve the implementation efficiency and enhance the comprehension of the discovered spatial knowledge.

The BP (Back-Propagation) neural network method has been applied to data mining, and it possesses the characteristics of high memory ability, high adaptability, accurate knowledge discovery, no restriction to the quantity of data and a fast speed of calculation. Liu et al. put forward a method that combines the learning algorithm of a BP neural network with a genetic algorithm to train the BP network and optimize the weight values of the network on a global scale. This method is featured as an example of global optimization, high accuracy and fast convergence. Using the genetic algorithm to optimize the BP network can effectively avoid the problem of local minima. Therefore, the genetic neural network based data-mining model has many advantages over other data mining models. The advantages have been fully embodied in data mining in the command centre of a police office.

The so-called "internal cardinal-direction relation" is discussed for spatial reasoning in the paper by Liu et al. This relation emphasizes the relations between two spatial objects, namely container and containee. Based on the authors' ICD model, the paper deduces the external (conventional) cardinal relations, qualitative distance relations and topological relations. Focusing on the ICD-9 model, the paper summarizes ICD-related qualitative reasoning. By using table tools, the composition of ICD relations is presented. The rule for complex ICD relations can be simplified. Because basic spatial relations can be induced from ICD relations, it is possible to tell that the container and the ICD relations together form a positioning framework for spatial knowledge.

Prediction is always a composite research procedure and its solution depends on the purpose of the modelling. Whatever the purpose, the scientist usually wants the model to behave in a manner similar to the true system; therefore, the scientist would want to select a model that has predictive power (Pontius and Malanson 2005). Miao and Liu present a model to forecast the demands of cultivated land. Their model, which is called the fuzzy Markov chain model with weights, ameliorates the traditional time homogeneous finite Markov chain model to predict the future value of cultivated land demand in land use planning. The model applies data mining techniques to extract useful information from an enormous volume of historical data and then applies a fuzzy sequential cluster method to set up the dissimilitude fuzzy clustering sections. The model regards the standardized self-correlation coefficients as weights based on the special characteristics of correlation among the historical stochastic variables. The transition probabilities matrix of the new model is obtained by using fuzzy logic theory and statistical analysis methods. The experimental results show that the ameliorated model, combined with the technique of data mining, is more scientific and practical than traditional predictive models.

One interesting application of data mining is the extraction of content from fundamental GIS. China has already established a series of National Fundamental Geographical Information System (NFGIS) databases. However, with the progress in national informatization, NFGIS has been confronted by many new requirements from GIS users, including its refinement and updating. The paper by Zhai et al. mainly concerns content refinement for the National Fundamental Geographical Information System (NFGIS) 1:50,000 DLG database employing statistical data mining technology. They propose a methodology employing a clustering strategy in database content refinement. This approach is very suitable for exploring the survey data and for obtaining useful information. The results of a user survey, conducted to collect users' requirements of the NFGIS database such as data content and attribute information, were analysed by clustering analysis.

ACKNOWLEDGEMENTS

We must first acknowledge all the members of the Programme Committee for their extraordinary efforts to return reviews on time. They are Donna Peuquet, Deren Li, Michael F. Goodchild, Martien Molenaar, Mike Worboys, Peter Fisher, Peter van Oosterom, Chuang Tao, Manfred Elhers, John L. van Genderen, Monica Wachowicz, Yongqi Chen, Wenzhong Shi, Junyong Chen, Qingxi Tong, Zongjian Lin, Jun Chen, Yu Fang, Jianya Gong and Lun Wu. We thank all the members of Local Organizing Committee for their efforts to ensure the success of STM'05.

Our gratitude extends to the National Basic Research Program (973 program, Project Name: Multi-resource and mass spatial information integration, fusion theory and real-time dynamic updating, Project No.: 2006CB701304) and the National High Technology Research and Development Program (863 Program, Project Name: Spatio-temporal database technique and its application in geographical information management, Project No.: 2006AA12Z214) for their support.

We also thank the Chinese Academy of Surveying and Mapping, Wuhan University, and the National Natural Science Foundation of China (NSFC) for their funding support.

Our special thanks go to David Tait who spent long hours in editing the contributions. Without his enormous editing experience and dedication this book would not have reached this stage.

We also would like to acknowledge the support and patience of Dr. Paul Aplin and Taylor & Francis in the publication of this book.

Finally, but not least, we would like to thank Ms. Xiaoming Gao, Mr. Hui Zhang and Mr. Liang Zhai for their hard work for this event.

REFERENCES

Bennett, D.A. & Tang, W. 2006. Modelling adaptive, spatially aware, and mobile agents: Elk migration in Yellowstone. *International Journal of Geographical Information Science* 20(9): 1039–1066.

Chen, J. & Jiang, J. 2000. An Event-Based Approach to Spatio-Temporal Data Modeling in Land Subdivision Systems. *GeoInformatica* 4: 387–402.

Chen, S.P, Yue, T.X. & Li, H.G. 2000. Studies on Geo-Informatic Tupu and its application. *Geographical Research* 9(4): 337–343 (in Chinese).

Egenhofer, M.J. & Franzosa, R.D. 1991. Point-set topological spatial relations. *International Journal of Geographical Information Systems* 5 (2): 160–174.

Egenhofer, M.J., Glasgow J., Gunther O., Herring, J.R. & Peuquet, D.J. 1999. Progress in computational methods for representing geographical concepts. *International Journal of Geographical Information Science* 13(8): 775–796.

Egenhofer, M.J. 2005. Spherical Topological Relations. *Journal on Data Semantics III, LNCS* 3534: 25–49.

Frank, A.U. 2001. Tiers of Ontology and Consistency Constraints in Geographic Information Systems. *International Journal of Geographical Information Science* 15(7): 667–678.

Goodchild, M.F., Egenhofer, M.J., Kemp, K.K., Mark, D.M. & Sheppard, E. 1999. Introduction to the Varenius project. *International Journal of Geographical Information Science* 13(8): 731–745.

Goodchild, M.F. 2006. Time, Space, and GIS. Past Place: *The Newsletter of the Historical Geography Specialty Group, Association of American Geographers* 14(2): 8–9.

Langran, G.E. & Chrisman, N.R. 1988. A framework for temporal geographic information. *Cartographica* 25(3): 1–14.

Langran, G.E. 1989. *Time in geographic information systems*. PhD Thesis, University of Washington.

Langran, G.E. 1992. *Time in geographic information systems. Technical Issues in Geographic Information Systems.* Taylor & Francis.

Li, X., Zhang, S.C. & Wang, S.L. 2006. Advances in Data Mining Applications. *Special Issue of IJDWM (International Journal of Data Warehouse and Mining)* 2(3): i–iii.

Oosterom, P. van & Lemmen, C.H.J. 2001. Spatial data management on a very large cadastral database. *Computer, Environment and Urban Systems* 25(4): 509–528.

Peuquet, D. & Duan, N. 1995. An event-based spatiotemporal data model (ESTDM) for temporal analysis of geographical data. *International Journal of Geographic Information Systems* 9(1): 7–24.

Peuquet, D. 1994. A Conceptual Framework for the Representation of Temporal Dynamics in Geographic Information Systems. *Annals of the Association of American Geographers* 84(3): 441–461.

Peuquet, D. 2002. Representations of Space and Time. New York: The Guilford Press.

Pontius, G.R. & Malanson, J. 2005. Comparison of the structure and accuracy of two land change models. *International Journal of Geographical Information Science* 19(2): 243–265.

Reitsma, F. & Albrecht, J. 2005. Implementing a new data model for simulating processes. *International Journal of Geographical Information Science* 19(10): 1073–1090.

Tang, X.M., Kainz, W. & Fang, Y. 2005. Reasoning about changes of land cover objects with fuzzy settings. *International Journal of Remote Sensing* 26(14): 3025–3046.

Verburg, Peter H. & Veldkamp, A. 2005. Introduction to the Special Issue on Spatial Modelling to Explore Land Use Dynamics. *International Journal of Geographical Information Science* 19(2): 99–102

Worboys, M.F. 1992. Object-Oriented Models of Spatiotemporal Information. *Proceedings of GIS/LIS '92 Annual Conference, San Jose, California, USA*: 825–834.

Worboys, M.F. 1994. Object-oriented approaches to geo-referenced information. *International Journal of Geographical Information Systems* 8(4): 385–399.

Worboys, M. 2005. Event-oriented approaches to geographic phenomena. *International Journal of Geographical Information Science* 19(1): 1–28.

Worboys M. & Duckham, M. 2006. Monitoring qualitative spatiotemporal change for geosensor networks. *International Journal of Geographical Information Science* 20(10): 1087–1108.

Yuan, M. 1997. Modeling Semantical, Spatial and Temporal Information in a GIS. In M. Craglia and H. Couleclis (eds), *Progress in Trans–Atlantic Geographic Information Research*. Taylor & Francis.

Spatial modelling

Advances in Spatio-Temporal Analysis – Tang et al. (eds)
© 2008 Taylor & Francis Group, London, ISBN 978-0-415-40630-7

An integrated model for topology and qualitative distance in discrete space

Jianhua He & Yaolin Liu
Resource and Environment Science Department of Wuhan University, Wuhan, China
Key Laboratory of Graphic Information Systems, Ministry of Education of China, Wuhan University, China

ABSTRACT: The topological relation of disjoint is ubiquitous in the physical world, whereas the traditional topological relationship models do not take it into consideration. Point set topology is exhaustive for the topological relation set and rational for topological computation; nevertheless, it is not good at spatial reasoning. On the other hand, RCC5 or RCC8 is competent for spatial reasoning, but does not handle discrete space. So the ultimate aim of this paper is to propose a topological model for representation and reasoning spatial entities in discrete space. Two steps are taken. Firstly, based on the parthood relation P, the tri-tuple $\{P(X, Y), P(X, \neg Y), P(Y, X)\}$ is introduced into the RCC topological relations model. Then the discrete space is transformed into continuous space by construction of the discrete object's k-neighbour with a Voronoi diagram. Thus, we can use an extended RCC model to represent and reason discrete spatial relations in a continuous manner.

1 INTRODUCTION

Qualitative spatial reasoning has been paid more and more attention. Nowadays, its applications have become the focus of spatial reasoning in GIS, robot navigation, computer vision and the interpretation of natural language (Laguna 1992, Frank 1996). The qualitative representation and reasoning of spatial relationships are vital for the field of qualitative spatial reasoning in GIS, which includes spatial topological relationships and metric relationships. In this paper, we centre our research on topological and distance relationships. A topological relationship is a relation that is invariant under homeomorphism transformations, which is significant for GIS modelling, analysing and reasoning. How to identify the topological relationships between spatial objects is a key point in GIS. During recent years, topological relationships have been investigated extensively and many achievements have been reported. There are three development clues for topological relationships: Allen et al. (1983) identified 13 topological relationships between two temporal intervals; Egenhofer et al. (1994) proposed the 4-intersection and 9-intersection models based on point set; eight topological relationships were derived from RCC theory based on classical logic (Randell et al. 1992). The 9-intersection approach decomposed the object into the interior Ao, boundary ∂A and exterior A-, and the topological relationships can be concluded from the 3*3 intersection matrix made up of the three parts of both objects. The eight topological relationships between two simple regions (no holes and connected), the 33 line-line topological relationships and the 19 point-line topological relationships have all been defined. The exhaustiveness and rationality of the model have been extensively agreed (Cobb 1995). But the definition of region-region topological relationships is too loose, so that many relations defined by the model are rare, even not viable, for the naive geographical world. On the contrary, the disjoint relations, which are plentiful in the physical world, cannot be discriminated well (Cao 2001), and so the 9-intersection model needs to introduce another approach to represent disjoint relations. Also, the model is based on the point set theory, which includes vast, complicated computations and the abilities of spatial reasoning are limited. This situation has been addressed partly with the composition table of topological relations developed by Egenhofer et al. whereas it is still incapable of qualitative reasoning with incomplete information. The 13 temporal relationships can be extended to distinguish 13*13 topological relationships between two objects, which has the merits of simplifying reasoning, computation and convenience. However, the direct dimensional extension causes the loss of some relationships (Cao 2001). So the model is not a formal model. Randell et al. claimed that the topological relationship could be deduced based on logic. Eight topological relationships were identified based on their RCC theory. Subsequently, Cohn et al. (1997) adopted the theory for spatial reasoning and improved its JEPD (Joint Exhaustive Pair Disjoint), which is necessary for a spatial reasoning model. Unfortunately, the RCC8 or RCC5 is competent for continuous space but not suitable for discrete phenomena. For metric relations, the research now focuses on qualitative distance relations. The reasoning of distance relations is based on spatial cognition, and the reasoning approach avoids complex computation. The goal is to furnish people with query and reasoning engines among dispersed objects, which accords with human cognition in naïve geographical space. Therefore, a representation and reasoning model in discrete space is urgently required.

In this paper, we design a topological relation model based on a deep analysis of the mechanisms of the 9-intersection approach and RCC5 theory. The tri-tuple (P(A,B),P(A,⌐B),P(B,A)) is developed with the primitive operator of parthood (P) and the complement (⌐) of regions in the model. Besides borrowing the logic rationality of the RCC theory, the qualitative distance relation is introduced into the topological model to remedy the limitation of the 9-intersection approach for disjoint relationships by translating the dispersed geographic phenomena into continuous space with the region's Voronoi diagram. The remainder of this paper is arranged as follows. Section 2 introduces the RCC theory and extends the RCC5 model based on the parthood (P) and the complement (⌐) operators. Section 3 proposes the topological relation model in discrete space based on fuzzy regions. Section 4 discusses the topological relation reasoning in discrete space. In section 5, we conclude the paper and point out the directions for future research.

2   EXTENSION OF RCC5

The region is considered as the spatial primitive in RCC5, which replaces the point set with the set of regions. There is some debate about whether the region or the point should be recognized as the primitive. In the domain of GIS, conventional GIS is based on the crisp set theory and the basic spatial operations are based on the computation geometry and classical algebra; so it is reasonable to select points with precise positions as the spatial primitives. On the other hand, qualitative spatial reasoning emphasizes the cognitive rationality of spatial representation, with the view that regional phenomena are all over the world, such as lakes, cordillera, even rivers and roads; so they claim it is certainly reasonable. We think that regions are the components of the physical world, while points are the result of human abstraction of the world, which is complicated. In addition, the field of qualitative spatial reasoning is the geographic space that accords with human's commonsense, which is intuitional and approximate, so the region is adopted as the spatial primitive in this paper.

2.1   *Topological relation representation with RCC5*

RCC theory is closely based on Clarke's system, so it is necessary to briefly introduce some fundamental notations. Clarke presented an extended account of logic axiomatization for a region-based spatial calculus. He defined the dyadic relation of C(X,Y) (means X connects with Y) as the basic operator of the system, where

$$C(X, Y) \equiv def\, \exists x \in X, \exists y \in Y \, \|x - y\| = 0 \tag{1}$$

and the five topological relation predicates of DC, PO, PP, PPi and '=' are identified. The RCC theory predefined that any region should have an untangential interior, which guarantees that the space is infinitely dividable. In order to represent the points, lines and polygons used in conventional GIS, we must extend the RCC system.

2.2   *The extended RCC space*

The classical GIS abstractly depicts the geographic world with points, lines and polygons; meanwhile, the RCC system requests that all objects be regional-like. In order to use RCC to represent those entities, we must first revise the notation of region. Being different from the traditional region, three discriminated notations of regions called 0-dimension region, 1-dimension region and 2-dimension region (denoted as $\dot{R}, \tilde{R}$ and R) are defined as follows:

Definition 1: supposing U is a geographic space, for any geographic object x, $x \in U$, if $A(x) = 0 \wedge$ Perim(x) = 0 then $x \in \dot{R}$

Here $A(x)$ and Perim(x) are the object's area and perimeter. Obviously, if both the area and perimeter of an object are zero, this means that the object is a traditional point. The object with 0-area and 0-perimeter may not exist in the physical world, but it is important for the conceptual world, just as the number 0 is vital for mathematical development.

Definition 2: object $x \in U$, if $A(x) = 0 \wedge$ Perim(x) $\neq 0$ then $x \in \tilde{R}$

From the definition, it can be seen that $\tilde{R}$ is just like the traditional line object.

Definition 3: object $x \in U$, if $A(x) \neq 0 \wedge$ Perim(x) $\neq 0$ then $x \in R$

This R is just like the traditional region.

Definition 4: the geographic space which is composed of the $\dot{R}$, $\tilde{R}$ and R is called the extended RCC space denoted by Ur. That is: $Ur = \vee i \dot{R} \cup \vee i \tilde{R} \cup \vee i R$

With these definitions, the following proposition exists for Ur:

Proposition 1: geographic space

$$U \subset U^r \to \phi, X \in U^r, \forall A_i \in U^r \to \vee_i A_i \in U^r, \forall U, V \in U^r \to U \wedge V \in U \qquad (2)$$

That is to say, Ur is a topological space. On the basis of Ur, we can represent the whole geographic space with the RCC model.

2.3 *Topological representation based on P(X, Y)*

Based on the spatial primitives of a region, this paper revises the basic operator of $C(X, Y)$. In order to use the existing functions of conventional GIS to represent spatial objects and reasoning the spatial relation, the operator of $C(X, Y)$ is replaced with the relation of $P(X, Y)$ (read as X is part of Y). $P(X, Y)$ can be defined as follows:

$$\text{Definition 5:} \quad P(X, Y) \equiv def[\ \forall x \in X \to x \in Y\ ] \qquad (3)$$

From the definition, we can be seen that $P(X, Y)$ is just like the operator of Contain(Y, X), which is available in conventional GIS, and so we can implement the spatial representation and reasoning with the existing GIS tools.

On the basis of the operator $P(X, Y)$, we are able to depict the five relations of RCC5(DC, PO, =, PP, PPi) with the tri-tuple $(P(X, Y), P(X, \neg Y), P(Y, X))$, where $\neg Y$ is the complement of region Y. The results are listed in Table 1.

Using the dyadic logic to express the result of $P(A, B)$ and $P(A, \neg B)$, there are $2^3 = 8$ topological relations identified by them; however, some situations depicted by these tuples (T, T, T), (T, T, F) and (F, T, T) are impossible in the real world.

For the extended RCC space, we can work out the possible topological relations between two regions considering the characters of 0-dimension, 1-dimension and 2-dimension regions. The results are listed in Table 2.

3 TOPOLOGICAL RELATION MODEL FOR DISCRETE SPACE

From the analysis in Section 2.3, we can see that the tri-tuple is competent for the representation of the topological relations in RCC space. Following the approach of the extension from RCC5 to RCC8, we can extend the dyadic logic (TRUE and FALSE) to ternary logic (TRUE, FALSE and MAYBE), so that the

Table 1. The model of RCC5 based on $P(X, Y)$.

RCC5		Tri-tuple of P			Figure
Predicate	Semantics	P(A, B)	P(A,¬B)	P(B, A)	
DC	Disjoint	F	T	F	
PO	Intersection	F	F	F	
=	Equal	T	F	T	
PP	Contain	T	F	F	
PPi	ContainedBy	F	F	T	

Table 2. The topological relations between the regions of RCC.

	0-dimension region $\dot{R}$	1-dimension region $\tilde{R}$	2-dimension region R
0-dimension region $\dot{R}$	DC, =	DC, PP	DC, PP
1-dimension region $\tilde{R}$	DC, PPi	DC, PO, =, PP, PPi	DC, PO, PP
2-dimension region R	DC, PPi	DC, PO, PPi	DC,PO, =, PP, PPi

tri-tuple can discriminate $3^3 = 27$ topological relationships. Certainly, the different definition of TRUE, FALSE and MAYBE will result in a variant set of topological relations; we make use of the following definitions in this paper.

$$P'(X, Y) = \begin{cases} T & P(X, Y') = T \\ M & P(X, Y) = T \wedge P(X, Y')F \\ F & P(X, Y') = F \end{cases} \quad (4)$$

$$P'(X, \neg Y) = \begin{cases} T & P(X, \neg Y) = T \\ M & P(X, \neg Y) = F \wedge P(X, \neg Y') = T \\ F & P(X, \neg Y) = F \end{cases} \quad (5)$$

where $P'(X, Y)$ is the operator of parthood using ternary logic; Y_o is the interior of the region Y, and $\neg Y_o$ is the complement of the interior of region Y.

By comparing every situation identified by the model, we found that only eight situations are possible in the physical world, which are just the RCC8, illustrated in Figure 1.

3.1 Fuzzy region

We can use tri-tuple to represent the topological relationships in RCC8 by extending dyadic logic to ternary logic. Meanwhile, we can extend the region from crisp to fuzzy, so that a continuous space can be formed with dispersed geographic phenomena.

According to the Decompose Theorem of fuzzy sets, supposing $\underset{\sim}{A} \subset \tilde{f}(x)$, $\tilde{f}(x)$ is a fuzzy set, A_λ is the λ cut set of $\underset{\sim}{A}$, so, we can get that $\underset{\sim}{A} = \bigcup_{\lambda \in [0,1]} \lambda A_\lambda$, and the membership function of λA_λ is:

$$\mu_{\lambda A_\lambda}(x) = \begin{cases} \lambda & x \in A_\lambda \\ 0 & x \notin A_\lambda \end{cases} \quad (6)$$

14

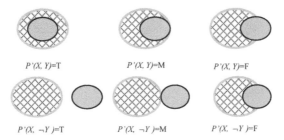

$$P'(X, Y)=T \qquad P'(X, Y)=M \qquad P'(X, Y)=F$$

$$P'(X, \neg Y)=T \qquad P'(X, \neg Y)=M \qquad P'(X, \neg Y)=F$$

Figure 1. Parthood relations based on ternary logic of $P(X, Y)$.

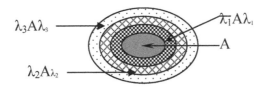

Figure 2. Fuzzy region.

Here, A_λ is a crisp set and A_1 is the kernel of the fuzzy set $\tilde{A}$.

Dividing the λ into 1, 2, 3 …, we can get the notation of fuzzy region as follows:

Definition 6: the fuzzy region $\tilde{A'}$ of crisp region A is composed of the kernel and the definite crisp regions (λ cut regions), $\tilde{A'} = \bigcup_{\lambda_i \in \lfloor 0,1 \rfloor, i \in N} \lambda_i A_{\lambda_i}$, which is to say that a fuzzy region is a set of encircled crisp regions. It can be illustrated as in Figure 2.

Here, λ_i is assigned with $\{1, 2, 3\}$ and $\lambda_1 > \lambda_2 > \lambda_3$. According to real situations, i can be assigned a much bigger value. Based on the above discussion, the following propositions can be deduced:

$$\text{Proposition 2:} \quad \forall i, j \in N \quad i < j \rightarrow P(\lambda_i A_{\lambda_i}, \lambda_j A_{\lambda_j}) = T \tag{7}$$

$$\text{Proposition 3:} \quad a \in R, \forall i, j \in N, i < j, P(a, \lambda_i A_{\lambda_i}) = T \rightarrow P(a, \lambda_j A_{\lambda_j}) = T \tag{8}$$

3.2 *Fuzzy region construction based on k-neighbourhood*

The topological relationship of disjoint is ubiquitous in the physical world, whereas the traditional topological relationship model does not take it into consideration, which is a severe fault for commonsense reasoning. This paper integrates the qualitative distance into topological relation representation to remedy the flaws of the traditional topological model. Many experts have paid attention to the qualitative distance relation (Frank 1992, Renz 2002), but the focus is how to transform quantitative distance into qualitative distance relations. The semantics of qualitative distance relations are complex, and are related to the object's size, figure, correlative position with associated objects and the reference frame (Frank 1992). A Voronoi diagram identifies the qualitative distance relation with no misinterpretations, and can consider the object's size, position and the other related surroundings. In this way, the k-neighbourhood is available for constructing the topological model, integrated with distance.

3.3 *Topological relation model based on fuzzy region*

According to the classification of qualitative distance, we can correspondingly construct the k-neighbourhood, for example the qualitative distance is graded into far, not-far and close. Then we can

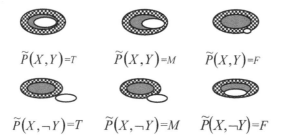

$$\widetilde{P}(X,Y)=T \qquad \widetilde{P}(X,Y)=M \qquad \widetilde{P}(X,Y)=F$$

$$\widetilde{P}(X,\neg Y)=T \qquad \widetilde{P}(X,\neg Y)=M \qquad \widetilde{P}(X,\neg Y)=F$$

Figure 3.   The part relation of tri-tuple based on fuzzy region.

construct 3-neighbourhood. For a simplified description, we use the 1-neighbourhood to construct fuzzy region $\underset{\sim}{A'}$. Then, TRUE, MAYBE and FALSE can be redefined as follows:

$$\tilde{P}(X,Y) = \begin{cases} T & P(X,Y_\perp) = T \\ M & P(X,Y_T) = T \wedge P(X,Y_\perp) = F \\ F & P(X,Y_T) = F \end{cases} \tag{9}$$

$$\tilde{P}(X,\neg Y) = \begin{cases} T & P(X,\neg Y_T) = T \\ M & P(X,\neg Y_T) = T \wedge P(X,\vee Y_\perp) = F \\ F & P(X,\neg Y\perp) = F \end{cases} \tag{10}$$

$$\tilde{P}(Y,X) = \begin{cases} T & P(Y_T,X) = T \\ M & P(Y_\perp,X) = T \wedge P(Y_T,X) = F \\ F & P(Y_\perp,X) = F \end{cases} \tag{11}$$

Here, X is a crisp region; Y is a fuzzy region formed with 1-neighbourhood; $Y_\perp$ is the farthest upper boundary of fuzzy region $\underset{\sim}{A'}$, which is just the kernel of $\underset{\sim}{A'}$. Y_T is the least inner boundary, so the parthood can be illustrated as in Figure 3:

Based on the operator $\tilde{P}(X,Y)$, we can use the tri-tuple $\{\tilde{P}(X,Y), \tilde{P}(X,\neg Y), \tilde{P}(Y,X)\}\}$ to identify the topological relation between a crisp region and a fuzzy region formed with 1-neighbourhood. The results are listed in Table 3.

By analysing the tri-tuple, we find that $3^3 = 27$ topological relationships can be discriminated theoretically. Considering the Definition 6, Proposition 2 and Proposition3, we can deduce that some situations are impossible such as (T, T, T), (T, M, T), and only the 10 cases listed in Table 3 are valid.

In the above, we only interpret the representation of topological relationships between a fuzzy region formed with 1-neighbourhood; in the same way, we can substitute ternary logic for many-value logic, and the 1-neighbourhood would be replaced with n-neighbourhood accordingly, so that the qualitative distance can be distinguished in more grades.

4   REASONING OF TOPOLOGICAL RELATION IN DISCRETE SPACE

The basic approach of spatial reasoning is constraint (Frank 1996, Cohn 1995, 2001), which needs to define the set of fundamental spatial topological relations first, and then deduce the composition table of topological relations. This reasoning approach presumes that any topological relation is composed of the basic relations, so that the set of basic relations must be Joint Exhaustive and Pair Disjoint (JEPD).

4.1   *Explanation of the model's JEPD*

There are ten basic topological relations between the crisp region and fuzzy region constructed with 1-neighbourhood (as listed in Table 3). From representation procedure, we can see the Pair Disjoint is obvious (any pair of relations tri-tuple is dissimilar).

Table 3. Topological relations between crisp and fuzzy regions.

Predicate	Semantics	$\tilde{P}(X,Y)$	$\tilde{P}(X,\neg Y)$	$\tilde{P}(Y,X)$	Figure
DD	Disjoint	F	T	F	
DO	k-1Disjoint & k Intersection	F	M	F	
DP	k-1Disjoint & k Contain	M	M	F	
OP	k-1Intersect & k Contain	M	F	F	
OO	Disjoint	F	F	F	
EP	k-1Equal & k Contain	T	F	M	
PP	Contain	T	F	F	
IP	k-1ContainBy k Contain	M	F	M	
IE	k-1Containby & k Equal	M	F	T	
II	ContainBy	F	F	T	

For the property of Joint Exhaustive, we can get interpretations from the RCC5 model. From Definition 6, we can conclude that the fuzzy region A' is the family of crisp region $\lambda iA\lambda i$. In the light of the basic principle of mereology, the fuzzy region $\underset{\sim}{A'}$ can be decomposed into n components; consequently, the computation of the topological relationship can be transformed into inferring the topological relationship between crisp region R and $\lambda iA\lambda i$. For the topological relationships of the crisp region and the fuzzy region constructed with 1-neighbourhood, the result can be gained by the intersection of crisp region R with both the kernel A of A' and the 1-neighbourhood $\lambda 1A\lambda 1$ of A. By using RCC5, there are five basic relations for each element of the intersection. Additionally, Proposition 1 shows that A is part of $\lambda 1A\lambda 1$, so the topological relationship between R and $\underset{\sim}{A'}$ can be expressed as follows:

$$a \in R, b \in \underset{\sim}{A'}, b = b_0 \cup b_1, P(b_0,b_1) = T \wedge P(\neg b_1, \neg b_0) = T \qquad (12)$$

$$Topo(a,b) = Topo(a,b_0) \circ Topo(a,b_1)$$

Here b_0 and b_1 are the kernel and the 1-neighbourhood of the region R, respectively; Topo(a, b) is the topological relation between region a and region b, and "$\circ$" is the composite operator of topological relations. Based on the composite table of RCC5 combined with Propositions 1 and 2, we can deduce the ten basic relations between a crisp region and a fuzzy region as listed in Table 3.

4.2 Predicates of topological relations and its reasoning

The topological model proposed in this paper can be regarded as the extended form of RCC5 in discrete space, which substitutes the C(X,Y) for P(X,Y). Consequently, we can get the predicates of this topological model by extending the basic predicates of RCC5. Synthesizing with the character of the fuzzy region, we define ten basic topological relation predicates as listed in Table 3. The set of topological relationship predicates is diverse with the different of k values of k-neighbourhoods. Each topological predicate consists

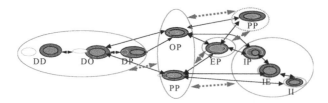

Figure 4.   Conceptual neighbour diagram of topologic relations.

of k + 1 basic predicates of RCC, and DC, PO, =, PP and PPi are simplified to D, O, E, P and I, for short. From the notation of the predicate, we can infer information from two sides; for example, DO(A, B) shows that the fuzzy region is formed with 1-neighbourhood and A is disjoint from B, A intersects with the 1-neighbourhood of B.

Corresponding with RCC5's logic reasoning approach by using axiomatic topological properties and the C(X,Y) operator, this model also can adopt the operator of P(X,Y) and the properties of $P(X, A) = T \wedge P(A, Y) = T \rightarrow P(X, Y) = T$ for reasoning and the computation of the tri-tuple $\{\tilde{P}(X, Y), \tilde{P}(X, \neg Y), \tilde{P}(Y, X)\}$. The detailed process of reasoning with this topological model will be discussed in another paper due to the limitation of this paper's topic. But we design the concept diagram of the set of basic topological relations based on the diagram of RCC5, which is useful for reasoning and predicting the topological relations between moving objects, which illustrated in Figure 4.

From all of the above, we can see that the semantics of this topological model are plentiful, and represent not only the topological relations but also the qualitative distance by using k-neighbourhood to replace the object itself. At the same time, this model makes it easy to encode these topological relationships in application systems so that we can reason and query spatial topological relationships under digital environments.

5   CONCLUSIONS AND DISCUSSIONS

This paper first reviews the existing topological models and analyses their merits and shortcomings for representing topological relations. Point set topology is exhaustive for a topological relation set and rational for topological computation; nevertheless, it is not good at dealing with qualitative spatial reasoning. Although RCC5 and RCC8 have the abilities of spatial reasoning and management of uncertainty, they are not fit for discrete space. In order to represent and reason the topological relations in discrete space, we propose a topological relation model for discrete space based on fuzzy regions. To illustrate this model, our investigations developed from the following aspects:

(1) Regard the region as the spatial primitive of spatial reasoning, and construct a topological space based on the definitions of a 0-dimension region, 1-dimension region and 2-dimension region, which represent the naive geographic world with regions and integrates with traditional GIS.
(2) Take P(X, Y) as the basic spatial operator, and define the tri-tuple {P(X,Y), P(X,¬Y), P(Y,X)} to identify the topological relationships between spatial objects; this extends RCC5 to fit with traditional GIS tools.
(3) Based on the fuzzy region formed with the objects k-neighbourhood, we propose a spatial topological relation model in discrete space and deduce the set of basic topological relations, which can represent the topological relations between discrete objects and integrate the qualitative distance into topological relations.

However, the representation and reasoning of spatial relations is a complex subject. Although we have adopted the RCC for discrete space, it is only the jumping-off point for spatial reasoning and representation in discrete space. Our future studies will be focused on these subjects:

(1) Based on the ontology that this paper proposes, study a consistent representation and reasoning model integrated with topological, directional and distance relationships.

(2) Study the representation and reasoning of spatio-temporal relationships between moving objects.
(3) Research the applications of this model and modify it according to the domains of the applications.

REFERENCES

Aurenhammer, F. & Edelsbrunner, H. 1984. An Optimal Algorithm for Constructing the Weighted Voronoi Diagram in the Plane. *Pattern Recognition* 17(2): 251–257.

Cao, H. & Chen, J. 2001. Qualitative Description and Reasoning of Directional and Topological Relationships. *Journal of Xi'an College of Petroleum (Science Editor)* 16(1): 68–72.

Clark, B.L. 1981. A Calculus of Individuals based on 'Connection'. *Notre Dame Journal of Formal Logic* 22(3): 559–607.

Cobb, M. 1995. *An Approach for the Definition. Representation and Querying of Binary Topological and Directional Relationships between Two–dimensional Objects.* Ph.D. Thesis, Tulane University.

Cohn, H. 2001. Qualitative Spatial Representation and Reasoning: An Overview. *Fundamenta Information* 43: 2–32.

Cohn, R. & Cui, Z. 1995. Taxonomies of Logically Defined Qualitative Spatial Relations. *International Journal of Human–Computer Studies: Special Issue on Formal Ontology in Conceptual Analysis and Knowledge Representation* 43(5–6): 831–846.

De Laguna, T. 1992. Point, Line and Surface as Sets of Solids. *Journal of Philosophy* 19: 449–461.

Egenhofer, M.J. 1994. Deriving the Composition of Binary Topological Relation. *Journal of Visual Languages and Computing* 5(2): 133–149.

Frank, A.U. 1992. Qualitative Spatial Reasoning about Distances and Directions in Geographic Space. *Journal of Visual Languages and Computing* 3(4): 343–371.

Frank, A.U. 1996. Qualitative spatial reasoning – cardinal directions as an example. *IJGIS* 10(3): 269–290.

Liu, D.Y. & Liu, Y.B. 2001. Reasoning of the Topological Relationships between Spatial Objects in GIS. *Journal of Software* 12(12): 1859–1865.

Renz, J. 2002. *Qualitative Spatial Reasoning with Topological Information.* Berlin: Springer-Verlag.

Goyal, R. & Egenhofer, M.J. 2000. Cardinal Directions between Extended Spatial Objects. *IEEE Transaction on Knowledge and Data Engineering,*
http://www.spatial.maine.edu/~max/max.html (accessed 12 June. 2003).

Smith, J. 1987. Close range photogrammetry for analyzing distressed trees. *Photogrammetria* 42(1): 47–56.

Advances in Spatio-Temporal Analysis – Tang et al. (eds)
© 2008 Taylor & Francis Group, London, ISBN 978-0-415-40630-7

Boolean operations on triangulated solids and their applications in 3D geological modelling

Honggang Qu, Mao Pan, Bin Wang, Yong Wang & Zhangang Wang
School of Earth and Space Science, Peking University, Beijing, China

ABSTRACT: In this paper, we present a new method for Boolean operations (intersection, union and difference) on triangulated solids and their applications in three-dimensional (3D) computerized geological modelling (3DGM). This method can be adapted to arbitrary geological solids. It includes three major steps. (1) Calculation of the intersection points between pairs of triangles. Our work is mainly concerned with coplanar triangles, while non-planar cases are dealt with by a previous algorithm (Möller 1997). (2) Retriangulation of the intersected triangles to keep the consistency between pairs of solids. (3) An inclusion test between triangles and solids. This work can be reduced to testing whether a point is inside a solid or not. We extend the "Crossing Number Method" from 2D to 3D, and deal with some particular cases when the ray passes vertices or edges of triangles. Several examples, such as intrusions, ore bodies and tunnels, bifurcated geological bodies, are presented to validate this method.

1 INTRODUCTION

Solid Modelling has broad applications in industrial design (Braid 1975, Krouse 1985), geological modelling (Bak & Mill 1989, Lemon & Jones 2003, Wu 2005) and other fields. Boundary representation (B-Rep), such as TIN (Triangulated Irregular Network), is one of the common representation methods for solid models. In this paper, we present a new method for Boolean operations on B-Rep solids (TIN). The solid can be arbitrary (convex or concave), but not self-intersected and cannot have holes. The types of Boolean operations include Union, Intersection and Difference (Fig. 1).

The broad applications of computers in geological fields such as oil and mining urgently need 3 DGM. Because of geological complexity, uncertainty and limitations of available useful data (Turner 1989, Kelk 1992, Simon 1994, Mallet 1997), 3 DGM is a complicated process. The complex spatial topological relationships of geological bodies have to be maintained, especially in multi-solid modelling. Therefore, the use of geological knowledge in 3 DGM is very important, even a pre-requisite in many cases.

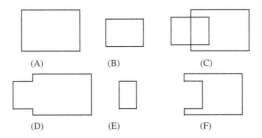

Figure 1. Boolean operations: (A) Body A; (B) Body B; (C) Spatial relationship of A and B; (D) Union (A ∪ B); (E) Intersection (A ∩ B); (F) Difference (A–B).

From the perspective of geology, geological modelling is a process of analysing and deducing the diverse geological history events and their characteristics from all kinds of "information" preserved in strata and current geological phenomena. It generally follows the rule of "uniformitarianism" and employs the method of "strata history comparison". This method of geological research can also be applied in 3 DGM. For example, we can first build the basic geological models ignoring the complex processes in geological history, and then build the complex body, "inserting" them into the basic geological bodies by the use of Boolean operations on body objects. This process is like the recurrence of the geological evolvement (Wang 2003). This paper presents the applications of this method in the modelling of intrusions, ore bodies, tunnels and bifurcated geological bodies.

2 ALGORITHMS AND SOLID BOOLEAN OPERATIONS

As introduced in the above sections, Boolean operations of solids include union, intersection and difference. The key algorithms of these three operations are similar, so we just take the difference operation as an example. Suppose that there are solids A and B. We will examine the Difference operation between A and B, namely A–B.

2.1 *Major steps*

This algorithm includes three major steps: (1) Calculate intersection points between triangles. For every triangle of one solid, we should compute its intersection points with all the triangles of other solids. If there is more than one intersection point, we should take the link lines of the points as constrained edges for triangulation in the next step. The whole process can be speeded up by filtration with the "bounding box technique" (Aftosmis et al. 1998, Zhou 1999). Here, we do not use the method of "neighbour triangle searching" (Lo & Wang 2004), because it cannot improve the efficiency when dealing with arbitrary solids that may have several intersected parts. (2) Retriangulate the intersected triangles to keep consistency between pairs of intersected solids and to make sure that there are no triangles crossing the intersection lines. With the constraint of new points and constrained edges, intersected triangles are retriangulated by the point-by-point insertion triangulation method (Lawson 1977, Tsai 1993). (3) Perform an inclusion test between triangles and solids. Delete the solid A's triangles that are inside B, and add the B's triangles that are inside A. This step can be reduced to the test between point and solid.

2.2 *Calculate intersection points between triangles*

There are two cases when calculating intersection points between two triangles: they are coplanar or non-planar. The latter case is dealt with by the algorithm proposed by Möller (1997), and our work is concerned with the coplanar cases.

2.2.1 *Non-planar triangles intersection (Möller 1997)*
Suppose that triangles T_1 and T_2 lie in planar π_1 and π_2 respectively, and π_1 intersects π_2 at line L. The process of calculation is as follows (Fig. 2):

(1) Filtering: compute plane equation of π_2, taking every vertex coordinate of T_1 into the equation; if the results have the same sign, all positive or negative, the three vertexes of T_1 must be on the same side of π_2 and it certainly does not intersect with π_2. If the results are equal to zero, T_1 and π_2 are coplanar. The relation between T_2 and π_1 can be computed in the same way.
(2) If the triangles are not filtered, they may intersect; then compute the intersection line L of π_1 and π_2.
(3) By projecting vertices of triangles to L, we can calculate that T_1 and π_2 intersect at $I_1^1 I_1^2$; T_2 and π_1 intersect at $I_2^1 I_2^2$.
(4) Judge whether $I_1^1 I_1^2$ and $I_2^1 I_2^2$ overlay. If they are intersected, calculate the intersection points.

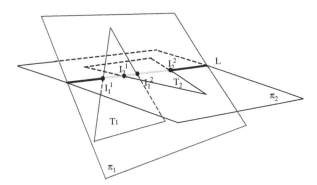

Figure 2. Calculate intersection points between non-planar triangles.

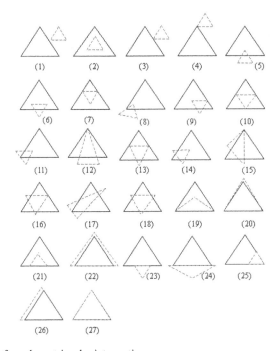

Figure 3. Classification of co-planar triangles intersection.

2.2.2 Co-planar triangles intersection

The coplanar case is complicated and can be classified by the number of intersection points; there are ten classes (Fig. 3)

- 0 intersection points: (1), (2)
- 1 intersection points: (3), (4)
- 2 intersection points: (5), (6), (7), (8)
- 3 intersection points: (9), (10), (11), (12)
- 4 intersection points: (13), (14), (15)
- 5 intersection points: (16), (17)
- 6 intersection points: (18)
- share one edge: (19), (20), (21), (22), (23), (24)
- share two edges: (25), (26)
- share three edges: (27)

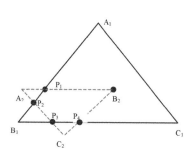

Figure 4.   Intersection of co-planar triangles.

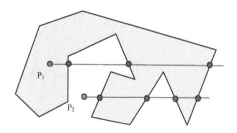

Figure 5.   Point inclusion in polygon test by "crossing number method" in 2D. The ray from Point P1 has 3 intersection points with the polygon, so it is in the polygon; P2 has 4 points and is outside the polygon.

Since it is troublesome to deal with these cases individually, we propose a method to deal with them uniformly. We found that the overlay part of intersection triangles is a convex region, which degenerates to a point or a line in a few cases, so we can calculate the convex region (Graham 1972) from the intersection points and other triangles' vertices that lie in the current operating triangle, and then every edge of the convex region will be a constrained edge for triangulation. Taking triangle 1($A_1B_1C_1$) and triangle 2($A_2B_2C_2$) in Figure 4 as an example. They are co-planar intersected and the former is the current operating triangle. We first compute their intersection points P_1, P_2, P_3 and P_4, and then find B_2, which is one vertex of triangle 2 but lies in triangle 1. Thirdly, we compute the convex region by these points, and then we get the convex region of P_1-P_2-P_3-P_4-B_2. Add points P_1, P_2, P_3, P_4, B_2 and constrained edges P_1P_2, P_2P_3, P_3P_4, P_4B_2, B_2P_1 to triangle 1; the newly added data will be used as constraints for the next step of triangulation.

2.3 *Retriangulation of intersected triangles*

Insert the constrained points and the edges added in the former step and retriangulate them using the "point-by-point insertion triangulation method" (Lawson 1977, Tsai 1993), making sure that the newly generated triangle edges will not cross the newly added constrained edge (Vigo & Pla 1995). Here, we do not deal with all of the triangles by the "edges swap method" (Lo & Wang 2005) at the intersection lines, because of the multi-value projection problem.

2.4 *Containment testing between triangles and solids*

Because intersected triangles have been dealt with, we can guarantee that the relationship between a triangle and a solid must be one of the following: inside the solid, on the solid or outside the solid. This work can be reduced to testing whether one point is inside a solid or not. If three vertices of a triangle are all inside the solid, it must be in the solid; if they are all outside the solid, it must be outside; we cannot be sure whether a triangle is on the surface or not if its three vertices are on the solid's surface for the case of concave solids. In this case, we have to compute the relationship between the inner point of the triangle and the solid, and then the triangle is on the surface if the random inner point is on the surface.

In 2D, one method of the "Jordan Curve theorem" is used to define whether a point is in a polygon: lead a ray from the test point, and the number of the intersection points between the ray and the polygon's boundary edges must be odd. As shown in Figure 5, P_1 has three intersection points, so it is in the polygon; P_2 has four intersection points, so it is outside. This method of point inclusion for polygon testing is commonly called the "crossing number method" (Shimrat 1962, Haines 1994).

We extend this method from 2D to 3D to test whether a point is in a solid or not. Leading an arbitrary ray from this point, such as a line along the positive Z-axis and having a length of two times the height of the bounding box, the number of intersection points between the ray and the solid boundary is counted. The test point will be in the solid if the number is odd and out if even. Special approaches are proposed when the ray intersects with triangles at vertices or on edges, because there would be several intersected points being the same in this case. To solve this problem, we project the intersection point and all the correlative triangles intersected at this point to the XOY plane and compute the convex region. If the point is in the region, the number of intersected points at this point is regarded as 1, or 0 if it is out. If one triangle is parallel with the Z-axis, ignore it and continue computing.

After the inclusion test, we must carry out the following operations:

(1) Delete solid A's triangles, which lie in B, including the ones that lie on B's boundary.
(2) Add the solid B's triangles which lie in A to solid A, including the ones lying on A's boundary, which are marked for the next step to decide whether we should add it in the final step.
(3) Delete the added triangles that should not be added to A. There are two steps: (a) Build the topological relationships of triangles of solid A (Hrádek et al. 2003), and delete the triangles which have the wrong topological relationship (three or more triangles share an edge) and have been marked in step2 (the B's triangles but lying on A's boundary and have been added to A). (b) Because the triangles directly linked with original solid A have just been deleted in the last step, there will be some separated triangles that should also be deleted. For this, we have to build topological relationships again, and find the triangles that have at least one edge without linked triangles, push them into a stack and mark with a deleting sign, then pop a triangle from the stack and find its neighbouring triangles, mark them and push them into the stack, recursively dealing with the triangles in the stack until it becomes empty. The final step is to delete all the triangles that have the deleting sign.

After all of above operations, we obtain the result of A–B.

3 APPLICATIONS IN 3D GEOLOGICAL MODELLING

The Boolean operations on solids, combined with geological knowledge, have broad applications in geological modelling. Generally, the process of application has 3 major steps:

(1) Reconstructing the basic geological bodies, such as the basic strata.
(2) Reconstructing the complex geological bodies, such as intrusions.
(3) Taking Boolean operations between complex and basic geological bodies and obtaining the final solids.

Three examples, intrusions, ore bodies and tunnels, and bifurcated geological bodies are presented below.

3.1 *Intrusions*

The magma intrudes the wall rock, alters the rock and finally forms the intrusion.

(1) As shown in Figure 6A, geological bodies A and B, the wall rock, are built first.
(2) Then the intrusion model C is reconstructed, whose relationships with A and B are shown in Figure 6B. It divides B into two parts and intrudes into A.
(3) Take the Difference operation A–C (delete A's triangles that lie in C and add C's triangles that lie in A to solid A) and B–C (delete B's triangles that lie in C and add C's triangles that lie in B to solid B). Figure 6C is the rendering result figure and Figure 6D is the wire frame figure.

3.2 *Ore bodies and tunnels*

This method can also be applied in geological engineering projects. For example, we can build ore bodies and tunnels independently, then apply the difference Boolean operation to them and find the practical representation of ore bodies with tunnels. Based on this model, we can calculate the volume, reserves and other parameters of the ore body.

(1) As shown in Figure 7A, ore body model A is reconstructed.
(2) The tunnel Model B is built (Fig. 7B).
(3) Figure 7C is the rendering figure after taking the A–B operation; Figure 7D is the wire frame figure.

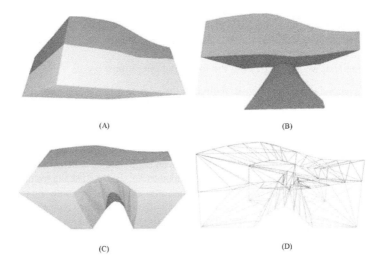

Figure 6. Reconstruction of intrusion models based on solid Difference operations: (A) Basic geological bodies A and B; (B) Intrusion C, and the relationships with A and B; (C) Results of A–C, B–C (rendering figure); (D) Wire frame figure.

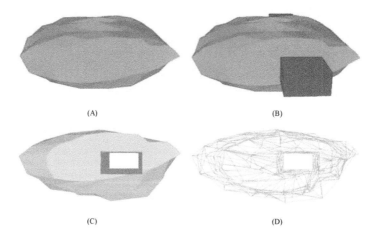

Figure 7. Ore and tunnel models based on solid Difference operations: (A) Ore body A; (B) Tunnel B and the relationship with A; (C) Result of A-B (rendering figure); (D) Wire frame figure.

3.3 *Bifurcated geological bodies*

Bifurcated geological bodies are common phenomena in geology, such as the furcation of rivers. Bifurcations lead to one geological body having different parts in different cross-sections. As shown in Figure 8, the geological body has one part in one section but three parts in the other.

(1) As shown in Figure 8A, three individual bodies (A, B and C) are built first.
(2) Figure 8B is the rendering figure after taking A ∪ B ∪ C operation (Unit A's triangles that lie outside B and C, B's triangles lie outside A and C, and C's triangles lie outside A and B); Figure 8C is the wire frame figure.

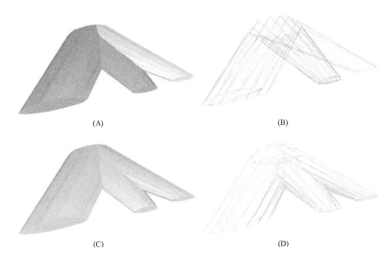

(A) (B)

(C) (D)

Figure 8. Building bifurcations models based on Union operations: (A) Basic geological bodies A, B, C and their spatial relationships; (B) Basic geological bodies shown in line frame; (C) Result of A ∪ B ∪ C (rendering figure); (D) Wire frame figure.

4 CONCLUSION

In this paper, a new method for Boolean operations (Union, Intersection and Difference) on solids is proposed. And the main steps and algorithms of the method are discussed in detail.

We also present the applications of this method in 3D geological modelling. This method, combined with geology knowledge, can reconstruct complex geological bodies and human engineering activities. Three examples, intrusions, ore bodies and tunnels, and bifurcated geological bodies, show that the method has the ability to build complex geological models.

ACKNOWLEDGEMENTS

This research paper is funded by the China National Science Foundation (40002024) and the Key Research Project of the Ministry of Education of China (99003).

REFERENCES

Aftosmis, M.J., Berger, M.J. & Melton, J.E. 1998. Robust and efficient Cartesian mesh generation for component-based geometry. *American Institute of Aeronautics and Astronautics Journal* 36 (6): 952–960.
Bak, P.R.G. & Mill, A.J.B. 1989. Three dimensional representation in a Geoscientific Resource Management System for the minerals industry. In J. Raper (ed.), *Three Dimensional Applications in Geographic Information Systems*: 155–182. New York: Taylor & Francis.
Braid, I.C. 1975. The synthesis of solids bounded by many faces. *Communications of the ACM* 18 (4): 209–216.
Graham, R.L. 1972. An Efficient Algorithm for Determining the Convex Hull of a Finite Point Set. *Information Processing Letters* 1 (4): 132–133.
Haines, E. 1994. Point in Polygon Strategies. In P. Heckbert (ed.), *Graphics Gems IV*: 24–46. Academic Press.
Hrádek, J., Kuchař, M. & Skala, V. 2003. Hash functions and triangular mesh reconstruction. *Computers & Geosciences* 29 (6): 741–751.
Kelk, B. 1992. 3-D Modelling with Geoscientific Information Systems: the problem. In A.K. Turner (ed.), *Three-dimensional Modeling with Geoscientific Information System*: 29–37. Dordrecht, the Netherlands: Kluwer Academic Publishers.
Krouse, J.K. 1985. Solid modeling catches on. *Machine Design* 57 (3): 60–64.

Lawson, C.L. 1977. Software for C1 Surface Interpolation. In J. Rice (ed.), *Mathematical Software III*: 161–194. New York: Academic Press.

Lemon, A.M. & Jones, N.L. 2003. Building solid models from boreholes and user-defined cross-sections. *Computers & Geosciences* 29 (5): 547–555.

Lo, S.H. & Wang, W.X. 2004. A fast robust algorithm for the intersection of triangulated surfaces. *Engineering with Computers* 20 (1): 11–21.

Lo, S.H. & Wang, W.X. 2005. Finite element mesh generation over intersecting curved surfaces by tracing of neighbours. *Finite Elements in Analysis and Design* 41 (4): 351–370.

Mallet, J.L. 1997. Discrete modeling for natural objects. *Mathematical Geology* 29 (2): 199–219.

Möller, T. 1997. A fast triangle-triangle intersection test. *Journal of Graphics Tools* 2(2): 25–30.

Shimrat, M. 1962. Algorithm 112, Position of Point Relative to Polygon. *Comm. ACM* 5(8): 434.

Simon, W.H. 1994. *3D Geoscience Modeling: Computer Techniques for Geological Characterization*: 27–36. Springer-Verlag.

Tsai, V.J.D. 1993. Delaunay Triangulations in TIN Creation: an overview and a Linear-time Algorithm. *International Journal of GIS* 7 (6): 501–524.

Turner, A.K. (ed.) 1992. *Three-Dimensional Modeling with Geoscientific Information Systems*: 443. Dordrecht, the Netherlands: Kluwer Academic Publishers.

Vigo, M. & Pla, N. 2000. Computing Directional Constrained Delaunay Triangulations. *Computers & Graphics* 24: 181–190.

Wang, Y. 2003. *Algorithm Research on Three Dimensional Vector Data Creation and Prototype Implementation*: 75–76. Dissertation, Peking University (in Chinese).

Wu, Q., Xu, H. & Zou, X.K. 2005. An effective method for 3D geological modeling with multi-source data integration. *Computers & Geosciences* 31 (1): 35–43.

Zhou, Y. & Suri, S. 1999. Analysis of a bounding box heuristic for object intersection. *Journal of the ACM (JACM)* 46 (6): 833–857.

Advances in Spatio-Temporal Analysis – Tang et al. (eds)
© 2008 Taylor & Francis Group, London, ISBN 978-0-415-40630-7

Towards uncertainty-based geographic information science – theories of modelling uncertainties in spatial analyses

Wenzhong Shi

Advanced Research Centre for Spatial Information Technology, Department of Land Surveying and Geo-Informatics, The Hong Kong Polytechnic University, Hong Kong, China

ABSTRACT: Within the framework of uncertainty-based geographic information science, this paper addresses modelling uncertainties in integrating multiple sources of data, modelling uncertainty in overlay analysis and in line simplification, uncertainty-based spatial data mining and spatial queries, theory and methods for controlling the quality of spatial data, modelling uncertain topological relations and the dissemination of uncertainty information of spatial data and analyses.

1 INTRODUCTION

In the first of two consecutive papers (Shi 2005a), we raised the point of view that one of the development trends in geographic information science is a move from determinate geographic information science to uncertainty-based geographic information science. In order to establish uncertainty-based geographic information science, we need to develop a number of corresponding supporting theories. These two consecutive papers describe two major supporting theories that we have developed so far: (a) modelling uncertainties in spatial data, and (b) modelling uncertainties in spatial analysis. Specifically, these include:

- Theories, models and methods for positional uncertainties in spatial data;
- Theories, models and methods for thematic uncertainties in spatial data;
- Theories, models and methods for uncertainties in digital elevation models (DEM);
- Quality improvement techniques and methods for satellite images;
- The modelling of fuzzy topological relationships between spatial objects;
- Theories and methods for uncertainties in spatial analyses and their propagation;
- Uncertainty-based spatial queries; and
- Methods for managing uncertainty information in spatial data and analyses.

The logical relations among the above theories are graphically described in the following diagram.

The previous paper addressed uncertainty modelling for spatial data and this paper will concentrate on uncertainty modelling for spatial analyses and other tasks. Specifically, this paper will cover modelling uncertainties in integrating multiple sources of data, modelling uncertainty in overlay analysis, modelling uncertainty in line simplification, uncertainty-based spatial data mining, uncertainty-based spatial queries, the theory and methods for controlling the quality of spatial data, modelling uncertainties in topological relations, and the dissemination of uncertainty information.

2 MODELLING UNCERTAINTIES IN INTEGRATING MULTI-DATA SOURCES

2.1 *The 'S-band' model – integrating attribute and positional uncertainties*

Compared with commonly used approaches where positional and thematic uncertainties are modelled separately, the philosophy behind the 'S-band' model is that the uncertainty within the fuzzy boundary

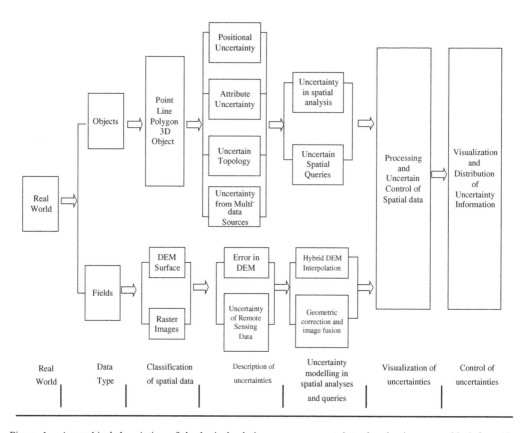

Figure 1. A graphical description of the logical relations among uncertainty theories in geographic information science.

region of a composed object is related to both positional and thematic uncertainties in the description of the object, as well as the correlation between them. The general 'S-band' model can be described by the following formula:

$$UPAT(X) = F(UP(X), UT(X), \rho(UP(X), UT(X))),$$

where UPAT(X) is the PAT uncertainty at X, and $X \in R2$ is a generic lattice or pattern data location in a two-dimensional Euclidean space. X varies over the index set $D \in R2$, which is the fuzzy boundary region; UP(X) is the positional uncertainty at X; UT(X) is the thematic uncertainty at X; and $\rho(UP(X), UT(X))$ is the correlation between UP(X) and UT(X) at X.

The general 'S-band' model illustrates that the uncertainty at a location X is related to its positional and thematic uncertainty and the correlation between them. UP(X) is quantitative in nature, UT(X) can be either qualitative or quantitative and the correlation $\rho(UP(X), UT(X))$ between them can vary from one case to another. Therefore, we cannot define a function F to cover every case. A concrete form of F can only be given when a practical problem is well defined.

Based on the general theoretical 'S-band' model, a practical 'S-band' model can be developed. For example, in a GIS-based spatial data analysis, two layers of data are used. On the first layer are geometric features such as administrative boundaries. The uncertainty of this layer results from positional errors, e.g. errors in digitizing the boundaries. The second layer is the result of the classification of a remote sensing image; the uncertainty of this layer is caused by an error in classification, i.e. mainly a thematic error. What will be the uncertainty of a new layer generated by the two basic data layers? A practical 'S-band' model is designed to solve this problem.

This problem can be defined as:

What is $P(HP \cap T)$, given $P(HP / ZP)$ and $P(HT / ZT)$?

Two methods have been developed to handle the combinational uncertainty resulting from both positional and thematic uncertainties (Shi 1994). One is based on probability theory and the other is based on a certainty factor model with a probabilistic interpretation.

2.1.1 Probability theory-based combination model

In the probability theory, we have the product rule of probabilities as follows:

$$P(A \cap B) = P(A)P(B/A) = P(B)P(A/B),$$

where $P(A \cap B)$ is the probability that both A and B will occur; $P(B/A)$ is the probability of B given the occurrence of A, i.e. the conditional probability of B given A; $P(A/B)$ is the probability of A given the occurrence of B, i.e. the conditional probability of A given B; $P(A)$ is the probability of A; and $P(B)$ is the probability of B.

2.1.2 Combination model based on the certainty factor concept

When considering the propagation of both certainty and uncertainty, the certainty factor model is used in MYCIN (Shortliffe & Buchanan 1975) for uncertainty-based reasoning. MYCIN is an expert system used for medical diagnosis. The certainty factor model can overcome the constraint of using a product rule based on the probability theory. Therefore, the certainty factor model can be used as an alternative.

We have therefore adopted the certainty factor model with the probabilistic interpretation, which was developed by Heckerman (1986) based on the original definition of the certainty factor. A detailed description of the solution can be found in Shi (1994).

2.2 Modelling uncertainties in change detection based on multi-temporal remote sensing data

An approach to determining uncertainties and their propagation in dynamic change detection based on classified remotely sensed images was proposed by Shi and Ehlers (1996). First, the uncertainties of a classified image using the maximum likelihood (ML) classification are determined. The probability vectors, which are generated during the maximum likelihood classification, are used as indicators of uncertainty. Secondly, the uncertainty propagation of classified multi-data images is described using two mathematical models. One is based on the product rule of probability theory and the other is based on a certainty factor model with the probabilistic interpretation. Thirdly, a visualization technique, using 3-D and colour, is used to visually present uncertainties in the detection of change. This model is different from the 'S-band' model. Here, instead of bundles of objects, we model the whole area covered by two raster images.

A case study was carried out, with the results indicating that the propagated change detection uncertainty is much higher than the uncertainty of the individual ML classifications. This increase is due to the propagation of uncertainty and indicates that the assessment of classification uncertainty alone is not sufficient to accurately quantify the uncertainties in the detection of change.

2.3 Rate of disFig.ment model for assessing the quality of multi-data source

In the paper (Shi 2005a), the rate of disFig.ment model is proposed to assess the uncertainties of attribute data. In fact, the rate of disFig.ment model can also be applied to assess the quality of an analysis based on multiple sources of data, which includes positional, thematic and other types of uncertainties. In this circumstance, there are a number of issues that need to be considered, such as determining the weight of each type of data, determining the sampling size and method(s), and so on.

Instead of modelling uncertainties in raster-based spatial analyses from previous studies, the uncertainty of modelling in vector-based spatial analyses is another area of research that is more generic and, at the same time, more complicated. Two approaches – analytical and simulating – were applied. Uncertainties and their propagation in overlay, buffer, line simplification and uncertainty-based spatial data mining will be addressed in the following sections. In the area of GIS, this is a step further from static error modelling for spatial data to uncertainty modelling for spatial data.

3 MODELLING UNCERTAINTY IN OVERLAY SPATIAL ANALYSES

Two methods are proposed to estimate the propagation of errors in vector overlays: an analytical error model derived from the error propagation law, and a simulation error model (Shi et al. 2004). For each of these two error models, it is proposed that the positional error in the original or derived polygons be assessed by three measures of error: (a) the variance-covariance matrices of the polygon vertices, (b) the radial error interval for all vertices of the original or derived polygons, and (c) the variance of the perimeter and that of the area of the original or derived polygons. Whereas the area and perimeter of the derived polygons have been introduced in previous studies, the x- and y-directional error intervals, as well as the radial error interval for the vertices of the derived polygons, are newly proposed.

The variance-covariance matrix of the vertices of an original or derived polygon is a relatively comprehensive description of error. However, such a matrix is neither practical nor easy to use as a measure of error and may also be large in size. Therefore, the radial positional error interval has been proposed as being more practical for describing the errors at the vertices of a polygon.

The results of a case study have demonstrated that the intersecting points of the sides of the original polygon have a higher accuracy than the vertices of the intersecting sides of the original polygon. Furthermore, it is also shown that the number of vertices and the error at the vertices of a polygon are two major factors that may affect the accuracy of parameters of the polygon, such as its perimeter and area. Increasing (or decreasing) both the number and the error at the vertices of a polygon will similarly influence the error of these polygon parameters. However, there is no consistent trend to the changes in the error of these parameters, if the number of vertices and the error at the vertices are not decreased or increased simultaneously.

4 MODELLING UNCERTAINTY IN BUFFER SPATIAL ANALYSES

A method of modelling the propagation of errors in a buffer spatial analysis for a vector-based GIS has been developed (Shi et al. 2003). Here, the buffer analysis error is defined as the difference between the expected and measured locations of a buffer. The buffer can be for the GIS point, line-segment, and linear and area features. The sources of errors in a buffer analysis include errors of the composed nodes of point or linear features, and errors in buffer width. These errors are characterized by their pdf.

Four indicators of error and their corresponding mathematical models in multiple integrals have been proposed for describing the propagated error in a buffer spatial analysis. These include the error of commission, error of omission, discrepant area, and normalized discrepant area. These indicators are defined for describing various error situations in a buffer analysis. Both the error of commission and the error of omission are suitable for describing a situation where the expected and measured buffers overlap. The definition of these two indicators is consistent with the error indicators used for assessing classified images in remote sensing. The discrepant area indicator has been defined by taking into consideration the possibility that the expected and measured buffers might not overlap with each other. This error indicator in fact provides a more generic solution for the cases where the measured and expected buffers either overlap with each other or do not. Furthermore, the normalized discrepant area has been proposed by considering the mathematical rigors of the definition – to meet the conditions of the definition of distance in algebra.

To apply the proposed mathematical models of the indicators in a practical manner, it is recommended that the Gaussian Quadrature numerical integration method be used to solve the models in multiple integrals.

One characteristic of this method is that it is highly accurate, and the authors have also proved in a previous study that it is capable of analyzing errors of spatial features.

5 MODELLING UNCERTAINTY IN LINE SIMPLIFICATION

Line simplification is performed in GIS in order to remove redundant points in a line or to reduce the volume of data for representing lines; however, such a process leads to positional uncertainties caused by (a) the uncertainty in an initial line and (b) the uncertainty due to the deviation between the initial and simplified lines. In our study (Cheung and Shi 2004), the uncertainties in a line simplification process are classified as a combination of the propagated uncertainty, the modelling uncertainty and the overall processing uncertainty. The propagated uncertainty is used to identify the uncertainty effect of the initial line and the modelling uncertainty represents the uncertainty arising from the line simplification process. The overall uncertainty in the simplified line is modelled by the overall processing uncertainty that integrates both the propagated and the modelling uncertainties in the line simplification process. Three uncertainty indices and corresponding mathematical solutions were proposed for each type of uncertainty by measuring its mean, median and maximum values. For the propagation uncertainty, we proposed the mean discrepancy, the median discrepancy and the maximum discrepancy; for the modelling uncertainty, we proposed the mean distortion, the median distortion and the maximum distortion; for the overall processing uncertainty, we proposed the mean deviation, the median deviation and the maximum deviation.

The mean and median uncertainty indices for each type of uncertainty are different in value although both types measure the central tendency of a particular type of uncertainty. The distributions of all of the types of uncertainty are positively skewed. The mean uncertainty index provides the general value of the type of uncertainty. If users want to minimize the average value of the type of uncertainty, the median uncertainty index will be given. In this study, the relation between the mean uncertainty index for the overall processing uncertainty and the threshold distance in the DP line simplification was studied. It was found that the mean uncertainty index is a monotonic increasing function of the threshold distance. Also, this function is used to determine the threshold distance such that an uncertain simplified line is close to the 'true' initial line to a predefined acceptable level of accuracy.

The maximum uncertainty index considers an extreme case in the type of uncertainty. This index for the overall processing uncertainty is important for describing an uncertainty distribution of the simplified line. The maximum deviation is a maximum distance measure for the 'true' initial line and the uncertain simplified line. It is considered a buffer width such that the buffer around the mean location of the simplified line contains the 'true' initial line. Therefore, points of the simplified line are expressed as a linear combination of their mean locations plus a random number evenly distributed in a square of the maximum deviation. The uncertainty distribution of the simplified line is important for assessing the uncertainty in the spatial results derived from the uncertain simplified line.

According to our experimental study, the uncertainty indices for the propagated and modelling uncertainties were not small when compared with the overall processing uncertainty. Since the propagated and modelling uncertainties assess the effect of the uncertainty in the initial line on the line simplification process, and the distortion between the measured initial line and the corresponding simplified line, respectively, it is more rational to measure the uncertainty in the simplified line by considering these two sources of uncertainty. Most previous studies have neglected the uncertainty in the initial line when modelling the uncertainty problem for a line simplification process. The uncertainty indices for the overall processing uncertainty proposed in this study are therefore more comprehensive in order to measure the uncertainty in the simplified line.

6 UNCERTAINTY-BASED SPATIAL DATA MINING

The idea of spatial data mining with uncertainty, and a corresponding theoretical framework, have been proposed (Shi et al. 2003). First, uncertainty in spatial data is presented in terms of its characteristics, perspectives and role in spatial data mining. Both randomness and fuzziness often appear at the same time. Secondly, a cloud model, a mathematical model that studies randomness and fuzziness in a unified way, is applied to model uncertainty in spatial data mining. Furthermore, the cloud model may act as an uncertainty transition between a qualitative concept and its quantitative data, which is implemented with the cloud generators. Finally, a case study on landslide-monitoring data mining was conducted. The results show that uncertainty is unavoidable when spatial data mining is carried out, and that the quality of the final discovered knowledge may be improved if the uncertainties are properly modelled.

7 UNCERTAINTY-BASED SPATIAL QUERIES

In a point-in-polygon analysis, a point and a polygon are uncertain due to random errors introduced in the process of capturing data or other processes. Existing research studies model the uncertainties of the point and polygon on the assumption that the uncertainties of all points located inside the polygon follow a circular normal distribution identically and independently. In our study of this field (Cheung et al. 2004), we have proposed a probability-based uncertainty model for a point-in-polygon analysis in a more generic case regarding existing research problems, including those in which (a) both the point and the polygon are uncertain, (b) the error ellipse of the point, which is more rigorous than the error circle model, should be used to describe the uncertainty of a point, and (c) where the uncertainties of the vertices of a polygon may be correlated and different from each other.

In order to provide the probability of an uncertain point located inside an uncertain polygon, the uncertainties of the point and of the vertices of the polygon are described in terms of probability density functions. The probability of an uncertain point located inside an uncertain polygon is then derived in terms of multiple integrals based on probability and statistical theories. Since this expression involves many variables of integration, we divide the point-in-polygon analysis into different cases depending on the intersection between the polygon and the error ellipse of the point. The mathematical expressions for the probability in the individual cases are also provided.

8 THEORY AND METHODS FOR CONTROLLING THE QUALITY OF SPATIAL DATA

In previous studies, the modelling of uncertainties in either spatial data or spatial analyses has focused mainly on describing the quantity of errors or uncertainties. A further question would be whether we can reduce the quantity of these errors or uncertainties. Studies on controlling the quality of spatial data are trying to address this issue.

A least squares based method, designed particularly for solving inconsistencies between areas of digitized and registered land parcels, has been proposed for adjusting the boundaries of area objects in a GIS. The principle of this approach is to take the size of the registered area of a land parcel as its true value and to adjust the geometric position of the boundaries of the digitized parcel. First, a generic area adjustment model is derived by incorporating the following two categories of constraints: (a) attribute constraints: the size of the true area of the parcel, and (b) geometric constraints: such as straight lines, right angles and certain distances. Secondly, the methods used to adjust the areas of the parcels for different cases are presented. It is demonstrated, via several case studies, that the proposed approach is able to maintain a consistency between the areas of the digitized and registered parcels. This study has solved one of the most critical problems in developing a land/cadastral information system, and this solution has been adopted in the processing of real world cadastral data in Shanghai and other cities in China (Tong et al. 2005).

The processing of cadastral areas is further studied, making the assumption that the digitized coordinates of a parcel contain errors, while the registered area of the parcel is regarded as a known value without error, and is less rigorous (Tong and Zhao 2002). The reason for this is that the registered area of a parcel is also obtained by surveying methods and thus contains errors. We should then treat the registered areas of the parcels as observations with errors. Therefore, it is necessary to solve the inconsistencies between the digitized and registered areas of land parcels under the condition that both the areas and coordinates are treated as observations with errors. The problem then comes to be defined as how to determine the prior knowledge of the weights of these observations, since these are two different types of observations, including different sources of data and units of measurement. In our study, a least squares adjustment based on the Helmert method is presented for estimating accurate weights between the area and coordinates. At the same time, the insistency between the registered area and digitized area of the parcel is adjusted through the least squares adjustment. The computational results show that the Helmert method can accurately estimate the weights of the observables, and the least squares adjustment based on the variance components of the unit weights can solve the insistency between the registered area and digitized area of the parcel more rigorously. As a result, the accuracy of the parcel area is improved via the surveying adjustment and quality control process.

9   MODELLING UNCERTAINTIES IN TOPOLOGICAL RELATIONS

The topological relation between spatial objects is one of the fundamental properties of GIS data, and can be used for spatial analysis, spatial queries and controlling the quality of data. In order to form a theoretical basis for spatial analysis, we extend topological relations to uncertain topological relations based on fuzzy topology and probability theory.

9.1   *Extended model of topological relations between crisp GIS objects*

The current models for describing topological relations between crisp objects in GIS (Egenhofer 1993, Cohn & Gotts 1996, Smith 1996) are first extended and followed by a new definition of the topological relations between two objects, which is an extension of the traditional definition based on empty and non-empty objects under homeomorphic mapping (Liu and Shi 2003). Based on this new definition, including the topology of the object itself and several topological properties, we have uncovered a sequence of topological relations between two convex sets.

There are a number of new findings from this study. The two major findings are: (a) that the number of topological relations between the two sets is not as simple as finite; actually, it is infinite and can be approximated by a sequence of matrices, and (b) that the topological relations between two sets are dependent on the shapes of the sets themselves.

9.2   *Modelling fuzzy topological relations in GIS*

The boundary of an object in GIS can be either crisp or vague/fuzzy. The classical set theory, which is based on a crisp boundary, may not be suitable for handling vague or fuzzy boundary problems. An incorrect modelling of fuzzy/vague GIS objects may lead not only to the loss of information, but also to incorrect descriptions of the reality in a GIS. For instance, a tide makes it difficult to determine the boundary of a sea, such as the boundary of the Pacific Ocean. Due to the tidal effect, some islands in the Ocean may appear and disappear from time to time. We may also have problems describing the topology of these islands based on a crisp and static GIS. Therefore, the classical set theory may not be a suitable basic tool for describing objects in GIS. Alternatively, the fuzzy set theory provides a useful solution to the description of uncertain objects in GIS. The fuzzy topology can be applied to investigate fuzzy topological relations among objects in GIS.

For modelling fuzzy topological relations among uncertain objects in GIS, the quasi-coincidence and quasi-difference are used (a) to distinguish the topological relations among fuzzy objects and (b) to indicate the effect of one fuzzy object on another, based on fuzzy topology (Shi & Liu 2004). Geometrically, features in GIS can be classified as point features, linear features and polygon or region features. In our study, we first introduced several basic concepts in fuzzy topology, then followed these by definitions of fuzzy points, fuzzy lines and fuzzy regions for GIS objects. Next, the level of one fuzzy object that affected the other is modelled based on the sum and difference of the membership functions, which are the quasi-coincidence and quasi-difference, respectively.

9.3   *Computable fuzzy topology in GIS*

The existing topological models can define conceptual definitions of interiors, boundaries and exteriors. However, in many cases, we need to compute the level of an interior, boundary and exterior. Therefore, we may need to propose new formulae to compute levels of interiors, boundaries and exteriors. Based on fuzzy topology, we developed a computable fuzzy topology for calculating the interior, boundary and exterior of a spatial object once the membership function is known.

Before we could determine a computable fuzzy topology, it was necessary to first define two new operators, the interior and closure operators, which are used to further define a computable fuzzy topology. We know that each interior operator corresponds to one fuzzy topology and that each closure operator corresponds to one fuzzy topology (Liu & Luo 1997). In general, if we define two operators, interior and closure, separately, they will define two fuzzy topologies, respectively. These two topologies may not cohere to each other. That is, the open set defined by the interior operator may not be the complement of the closed set, which is defined by the closure operator. In order to define a computable fuzzy topology, we want these two operators to be able to cohere with each other so that they can define the same fuzzy topology, in which the open sets are the image of the interior operator and the closed sets are the image of the closure operator. At the same time, we set the complement of an open set to be a closed set. The interiors, boundaries and closures of fuzzy spatial objects in GIS can thus be computed (Liu & Shi 2005).

9.4 *Qualitative and quantitative fuzzy topological relations under several invariant properties*

The above fuzzy topological relations are elementary relations in the study of topological relations between spatial objects in GIS. Many researchers have developed their models in this field based on these relations (Clementini & Di Felice 1996, Shi & Guo 1999, Tang & Kainz 2001).

The properties of topological spaces that are preserved under homeomorphic mappings are called the topological invariants of the spaces. To study the topological relations, we first need to investigate the properties of a fuzzy mapping, especially a homeomorphic mapping. Based on the newly developed computational fuzzy topology, methods for computing the fuzzy topological relations of spatial objects have been proposed in this study. Specifically, the following areas are covered: (a) the homeomorphic invariants of the fuzzy topology have been proposed; (b) the connectivity based on the new fuzzy topology has been defined; (c) for modelling the topological relations between spatial objects, the concepts of a bound on the intersection of the boundary and interior, and the boundary and exterior have been defined based on the computational fuzzy topology, which are $(A_\alpha \wedge \partial A)(x) < 1 - \alpha$ and $((A^c)_\alpha \wedge \partial A)(x) < 1 - \alpha$. With these, we can guarantee the properties that remain unchanged in a GIS transformation, such as the maintenance of topological consistency when transferring a map from one system to another. Moreover, among these topological relations, we can extract useful topological relations, as this activity commonly occurs in GIS (Shi & Liu 2005a).

9.5 *Shape-dependant topological relations*

Topology is normally considered to be independent of the shapes of spatial objects. This may not necessarily be true when describing relations between spatial objects in GIS. We have proven (Shi & Liu 2005b) that the topological relations between spatial objects are dependent on the shapes of spatial objects. That is, the topological relations of non-convex sets cannot be deformed to the topological relations of convex sets. The significant theoretical value of this research is in its finding that the topologies of spatial objects are shape-dependent. This indicates a necessary consideration of both topologies and shapes of objects when describing topological relations among spatial objects in GIS. As a result, some of the spatial data modelling, queries and analyses based on the existing understanding of the topologies of spatial objects may need to be re-assessed.

10 DISSEMINATION OF UNCERTAINTY INFORMATION OF SPATIAL DATA AND ANALYSIS

Two technologies have been developed for disseminating uncertainty-related information for GIS: (a) to manage error metadata in a database, and (b) to distribute uncertain information by Web-service technology.

An object-oriented error metadata database—EMMS—has been designed and developed for managing metadata at the object level rather than at the map or feature level (Gan & Shi 2002). It captures quality

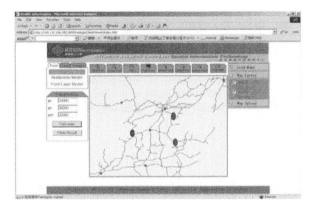

Figure 2. An interface of a Web service-based information system for disseminating uncertainty-related information for users.

information and temporal information, information about sources of data, processing steps and carto-graphic generalizations on the dataset. Such information relating to data sources and data validation will be captured automatically during the process of being manually input by operators. In the example of the Hong Kong 1:20,000 Digital Topographic Map Series, the EMMS is used to display metadata, especially when data are conflated. Metadata for other datasets such as 1:10,000 and 1:50,000 maps in Hong Kong or elsewhere can also adopt the EMMS technology.

Metadata for individual objects may not yield instant benefits to data producers. However, in the long run, the error metadata can support users in making decisions regarding the appropriateness-for-use of a dataset. It also facilitates the determination of positional errors of GIS objects in the geometric form of point, line and polygon. Quality control and quality assurance can also be implemented through the installation of the EMMS for data providers. Data users can ultimately select appropriate GIS data for a particular application based on the metadata in the EMMS.

10.1 *Web service-based uncertainty information distribution techniques*

Based on Web service technology, an internet-based distributed computing environment has been built to provide a solution to the Web service-based uncertainty information service and metadata distribution (Shi 2005b). Specifically, it can be used to design the Web service-based data quality information system, pro-vide its technical framework and provide a practical solution to the distribution of uncertainty information. Figure 2 is an example of the user interface of a Web service-based information system for disseminating the error information of points, where the error of each point feature is illustrated by its error ellipse.

11 CONCLUDING REMARKS

This paper addresses a number of theories or solutions in modelling uncertainties in spatial analyses, which is one of the two major supporting theories for uncertain geographic information science. At the same time, it also takes the research on modelling uncertainties in spatial data a step further. Progress has been made in the following areas:

- From the modelling of static uncertain data to dynamic uncertain spatial analysis;
- From modelling errors for single types of data to multiples sources of data;
- From *measuring* the quality of data to *controlling* the quality of data, and possibly *reducing* overall errors in GIS data via least squares or other possible solutions, such as image fusion and geometric corrections to remote sensing images;

- From traditional positional and attribute error modelling to spatial data mining, uncertain topological relations; and
- To disseminating uncertain information using the latest information technology—Web service.

This and the previous paper (Shi 2005a) only address two major supporting theories of uncertainty-based geographic information science. This is just a beginning in the development of uncertainty-based GIS. We expect more developments, which will enrich uncertainty-based GIS in the near future.

ACKNOWLEDGMENTS

The work described in this paper was supported by grants from the Hong Kong Polytechnic University (PolyU 5071/01E), and Hong Kong Research Grants Council (Project No. PolyU 5167/03E).

REFERENCES

Cheung, C.K. & Shi, W.Z. 2004. Estimation of the positional uncertainty in line simplification in GIS. *The Cartography Journal* Vol. 41, No.1: 37–45.

Cheung, C.K., Shi, W.Z. & Zhou, X. 2004. A probability–based uncertainty model for point–in–polygon analysis. *GeoInformatica* Vol. 8, No. 1: 71–98.

Clementini, E. & Felice, D.P. 1996. An algebraic model for spatial objects with indeterminate boundaries. In P.A. Burrough & A.U. Frank (eds), *Geographic Objects with Indeterminate Boundaries*: 155–169. London: Taylor & Francis.

Cohn, A.G. & Gotts, N.M. 1996. The "egg–yolk" representation of regions with indeterminate boundaries. In P.A. Burrough & A.U. Frank (eds), *Geographic Objects with Indeterminate Boundaries*: 171–187. London: Taylor & Francis.

Egenhofer, M. 1993. A model for detailed binary topological relations. *Geomatica* 47 (3) & (4): 261–273.

Gan, E. & Shi, W.Z. 2002. Error metadata management system. In W.Z. Shi, P.F. Fisher & M.F. Goodchild (eds), *Spatial Data Quality*: 251–266. Taylor & Frances.

Heckerman, D. 1986. Probabilistic interpretation of MYCIN's certainty factors. In L.N. Kanal & J.F. Lemmer (eds), *Uncertainty in Artificial Intelligence*: 167–196.

Liu, K.F. & Shi, W.Z. 2005. A fuzzy topology for computing the interior, boundary, and exterior of spatial objects quantitatively in GIS. *Photogrammetric Engineering and Remote Sensing* (submitted).

Liu, K. & Shi, W.Z. 2003. Analysis of topological relationships between two sets. In W.Z. Shi, M.F. Goodchild & P.F. Fisher (eds), *Proceedings of the Second International Symposium on Spatial Data Quality, March 19th 20th, Hong Kong*: 61–71.

Liu, Y.M. & Luo, M.K. 1997. Fuzzy Topology. World Scientific.

Shi, W.Z. 1994. *Modeling Positional and Thematic Uncertainties in Integration of Remote Sensing and Geographic Information Systems*. Enschede: ITC Publication.

Shi, W.Z. & Guo, W. 1999. Topological relations between uncertainty spatial objects in three–dimensional space. *Proceedings of International Symposium on Spatial Data Quality, Hong Kong*: 487–495.

Shi, W.Z., Tong, X.H. & Liu, D.J. 2000. An approach for modeling error of generic curve features in GIS. *ACTA Geodaetica et Cartographica Sinica* Vol. 29, No. 1: 52–58.

Shi, W.Z., Cheung, C.K. & Zhu, C.Q. 2003. Modeling error propagation in vector–based buffer analysis. *International Journal of Geographic Information Science* Vol. 17, No. 3: 251–271.

Shi, W. Z., Wang, S.L., Li, D.R. & Wang, X.Z. 2003. Uncertainty–based spatial data mining. *Proceedings of Asia GIS 2003 Conference, October 16–18, 2003, Wuhan, China.*

Shi, W.Z., Cheung, C.K. & Tong, X.H. 2004. Modeling error propagation in vector–based overlay spatial analysis. *ISPRS Journal of Photogrammetric and Remote Sensing* Vol. 59, Issues 1–2: 47–59.

Shi, W.Z. & Liu, K.F. 2004. Modeling fuzzy topological relations between uncertain objects in GIS. *Photogrammetric Engineering and Remote Sensing* Vol. 70, No. 8: 921–930.

Shi, W.Z. & Liu, K.F. 2005a. Computing the fuzzy topological relations of spatial objects based on induced fuzzy topology. *International Journal of Geographic Information Science (submitted)*.

Shi, W.Z. & Liu, K.F. 2005b. Are topological relations dependent on the shape of spatial objects? *Progress in Nature Science (Accepted)*.

Shi, W.Z. 2005a. Towards uncertain geographic information science (part A) — modeling uncertainties in spatial data. *Proceedings of International Symposium on Spatial Data Quality, August 25–25, 2005, Beijing.*

Shi, W.Z. 2005b. *Principle of modeling uncertainties in spatial data and analysis.* Beijing: Science Press.

Shortliffe, E.H. & Buchanan, B.G. 1975. A model of inexact reasoning in medicine. *Mathematical Biosciences* 23: 351–379.

Smith, B. 1996. Mereotopology: A theory of parts and boundaries. *Data & Knowledge Engineering* 20: 287–303.

Tang, X.M. & Kainz, W. 2001. Analysis of topological relations between fuzzy regions in general fuzzy topological space. *Proceedings of the SDH Conference 02', Ottawa, Canada.*

Tong, X.H., Shi W.Z. & Liu, D.J. 1999. Uncertainty model of circular curve features in GIS. *ACTA Geodaetica et Cartographica* Vol. 28, No. 4: 325–329.

Tong, X.H., Shi, W.Z. & Liu, D.J. 2005. A least squares–based method for adjusting the boundaries of area objects. *Photogrammetric Engineering and Remote Sensing* Vol. 71 No.2: 189–195.

Advances in Spatio-Temporal Analysis – Tang et al. (eds)
© 2008 Taylor & Francis Group, London, ISBN 978-0-415-40630-7

The expression of spherical entities and generation of voronoi diagrams based on truncated icosahedron DGG

Xiaochong Tong, Jin Ben, Yongsheng Zhang & Li Meng
Institute of Surveying and Mapping, Zhengzhou, China

ABSTRACT: Based on the analysis of current spherical spatial data structures, the paper discusses the concepts of the spherical spatial subdivision mode, which uses an inverse Snyder polyhedron equal area projection on the surfaces of a truncated icosahedron, and then obtains global multi-resolution overlays through hierarchical subdivision on the initial unfolded projection plane using hexagonal grids. The paper puts forward essential advice for managing the three-leaf nodes and coding tiles on the basis of a hexagonal grid, and establishes the hexagonal grid expression modes of different spherical entities. Then, based on these clues and the spherical hexagonal overlay, the paper puts forward the algorithm for generating the Voronoi diagram and verifies the correctness and efficiency of the algorithm through experiments. Finally, after comparison with some similar algorithms, the paper summarizes the advantages of the algorithm and also gives further research directions.

1 INTRODUCTION

With the development of various data acquisition methods, we can obtain all sorts of spatial data, especially global remote sensing and mapping data. In order to manage, extract and analyse the spatial data effectively, it is important to adopt a new mode to supervise global spatial data. In the 1980s, and in order to effectively manage extensive or even global multi-resolution geographical spatial data, many academicians brought forward global discrete grid data models based on the subdivision of regular polyhedrons. This involved splitting spherical surfaces into tiles with approximately equal shape or area and, instead of geographical coordinates, using hierarchical recursive subdivision and the corresponding address codes of each tile to carry out all kinds of operation. Because the address codes not only show the position but also express the scale and precision, it has the potential to manage multi-scale data. Moreover, thanks to the special mathematical characteristics of Voronoi diagrams, it became the most hopeful method to solve dynamic GIS (Chen 2002). How to effectively combine them both? This, however, has become the focus for many academicians.

The Voronoi diagram is a widely researched issue in computational geometry and is an important geometric structure, which is only inferior to convex hull. It has many egregious funny mathematical characteristics, and is a powerful tool to research and to solve some issues in geoscience, computer science, mathematics and so on. As the Earth is an approximate sphere, it is very appropriate to establish a spherical Voronoi diagram for the management of global spatial data and dynamic maintenance of spherical spatial relations.

At present, there is little research into the theories of spherical Voronoi diagrams, but several typical algorithms have emerged. For example, Augenbaum (1985) gave the spherical Voronoi diagram algorithm of n points using an incremental method, and the time complexity is $O(n^2)$; Robert (1997) put forward the divide and conquer method to generate spherical Voronoi diagrams, and the time complexity is $O(nlogn)$; Zhao (2002) put forward the QTM-based algorithm, etc. The first two methods are vectorial algorithms aimed at spherical point sets, and the third aims at all kinds of spherical aggregation on the basis of a specific spherical triangular mesh. Until now, all the vectorial algorithms of spherical Voronoi diagrams have only been effective with point sets, and do not have operability for curve sets and area sets (Chen 2002, Zhao 2002). For grids, the paper also adopts the basic generating theories of Voronoi diagrams, which research

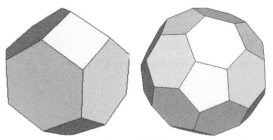

a. Truncated Octahedron   *b.* Truncated Icosahedron

Figure 1.   Truncated Solids.

spherical Voronoi diagrams using the spherical grid as the basic subdivision cell, and adopts grid interval transformation. However, instead of a triangular mesh, a hexagonal mesh is adopted as the basic spherical subdivision cell. Compared with triangles, the hexagons have more advantages. For instance, they have more directions, provide the greatest angular resolution (Golay 1969), have a smaller sampling interval, and all directions are the same, etc. Therefore, in this paper we put forward spherical hexagonal grid expression modes of different spherical entities and a corresponding generating algorithm for Voronoi diagrams.

2   BASIC SUBDIVISION GRID AND THE DEFINITION OF A SPHERICAL VORONOI DIAGRAM

2.1   *The establishment of basic subdivision grid*

As the basis for planar discrete grids, the hexagon has received a great deal of attention. Among the three regular flat surface bedding polygons (triangles, squares, and hexagons), the hexagon is the most compact one, because it can quantize the plane with the smallest average error (Conway & Sloane 1988) and provide the greatest angular resolution (Golay 1969). However, attention must be paid to the fact that it is impossible to completely tile a sphere with hexagons. When a polyhedron is tiled with hexagon-subdivided triangle faces, a non-hexagon polygon will be formed at some of the polyhedron's vertices. The number of such polygons relates to the number of polyhedron vertices, and has nothing to do with the grid resolution. For example, an octahedron has eight squares, while an icosahedron has 12 pentagons. They are all called Truncated Polyhedrons and are also called Archimedean Solids, e.g. Figure 1.

Generally speaking, perfect entities with small faces will reduce the distortion, when they are transformed between the face of a polyhedron and the corresponding spherical surface (White et al. 1998). The tetrahedron and cube have the largest face size and, compared with other entities, they are most unlike a sphere. However, because the faces of a cube can be easily subdivided into square quadtrees, it was chosen as the basic platonic entity by Alborzi and Samet (2000). The icosahedron has the smallest face size and, as a result, the DGGSs defined on it tend to have the smallest distortion. So the icosahedron is the best choice for a basic platonic entity. Geodesic DGGSs, which are based on the icosahedron, include those of Williamson (1968), Sadournay et al. (1968), Baumgardner and Frederickson (1985), Sahr and White (2003), White et al. (1998), Fekete and Treinish (1990), Thuburn (1997), White (2000), Song et al. (2002), and Heikes and Randall (1995a, b). Based on these considerations, we discuss mainly the icosahedron subdivision and hexagonal subdivision (also called Truncated Icosahedron Subdivision) in this paper.

How to project the discrete grid on a spherical surface is also a problem that needs to be discussed. It can be considered in two ways. One is direct spherical subdivision, the maximal bug of which, however, is that it is hard to define the subdivision, and it is difficult to ensure that the subdivision grids are of the same area. Moreover, it is difficult to ensure that the subdivision edges are major spherical arcs. Therefore, this first method tends to have a low precision and is hard to operate. The other possibility is the projection method, which can project discrete grid maps to the spherical surface. Compared with the first method, the second one is easier to operate, has controllable deformation and a rigorous mathematical reasoning.

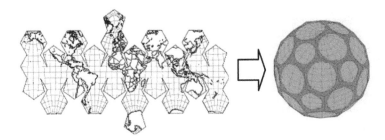

Figure 2.   The Snyder projection of truncated icosahedron.

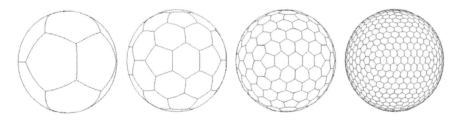

Figure 3.   Spherical hierarchical subdivision based on truncated icosahedron using ISEA projection.

We adopted the second method and established the direct relationship between the unfolded polyhedron and the spherical surface by ISEA Snyder equal area projection (Snyder 1992), e.g. Figure 2. Then we extended it to a multi-layer subdivision grid. The details can be found in Ben (2005) and Tong (2006).

Based on the above discussion, we chose the truncated icosahedron as the basic spherical subdivision cell, and used a hexagonal grid to conduct detailed subdividing. For the practical dispersed method, the aperture adopted is four (Ben 2005). Figure 3 shows the image of a discrete grid map on a spherical surface, projected by ISEA.

2.2   *The definition of a spherical Voronoi diagram*

Spherical surface S is not the Euclidian space; it is different from the plane, because it does not have the same embryo, and so we must give the definition of a spherical Voronoi diagram before it is researched.

In order to express it more tersely, we establish the Voronoi diagram on the surface of a unit sphere. First, we define the Voronoi diagram for a point set. Given a point set $P = \{p_1, p_2, \ldots, p_n\}$ of n points on a spherical surface, for every point p_i in P, the locus of x is the one containing all points closer to a point p_i in P than any another point in P.

$$\left\{ x \in S \,\middle|\, \left| x\overset{\cap}{p_i} \right| \leq \left| x\overset{\cap}{p_j} \right|, j = 1, 2, \cdots, n; i \neq j \right\} \tag{1}$$

where $|x\overset{\cap}{}y|$ denotes the shortest distance between x and y on S, i.e. the length of a minor arc on the spherical surface. If the central angle of $|x\overset{\cap}{}y|$ is τ, then $|x\overset{\cap}{}y| = 1 \cdot \tau = \tau$. The polygon region $V(p_i)$ that satisfies Formula 1 is called the spherical Voronoi region of p_i. The collection of all n Voronoi regions, one for each point in P, constitutes the Spherical Nearest-point Voronoi diagram of the point set, or simply the Spherical Voronoi diagram of the point set.

Then we define the spherical Voronoi diagram of an arbitrary figure G on spherical surface S. If the figure G on the spherical surface satisfies $G = \{g_i \in S | i \in N; i \geq 2\}$, then its spherical Voronoi diagram can be expressed as follows.

$$V = \left\{ v \in S \,\middle|\, \exists g_m, g_n \in G; g_m \neq g_n; \forall g_i \in G, \left| v\overset{\cap}{g_m} \right| = \left| v\overset{\cap}{g_n} \right| \leq \left| v\overset{\cap}{g_i} \right| \right\}. \tag{2}$$

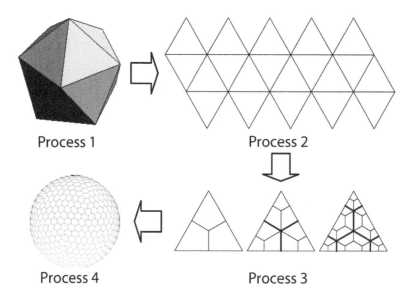

Figure 4.   The process of spherical hierarchical subdivision.

The arbitrary figure G on the spherical surface can be separated into a point set, curve set and area set. The generating algorithm for a spherical Voronoi diagram for a point set already exists, but for a curve set and area set the generating algorithms are not complicated, so we adopt the method based on a grid overlay. It has many advantages. For example, it is brief in concept, well defined in hierarchy and easily expanded in high dimensions. Thus it gives a new way to generate the spherical Voronoi diagram. The paper adopts the Voronoi diagram generating algorithm based on spherical subdivision of a truncated icosahedron hexagonal discrete global grid.

3   HIERARCHICAL MANAGEMENT AND MODE OF EXPRESSION FOR A SPHERICAL ENTITIES BASED ON DGG

3.1   *Multi-resolution subdivision method of spherical icosahedron*

Spherical subdivision was advanced by the German cartographer Fuller in 1944, when he was studying map projections. And from then on, some academicians have delved into different methods, in order to manage and analyse global data. The paper adopts an icosahedron as the basic spherical grid, and uses the method of Snyder equal area projection (ISEA) to get spherical surfaces mapped on a plane. It then uses a hexagon whose aperture is four to subdivide every projection plane, and finally uses the inverse Snyder equal area projection to obtain a planar grid map on the spherical surface. Figure 4 shows the process.

The idiographic multi-resolution subdivision is operated in Process 3. But the use of multi-resolution, hexagon-based discrete grid systems has been hampered by the fact that congruent discrete grid systems cannot be built by hexagons, for it is impossible to exactly decompose a hexagon into smaller hexagons (or, conversely, to aggregate small hexagons to form a larger one). This problem heavily confined the application of the multi-resolution hexagonal grid. On the basis of the subdivision characteristics of aperture 4, we put forward a new retrieving algorithm for multi-resolution grid nodes. Unlike the traditional retrieving methods of quad tree tiles, this algorithm can retrieve and manage multi-resolution hexagonal grids more effectively. We will now introduce the basic concepts of this method.

3.2   *Management of three-leaf nodes and encoding of tiles*

The discrete grid cannot use the managing method of hierarchical tiles, since it adopts the hexagonal subdivision mode on a triangular plane whose aperture is four. However, if we examine the mode more

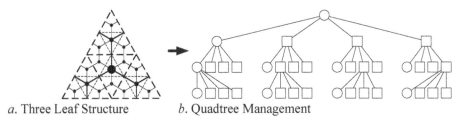

a. Three Leaf Structure   *b*. Quadtree Management

Figure 5.   The quadtree management of three leaf nodes (○: Reserved Nodes, □: New Nodes).

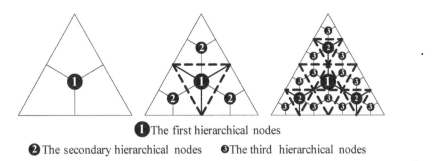

❶The first hierarchical nodes

❷The secondary hierarchical nodes   ❸The third hierarchical nodes

Figure 6.   The Voronoi diagram of different hierarchical nodes (dashed lines represent the Voronoi diagram).

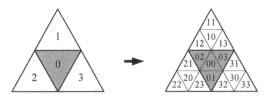

Figure 7.   Order of the QTM code.

carefully, we can find that all nodes of the last hierarchy are kept down in the next hierarchy, and there are three new nodes growing up based on the old ones, and the new nodes can compose a new structure called the Three-Leaf Node Structure (see Fig. 5a). Then, the three new nodes and the last hierarchical old node can make up a new node quad tree (Fig. 5b). Therefore, we can manage the hexagonal tile data via managing the leaf nodes of a quad tree.

An arbitrary point P on the spherical surface is mapped on one of the triangular planes and for the N nodes in the nth arbitrary subdivision hierarchy, the point must be in one of the regions (L_i), i.e. Formula 3.

$$|PL_i| \leq |PL_j| \quad j = 1, 2, 3, \cdots, N; i \neq j. \tag{3}$$

From Formula 3, we find that the aggregate of all P points compose the Voronoi region of the node L_i, and the collection of all N Voronoi regions, one for each point L_i in L, constitutes the Voronoi diagram of point set L, e.g. Figure 6.

The Voronoi diagram of nodes constitutes the overlay of the regular triangles on a single triangular plane, similar to the traditional QTM (Quaternary Triangular Mesh) subdivision method. In order to accomplish quick indexing and management of the hexagonal tiles, these are managed with the assistance of a node quad tree; so we need to manage nodes through encoding. As the Voronoi diagram of a node set makes up QTM subdivision, we record the nodes' coordinates through typical QTM coding. Detailed discussion of the methods above can be found in Zhao (2003). Figure 7 is the order of the QTM coding sequence in different directions.

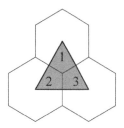 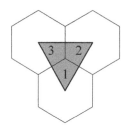 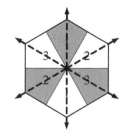

Figure 8. The order of identification code. Figure 9. Conjunct image of a hexagonal tile.

How can we convert an arbitrary point on a spherical surface to the Voronoi diagram space of node set? Here we put forward the Three Orientation Translating Algorithm and establish the three orientation geographic grid coordinates, detailed discussion of which will be presented in another article. We can convert an arbitrary point on a spherical surface to the Voronoi diagram space of the node set through the Three Orientation Translating Algorithm, and establish nodes' code indexes, whose form is the same as the code indexes of the QTM tree structure, i.e. $a_0 a_1 \ldots a_n$, among which, a_0 stands for the basic subdivision hierarchy. For an icosahedron, i.e. $a_0 \in [00, 19]$, n is the number of the subdivision hierarchy, $a_1 \rightarrow a_n$ is quaternary code(0–3) and the maximum of n is 30 (Zhao 2003). Because the encoding above only accomplishes the nodes' coding of an arbitrary point, we must add one identification code to the above codes to accomplish the conversion between a hexagonal tile and a spherical point, i.e. $a_0 a_1 \ldots a_n b$, where $b \in [1, 3]$. The identification code b shows a hexagonal tile to which the spherical arbitrary point belongs (Fig. 8).

We can position an arbitrary spherical point in the hexagonal tile through its identification code, which means a hexagonal tile is divided into six regular small tiles. For the points on different small tiles of the same hexagon, although their codes are different, they belong to the same hexagonal tile. We adopt a new, quick, indexing method to search for the codes of the six small tiles that are generated from the same node, and solve the issue in a better way. The indexing method is also a good choice for solving another problem of tile splicing of a different hierarchy. This will be described in another article. Figure 9 shows the hexagonal tile, which is made up of six small tiles (the numbers in the figure are the identification codes).

3.3 *Expression mode of spherical entity based on hexagonal grid*

In order to operate spherical entity more simply, the mode that we designed must not only be standard, but also of great advantage in designing databases of different resolution spatial data and dynamic operation of local data. In this paper, we combine the spherical entity with spherical overlay, and use the code of the hexagonal tile to manage the spherical point set, curve set and area set.

Point set: It is easy to express the point set by a hexagonal address code: however, the address code of a hexagon is not unique, as it is made up of six small tiles. The coordinates of the point must be in one tile, so the hexagonal address code of the point is specified by the address code of the small tile. In this way, we can search the other five, expressing modes through a new quick indexing method, and different mode expressions stand for different directions. Although not very useful in dynamic analysis, it is very useful to record the points' directions, and the analysis will be much easier and more effective.

Curve set: Spherical curve entity is expressed by connecting a series of points, just the same as the point set. And the coordinates are expressed by the hexagonal address code. One point in an arbitrary curve has two different directions, i.e. spherical normal direction and spherical tangent direction. We adopt two different encoding modes to record spherical curve entity and mark its directions, i.e.

(1) In the analysis of GIS, the tangent direction of every point in the curve is very important. And in the process of converting data from raster to vector format, such as the process of extracting the line symbol and sampling the point vectorization, the direction characteristics are even more important. Owning to

46

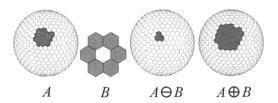

$$A \qquad B \qquad A \ominus B \qquad A \oplus B$$

Figure 10.   Erosion and dilation operation based on a spherical hexagonal grid.

the limitation of the traditional vector and raster data modes, many scholars use chain-coding, curvature scale, etc. to record the tangent direction of the points in a curve as an assisting tool. But in this way, complexity increases and efficiency decreases. Besides, it cannot record the coordinates and direction at the same time;

(2) In some special domains, such as the spherical Voronoi diagram generating process, which will be mentioned in the following paragraph, it is more important to record the spherical normal direction of a curve. So we use the small tile's normal direction code instead of the hexagonal tile's address code, and show the dilative direction of every point. It is of great advantages to the following process.

Each of the two modes of expression for spherical curve sets has its own use. We can also design different encoding modes based on specific demands. All these different encoding modes can be quickly converted one to another by the indexing method. We will discuss the issue in another article.

Area set: Spherical area entity is represented as a clockwise surrounding boundary line and a series of hexagonal cell address codes. Compared with the traditional vector data expression, it is easy to infer spatial relationships. This kind of two dimensional entity expressing method not only possesses the traditional surrounding boundary and regular grid expressing modes' advantages, but also abides by the spherical geometric characteristics of digital area (Zhao 2002). The area boundary adopts the encoding mode of curve entity, choosing different directions according to specific purposes, and the paper adopts encoding mode two above. The inner regular grid adopts the encoding mode of the point set.

4   GENERATION AND ANALYSIS OF SPHERICAL VORONOI DIAGRAMS BASED ON HEXAGONAL DGG

4.1   *Generation algorithm of spherical Voronoi diagrams*

The generating concept for spherical Voronoi diagrams is based on the hexagonal grid, according to the principles of regional dilation in mathematical morphology (Chen 2002, Zhao 2002, Chui 2002). It defines two basic operations on the spherical hexagon. If A is a primary region of the spherical hexagonal grid and B is a structure element, Figure 10 shows the operating definition of erosion and dilation on a spherical hexagonal grid.

$$\text{Erosion} \qquad A \ominus B = \cup b \in BA_b. \tag{4}$$

$$\text{Dilation} \qquad A \oplus B = \cap b \in BA_b.$$

The concrete algorithm for generating the spherical Voronoi diagram is as follows:

(1) Extract the spatial vector data from the spherical surface, and then convert them into address codes. For points, lines or areas, the ultimate operation is an aggregate of lines. In this case, we adopt the second encoding method to record the lines' normal direction.

(2) Using these normal directions, we can easily find the address codes of the hexagon tiles in the grads direction. Actually, these codes are different from the identification codes. So what we should do next is to convert them to the normal direction codes.

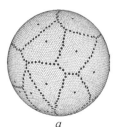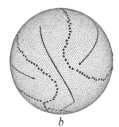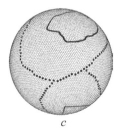

a   b   c   d

Figure 11.   The Voronoi diagram of different spherical sets based on a hexagonal grid. *a*: V diagram of point-set; *b*: V diagram of curve-set; *c*: V diagram of area-set; *d*: V diagram of composite set.

Table 1.   Experimental data.

	Points	Lines	Areas
Point V Diagram	19	0	0
Curve V Diagram	0	4	0
Area V Diagram	0	0	5
Composite V Diagram	1	1	7

(3) Delete the hexagon tiles that have the same codes and then a new spherical surface and a new verge will be produced through inflating or decaying. The new verge has the same distance as the original one.

(4) Repeat the steps above until the inflation (or decay) area of different spherical surface entities intersect with the others. This step will stop once the codes (except the identification codes) of different entities become the same, or different codes belong to the same hexagon tile.

(5) Repeat the steps from (1) to (4), until the whole sphere has been completely searched and the intersecting hexagon tiles produce a new verge. And then use the first method to generate the spherical Voronoi diagram.

4.2   *Experiment and analysis*

Based on the principles above, we developed a relevant system on the three-dimensional visual platform of OpenGL, using VC++. The computer's configuration was: PIV 1.5 GHz, memory 512 MB. We adopted the hexagonal tiles address coding method to calculate the Voronoi diagram of different entities. In the experiments, we used a mixed collection, as shown in Table 1. The level n is 6 ($n = 6$). The experimental results are shown in Figure 11. The experiments prove that the spherical Voronoi diagram based on hexagonal grids satisfies the precision of the theoretical spherical Voronoi diagram. The time required is related to the initial aggregate and the levels of the grid. So for the same precision, we can adjust the grid levels to control the time required.

As a prime algorithm for spherical Voronoi diagrams based on a hexagonal grid, the concept is easy to understand, the structure is simple and the levels are legible. After having been operated, the calculation result can construct the dynamic data format (Tong, 2006).

Compared with the generation algorithm for Voronoi diagrams based on the QTM triangular grid, this algorithm has many advantages.

First of all, thanks to the isotropy of the hexagonal grid, the inflation (or decay) result is closer to the ideal model, and it is easy to operate.

Secondly, by using the Snyder equal area projection, all the hexagonal tiles have the same area. And by using the morphologic mathematical operation method, the result is more accurate. Theoretically, the error of the Voronoi diagram can be strictly controlled in a hexagonal grid.

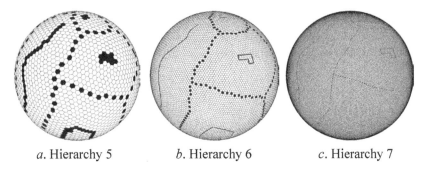

a. Hierarchy 5   *b*. Hierarchy 6   *c*. Hierarchy 7

Figure 12. The Voronoi diagrams of different subdivision hierarchies (real line and isolated point are the original sets; dashed is the generating spherical Voronoi diagram).

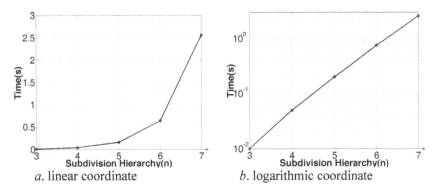

a. linear coordinate   *b*. logarithmic coordinate

Figure 13. The time required to generate Voronoi Diagrams from different hierarchical subdivisions.

In order to analyse the influence of hexagonal grids of different size on the generating speed, we adopted the composite aggregate (e.g. 1 point, 1 arc, 7 close curved faces) to operate at different hexagon levels. The subdivision hierarchy n was 3, 4, 5, 6, 7, respectively, taking three hierarchies as examples, and the spherical Voronoi diagrams for different subdivision hierarchies are shown in Figure 12. The time required is shown in Figure 13.

The dynamic stability of the Voronoi diagram makes it a useful tool to solve the global dynamic data model. In this paper, we give the generating algorithm for spherical Voronoi diagrams of different sets based on the truncated icosahedron hexagonal discrete global grid. After analysing the experimental results, we come to the following conclusions:

(1) The concept of the algorithm for the spherical Voronoi diagram based on a hexagonal grid is very intuitive. In addition, it is easy to operate and extend. Regardless of whether points, lines or areas are used, it consumes the same calculating time.

(2) The calculating time of the algorithm is mainly decided by the chosen subdivision hierarchy. With an increase in hierarchy, the consumed time follows an exponentially increasing trend (Fig. 13). For low precision spherical surface data, we can adopt a lower subdivision grid hierarchy to reduce the calculating time and the precision can still be satisfactory. For high precision spherical surface data, we can adopt a multi-layer hexagonal grid structure (that is, from high resolution to low resolution) to gradually approximate spherical Voronoi diagrams of different sets.

(3) When we operate with a large quantity of data, including data with complex original aggregates and those with excessive levels, the algorithm mentioned in this paper has some limitations. The

prime reason is that, with the increase in the scale of the aggregates, the consumed time takes on an exponentially increasing trend, which is disadvantageous to the generation of the spherical Voronoi diagram of large quantity data, and which is just what we have to study in the near future.

REFERENCES

Augenbaum, M. 1985. On the construction of the Voronoi mesh on a sphere. *Computational Physics* 59: 177–192.

Baumgardner, J.R. & Frederickson, P.O. 1985. Icosahedral discretization of the two–sphere. *SIAM Journal of Numerical Analysis* 22/1: 1107–1114.

Ben Jin, 2005. *A Study of the Theory and Algorithms of Discrete Global Grid Data Model for Geospatial Information Management.* PhD thesis. Zhengzhou: Institute of Surveying and Mapping.

Chen Jun, 2002. *Voronoi Dynamic Spatial Data Mode.* Beijing: Editorial Department of Acta Geodaetica et Cartography Sinica.

Chui Yi, 2002. *The Method and Application of Mathematical Morphology.* Beijing: Editorial Department of Science.

Conway, J.H. & Sloane, N.J.A. 1998. *Sphere packings, lattices, and groups.* New York: Springer–Verlag.

Randall, D.A., Ringler, T.D. & Heikes, R.P. 2002. Climate modeling with spherical geodesic grids. *Computing In Science & Engineering* Sep/Oct: 32–41.

Fekete, G. & Treinish, L. 1990. Sphere quadtrees: A new data structure to support the visualization of spherically distributed data. *SPIE, Extracting Meaning from Complex Data: Processing, Display, Interaction* 1259: 242–250.

Frisch, U., Hasslacher, B. & Pomeau, Y. 1986. Latticegas automata for the Navier–Stokes equations. *Physics Review Letters* 56: 1505–8.

Golay, J.E. 1969. Hexagonal parallel pattern transformations. *IEEE Transactions on Computers* C–18(8): 733–9.

Heikes, R. & Randall, D.A. 1995a. Numerical integration of the shallow–water equations on a twisted icosahedral grid. Part I: Basic design and results of tests. *Monthly Weather Review* 123(6): 1862–1880.

Heikes, R. and Randall, D.A. 1995b. Numerical integration of the shallow–water equations on a twisted icosahedral grid. Part II: A detailed description of the grid and an analysis of numerical accuracy. *Monthly Weather Review* 123(6): 1881–1887.

Hui, L. & Baoyan, D. 1999. The Voronoi Diagram of Random Plane. *Journal of Xidian University* 26(1): 118–123.

Renka, R.J. 1997. Delaunay Triangulation and Voronoi Diagram on the Surface of a Sphere. *ACM Transactions on Mathematical Software* 23(3), Sep: 416–434.

Snyder, J.P. 1992. An Equal-Area Map Projection for Polyhedral Globes. *Cartographica* 29(1): 10–21.

Sahr, K., White, D. & Kimerling, A.J. 2003. Geodesic Discrete Global Grid Systems. *Cartography and Geographic Information Science* 30(2): 121–134.

Sahr, K., White, D. & Kimerling, A.J. 2003. Geodesic Discrete Global Grid Systems. *Cartography and Geographic Information Science* 30(2): 121–134.

Song, L., Kimerling, A.J. & Sahr. K. 2002. Developing an equal area global grid by small circle subdivision. In M.F. Goodchild & A.J. Kimerling (eds), *Discrete global grids: A web book.* Santa Barbara: University of California. http://www.ncgia.ucsb.edu/globalgridsbook.

Thuburn, J. 1997. A PV–based shallow–water model on a hexagonal–icosahedral grid. *Monthly Weather Review* 125: 2328–2347.

Tong Xiaochong. 2006. *The Construct of Digital Space of Global Multiresolution Grid System and the Study of Index Mechanism.* PhD thesis. Zhengzhou: Institute of Surveying and Mapping.

White, D., Kimerling, A.J., Sahr, K. & Song, L. 1998. Comparing area and shape distortion on polyhedralbased recursive partitions of the sphere. *International Journal of Geographical Information Science* 12: 805–27.

White, D. 2000. Global grids from recursive diamond subdivisions of the surface of an octahedron or icosahedron. *Environmental Monitoring and Assessment* 64(1): 93–103.

Xu Yanfeng & Yang Boting. 1995. The Extend of Spherical Voronoi Diagram. *Journal of Engineering Mathematics* 12(2): 93–96.

Zhao Xuesheng & Chen Jun. 2002. QTM–based Algorithm for the Generating of Voronoi Diagram for Spherical Objects. *Acta Geodaetica et Cartographica Sinica* 31(2): 158–163.

Zhao Xuesheng & Chen Jun. 2003. Fast Translating Algorithm between QTM Code and Longitude/Latitude Coordination. *Acta Geodaetica et Cartographica Sinica* 32(3): 272–277.

Advances in Spatio-Temporal Analysis – Tang et al. (eds)
© 2008 Taylor & Francis Group, London, ISBN 978-0-415-40630-7

A fuzzy spatial region model based on flou set

Qiangyuan Yu & Dayou Liu
College of Computer Science and Technology, Jilin University, ChangChun, China
Key Laboratory of Symbolic Computation and Knowledge Engineering of Ministry of Education,
Jilin University, ChangChun, China

Jianzhong Chen
Department of Computing, Imperial College London, London, UK

ABSTRACT: Uncertainty modelling for geometric data is currently an important problem in spatial reasoning, geographic information systems (GIS) and spatial databases. In many geographical applications, spatial regions do not always have homogeneous interiors and sharply defined boundaries but frequently their interiors and boundaries are fuzzy. This paper provides a new fuzzy spatial region model based on flou sets. To the authors' knowledge, there has been no similar work based on flou sets. This model, which is different from the traditional fuzzy model that uses membership value, uses the relative relations between point sets and represents a fuzzy region as a flou set. The model uses operational properties of flou sets to obtain the properties and relations of fuzzy regions. The fuzzy region model based on flou sets is valuable in fields such as GIS, geography and spatial databases.

1 INTRODUCTION

A spatial region is a kind of geographic entity in many fields such as geography, geology, environment and soil, etc. These entities may often be fuzzy when they are represented as spatial objects. This kind of entity is continuous in space, and two fuzzy spatial regions may share a common gradual boundary, e.g. the transition zones between a desert and a prairie, between different soil strata, or between a hill and a valley. It is hard to decide the positions of boundaries for fuzzy spatial entities, such as forests, prairies, climatic regions and habitats of animals. This kind of spatial object is usually called a *fuzzy object*.

At present, there are two common perspectives in which spatial objects can be considered as being spatially fuzzy (Kulik 2003). One view is called *ontic fuzziness*, where the objects themselves are fuzzy. The other view is called *semantic fuzziness*, where the concepts or representations of the objects are fuzzy. Spatial fuzziness is considered as a variant of semantic fuzziness in this paper, i.e. a term, such as "forest", does not refer to a single fuzzy object in space but its spatial location will be represented as a fuzzy region.

In most current geographic information systems and spatial databases, the spatial extents of geographic objects are often modelled as sharp regions that have a unique boundary, and therefore are unable to represent and deal with fuzzy objects that have fuzzy boundaries. It has been one of the fundamental problems to define and represent fuzzy spatial regions in geographic information systems with fuzzy objects. The formalization of fuzzy regions also plays an increasingly important role in geographic information systems. In computer science and geographic information science, spatial fuzziness is often modelled by employing fuzzy approaches. Based on fuzzy set theory, fuzzy approaches use the concept of different degrees of membership to describe fuzziness.

This paper presents a fuzzy spatial region model based on flou set. It can formalize fuzzy spatial regions well and can be used conveniently in practical applications. It is different from the traditional fuzzy model that uses membership values. The flou-set model uses the relative relation between sets, and can obtain the property and relation of fuzzy spatial regions using the operational properties of flou sets. The rest of the paper is organized as follows. First, we compare the method with relevant work in Section 2. Then, the fundamental concepts and operations of flou sets are given in Section 3. In Section 4, we illuminate the detailed representation and properties of a fuzzy region model based on a flou set. A practical case is given in Section 5 and conclusions and future work are presented in Section 6.

2 RELATIVE WORK

In computer science and geographic information science, spatial fuzziness is often modelled by employing fuzzy approaches. Based on fuzzy set theory, fuzzy approaches use the concept of different degrees of membership to describe fuzziness. Most related work model fuzzy regions as a classical fuzzy set and, based on membership functions and cut sets, the relationships between fuzzy regions can be worked out.

Burrough considers fuzzy set theory as an appropriate method to deal with spatial fuzziness in mathematical or conceptual models of empirical phenomena (Burrough 1996). The fuzzy model proposed by Zhan uses fuzzy sets to represent indeterminate regions where every point is assigned a membership value within the interval [0, 1] and every α-cut level region is a determinate region (Zhan 1998). Brown applies fuzzy set theory to represent vegetation areas as a continuous spatial area (Brown 1998). Altman represents fuzzy regions and calculates the spatial relation based on operations of fuzzy sets (Altman 1994). Tinghua defines the boundary of a fuzzy region as a wide one where the degree of a certain point belonging to the thematic region is defined by a fuzzy membership function (Tinghua 1998). Molenaar and Cheng try to model fuzzy objects using a classical fuzzy set where the similarity degree is calculated based on the intersection of two fuzzy sets (Molenaar & Cheng 2000). Based on the fact that the spatial objects in a GIS thematic map as used in practice are usually plotted according to predefined thresholds of attribute membership degree, Wenbao and Min built up the conformation description mode, i.e. the boundary, interior and exterior of a fuzzy region (Wenbao & Min 2002). Guesgen and Albrecht use fuzzy techniques to model imprecise and qualitative spatial relations between geographic objects (Guesgen & Albrecht 2000). Schneider introduces fuzzy points, fuzzy lines, and fuzzy regions (Schneider 1999) and studies their algebraic and topological properties on a discrete geometric structure (Schneider 2003).

In many practical applications, it is difficult to acquire precise data, and the membership value of objects to the fuzzy concept is also not easy to ascertain. So in some cases the traditional fuzzy model is not suitable. The fuzzy region model based on flou sets discussed in this paper is suitable for the fuzzy character of spatial data in applications. The idea and result of this model is compatible with the cognition of human beings. It can model and analyse classical fuzzy regions and crisp regions uniformly.

3 FLOU SET

Flou set stems from some linguistic considerations of Yves Gentilhomme about the vocabulary of a natural language (Gentilhomme 1968). A flou set is now often used to model fuzziness and uncertainty. The fundamental concepts, properties and operations of the flou set are introduced as follows (Kerre 1999).

Definition 1: A flou set in a universe U is a pair (E, F) of subsets of U such that $E \subseteq F$. E is called the certain zone, F is called the maximal zone and $F \setminus E$ is called the flou zone (see Fig. 1). The class of flou sets in a universe U will be denoted by $FS(U)$, namely $FS(U) = \{ (E, F) | E \subseteq U, F \subseteq U, E \subseteq F \}$.

A flou set $A = (E, F)$ can be explained as follows: E is the set of centre elements in A, $F \setminus E$ is the set of surrounding elements, and the elements in E are said to be more belonging to A than the elements in $F \setminus E$.

For example, let U be the set of all possible heights expressed in cm for a person, e.g. an interval [0, 300]. The class of high persons could then be represented as a flou set ([180, 300], [160, 300]) indicating that every height ≥ 180 can be accepted as tall, every height < 160 is considered as not tall, and every $160 \leq$ height < 180 is classified as doubtful with respect to the label tall.

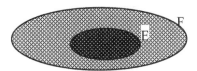

Figure 1. Illustration of a flou set.

Two flou sets $A = (E, F)$ and $B = (E', F')$ in $FS(U)$ may be provided with the following binary (AND and OR) and unary (NOT) operations:

$$A \cap B = (E, F) \cap (E', F') = (E \cap E', F \cap F')$$
$$A \cup B = (E, F) \cup (E', F') = (E \cup E', F \cup F')$$

$$\overline{A} = \overline{(E, F)} = (\overline{F}, \overline{E})$$

One can easily verify that $(FS(U), \cap, \cup, ^-)$ is a Morgan algebra, in which the smallest element is $(\emptyset, \emptyset)$ and the largest element is (U, U).

The corresponding order relation on $FS(U)$ is defined as

$$(E, F) \subseteq (E', F') \Leftrightarrow (E \subseteq E', F \subseteq F').$$

Definition 2: An *m-flou set* in a universe U ($m \geq 2$) is an m-tuple $(E_1, E_2, \cdots, E_m)$ of subsets of U such that $E_1 \subseteq E_2 \subseteq \cdots \subseteq E_m$. The class of m-flou sets in a universe U will be denoted by $FS_m(U)$.

The class $FS_m(U)$ of all m-flou sets in U can be endowed with the following binary and unary operations:

$$A \cap B = (E_1, E_2, \cdots, E_m) \cap (F_1, F_2, \cdots, F_m) = (E_1 \cap F_1, E_2 \cap F_2, \cdots, E_m \cap F_m)$$
$$A \cup B = (E_1, E_2, \cdots, E_m) \cup (F_1, F_2, \cdots, F_m) = (E_1 \cup F_1, E_2 \cup F_2, \cdots, E_m \cup F_m)$$

$$\overline{A} = \overline{(E_1, E_2, \cdots, E_m)} = (\overline{E_m}, \cdots, \overline{E_2}, \overline{E_1})$$

It is obvious that $(FS_m(U), \cap, \cup, ^-)$ is also a Morgan algebra. The corresponding order relation on $FS_m(U)$ is defined as

$$(E_1, E_2, \cdots, E_m) \subseteq (F_1, F_2, \cdots, F_m) \Leftrightarrow \forall i \in \{1, \cdots, m\}(E_i \subseteq F_i).$$

4 FUZZY REGION MODEL

4.1 *Fuzzy region representation*

In some applications, for a fuzzy instance such as a forest, which can be characterized by several different criteria, we can get different sharp regions to represent these regions. Any sharp region may be regarded as a crisp description for the fuzzy concept of a forest. Thus, a forest fuzzy region is characterized as a family of sharp regions.

Here, we use a flou set to represent a fuzzy region, and the subsets (sharp regions) of a flou set must satisfy nested relations from smaller to larger. When some sharp regions of a fuzzy region are not included in each other and all regions share at least a common intersection, we can construct a new family of nested sharp regions that will form a flou set.

If a fuzzy region is described as finite sharp regions D_i, $i \in \{1, 2, \cdots, m\}$ that share a common intersection, we can construct the same number of nested regions E_i, $i \in \{1, 2, \cdots, m\}$ in the following manner.

$$E_i = \bigcap_{\{j_1, \cdots, j_i\} \subseteq \{1, \cdots, m\}} \left(\bigcup_{k=1}^{i} D_{j_k} \right) \tag{1}$$

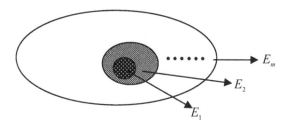

Figure 2.   A fuzzy spatial region based on a flou set.

For example, in the case m = 3, we would obtain three new regions from three sharp regions D_1, D_2, D_3 that have a common subset.

$$E_1 = D_1 \cap D_2 \cap D_3$$
$$E_2 = (D_1 \cup D_2) \cap (D_1 \cup D_3) \cap (D_2 \cup D_3)$$
$$E_3 = D_1 \cup D_2 \cup D_3$$

These new nested subsets form a flou set $A = (E_1, E_2, \cdots, E_m)$, which can be used to represent a fuzzy region of a fuzzy concept as shown in Figure 2. In the following, we assume the *core* of a fuzzy region A is $Core(A) = E_1$ and the *hull* of A is $Hull(A) = E_m$. After fuzzy regions are modelled as flou sets, we can obtain new fuzzy regions from the original fuzzy regions using the operations of intersection, union and complement on the flou set. We can also deduce the spatial relationships between fuzzy regions through the relations on the flou set. Cohn's Egg-Yolk model (Cohn & Gotts 1996) can be regarded as a special case of the flou set model, where it is a fuzzy spatial region model based on a flou set when m equals 2.

The alpha-cut of fuzzy set theory is a geometric structure specified by the above definitions of fuzzy regions. Let $\tilde{A}$ be a fuzzy set on universe U, $\alpha \in [0, 1]$, then the alpha-cuts of $\tilde{A}$ is $A_\alpha = \{u | \mu_{\tilde{A}}(u) \geq \alpha, u \in U\}$.

For $1 = \alpha_1 > \alpha_2 > \cdots > \alpha_n = 0$, it is obvious that $A_{\alpha_1} \subseteq A_{\alpha_2} \subseteq \cdots \subseteq A_{\alpha_n}$, then $A = (A_{\alpha_1}, A_{\alpha_2}, \cdots, A_{\alpha_n})$ is a fuzzy region that is described by a flou set. The sharp regions A_1 and A_0 are the core and hull of fuzzy region A, respectively.

4.2   Properties of fuzzy region

In this subsection, some properties and relations of fuzzy regions based on a flou set will be presented.

Definition 3: For two fuzzy regions $A = (E_1, E_2, \cdots, E_m)$ and $B = (F_1, F_2, \cdots, F_m)$, the equality relation can be described as $\forall i \in \{1, 2, \cdots, m\} [E_i = F_i] \Rightarrow A = B$, namely a fuzzy region is uniquely determined by the corresponding flou set.

Definition 4: For two fuzzy regions $A = (E_1, E_2, \cdots, E_m)$ and $B = (F_1, F_2, \cdots, F_m)$, the contained relation is defined as $\forall i \in \{1, 2, \cdots, m\} [E_i \subseteq F_i] \Rightarrow A \subseteq B$, namely the contained relation of two fuzzy regions is equivalent to that of the corresponding flou sets.

In some applications of fuzzy regions, the characterization of fuzzy regions by sharp regions allows us to compare points in a relative manner regarding their degree of membership of a fuzzy region. It is not necessary to decide an absolute membership value of an individual point but a relative value. Here, the relative membership values can be easily obtained by comparing the relative membership relations of two points in a fuzzy region.

Definition 5: The *relative membership value* of two points P and Q in a fuzzy region $A = (E_1, E_2, \cdots, E_m)$ can be defined as follows:

A point P belongs to a fuzzy region A at least to the same degree as a point Q, if $\forall i \in \{1, 2, \cdots, m\} [Q \in E_i \Rightarrow P \in E_i]$, which is denoted by $P \geq_A Q$.

A point P belongs to a fuzzy region A at most to the same degree as a point Q, if $\forall i \in \{1, 2, \cdots, m\} [P \in E_i \Rightarrow Q \in E_i]$, which is denoted by $P \leq_A Q$.

Point P and Q belongs to the same degree to a fuzzy region A, if $\forall i \in \{1, 2, \cdots, m\} [P \in E_i \Leftrightarrow Q \in E_i]$, which is denoted by $P \approx_A Q$.

A point P belongs more to a fuzzy region A than a point Q, if $(P \geq_A Q) \wedge \neg (Q \geq_A P)$, which is denoted by $P >_A Q$.

For the relative membership value of points in a fuzzy region, we introduce the following axioms.

$$\forall A \forall P \forall Q [P >_A Q \Leftrightarrow \exists i \in \{1, 2, \cdots, m\}[P \in E_i \wedge Q \notin E_i]]$$

$$\forall A \forall P \forall Q [P \geq_A Q \vee Q \geq_A P]$$

$$\forall A \forall P \forall Q [P \in Core(A) \Rightarrow P \geq_A Q]$$

$$\forall A \forall P \forall Q [P \notin Hull(A) \Rightarrow Q \geq_A P]$$

For some tasks, such as the visualization of a fuzzy region, an absolute representation of fuzziness is necessary. An absolute representation assigns a fixed value to every point of a fuzzy region. The absolute membership values of individual points in a fuzzy region can be calculated with the properties of a flou set. Here we assume that a fuzzy region consists of a set of finite sharp regions, namely the dimension m of the flou set is a finite number.

In the following, two alternatives of obtaining an absolute representation of fuzziness are presented. In the first method, we assume that a fuzzy region consists of a set of finite sharp regions. The degree of membership of a point P associated with a fuzzy region $A = (E_1, E_2, \cdots, E_m)$ (denoted as $\mu_A(P)$) can be characterized by the number of subsets of a flou set that contain the point, relative to the number of all subsets of a flou set.

$$\mu_A(P) = ||E_i|P \in E_i, i \in \{1, 2, \ldots, m\}||/m \tag{2}$$

Using the above equation, we have

$$\begin{cases} \mu_A(P) = 1 & \text{if } P \in E_1; \\ \mu_A(P) = 1/m & \text{if } P \in E_m \wedge P \notin E_{m-1}; \\ \mu_A(P) = 0 & \text{if } P \notin E_m \end{cases}$$

Alternatively, another method is using the Hausdorff metric of sets to calculate the membership value of a point in a fuzzy region. This metric is zero if and only if the two sets are identical. To characterize the Hausdorff metric, we first introduce the ε-neighbourhood. The ε-neighbourhood of a set T, abbreviated as $N_\varepsilon(T)$, is the set of all points with a distance smaller than ε, namely

$$N_\varepsilon(T) = \{P | \exists Q \in T[dist(P, Q)] < \varepsilon\},$$

where $dist$ () is a distance function between two points.

The Hausdorff metric H of two sets T_1 and T_2 is defined as the following equation:

$$H(T_1, T_2) = \inf\{\varepsilon | T_1 \subset N_\varepsilon(T_2) \wedge T_2 \subset N_\varepsilon(T_1)\}$$

The absolute membership value of a point P in a fuzzy region $A = (E_1, E_2, \cdots, E_m)$ is defined by the following equation:

$$\mu_A(P) = 1 - \inf\left\{\frac{H(E_i, E_1)}{H(E_m, E_1)} | P \in E_i, i \in \{1, 2, \ldots m\}\right\}$$

For a fuzzy spatial region, the result of the second alternative may be different to that of the first alternative.

Table 1. The values of copper and iron.

P	C (mg)	I (mg)
1	[267, 272]	[955, 1020]
2	[221, 225]	[1280, 1378]
3	[228, 234]	[913, 998]
4	[240, 245]	[1000, 1080]
5	[175, 180]	[1206, 1398]
6	[257, 262]	[2188, 2256]
7	[222, 227]	[239, 336]
8	[231, 236]	[1256, 1372]
9	[216, 221]	[958, 1052]
10	[218, 222]	[1080, 1116]
11	[212, 218]	[1138, 1216]
12	[107, 114]	[3166, 3289]
13	[213, 219]	[332, 424]
14	[167, 172]	[1732, 1812]
15	[207, 212]	[744, 868]
16	[224, 230]	[1198, 1286]
17	[235, 239]	[1244, 1368]
18	[261, 267]	[976, 1064]

5 A PRACTICAL CASE

We have applied the fuzzy region model in a system that synthesizes information for minerals prognosis. This system extracts and synthesizes the characteristic information of multi-source data in certain areas, which include geography, geology, geophysics, geochemistry, remote sensing and so on.

In this application, the observed spatial data are not often precise, but rather vary in an interval range. The data may also contain errors to some extent, such as in data of chemical elements, aeromagnetic and gravitational surveys. These make the degree of the spatial data satisfy the fuzzy concepts to be an interval value, and the region determined by the fuzzy concepts is an undetermined fuzzy region as well.

In the process of minerals prognosis, data on the chemical elements copper and iron are sometimes interval values. The content of copper and iron in the soil may imply the mineral. Let us illustrate a fuzzy region defined by a fuzzy concept as "target area", according to the content of copper and iron. The target area will exhibit relatively high contents of copper and iron. The sample points (P) are $\{P_1, P_2, \cdots, P_{18}\}$ and the values of copper (C) and iron (I) are given in Table 1.

Because the concept "Target area" is fuzzy, three sharp regions of target area are estimated by three individual experts, respectively, as follows and as shown in Figure 3:

$$D_1 = \{3, 8, 9, 10, 11, 15, 16\}$$
$$D_2 = \{2, 3, 4, 5, 8, 9, 10\}$$
$$D_3 = \{4, 9, 10, 11, 15, 16, 17\}.$$

Now we can further represent the target area as a fuzzy region by transforming the above three sharp regions according to equation (1). Then, we can get three new sharp regions that will form a flou set.

$$E_1 = D_1 \cap D_2 \cap D_3 = \{9, 10\}$$
$$E_2 = (D_1 \cup D_2) \cap (D_1 \cup D_3) \cap (D_2 \cup D_3) = \{3, 4, 8, 9, 10, 11, 15, 16\}$$
$$E_3 = D_1 \cup D_2 \cup D_3 = \{2, 3, 4, 5, 8, 9, 10, 11, 15, 16, 17\}$$

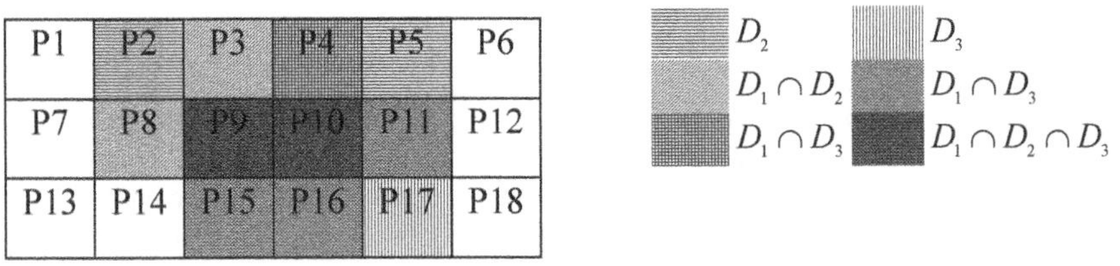

P1	P2	P3	P4	P5	P6
P7	P8	P9	P10	P11	P12
P13	P14	P15	P16	P17	P18

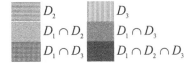

Figure 3. Three sharp regions estimated by three individual experts.

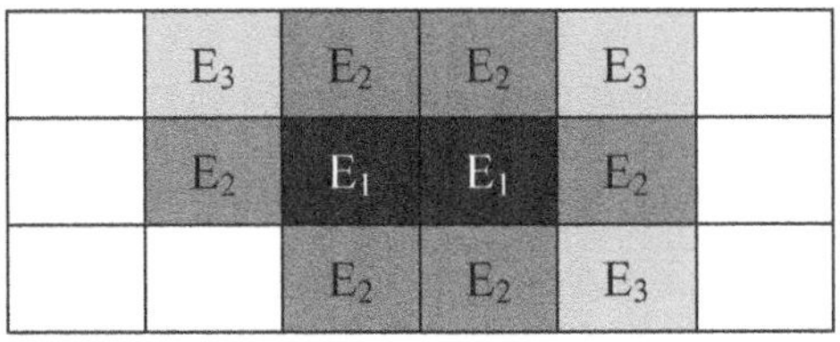

	E_3	E_2	E_2	E_3	
	E_2	E_1	E_1	E_2	
		E_2	E_2	E_3	

Figure 4. The fuzzy region of the target area represented with a flou set.

We thus use a flou set (E_1, E_2, E_3) to represent the fuzzy target area, as in Figure 4. Its core is E_1 and its hull is E_3. For the visualization task, we may compute the absolute membership values for sample points of the fuzzy concept "Target area" using equation (2). The absolute membership result for the eighteen sample points is

$$\{0, 1/3, 2/3, 2/3, 1/3, 0, 0, 2/3, 1, 1, 2/3, 0, 0, 0, 2/3, 2/3, 1/3, 0\}$$

In the synthesis of information for minerals prognosis, we represent the fuzzy regions as flou sets and we can use these flou sets to obtain the property and relation of fuzzy regions using the operational property of the flou set.

In the traditional fuzzy models, the membership functions associated with fuzzy concepts are often hard to determine and the numerical values for individual points are not necessary in some cases. In the above case, we can see that rather than using the membership values directly, the flou set model employs the relative relations between point sets to represent a fuzzy region as a flou set. One advantage of this model is that it is easy to model a fuzzy region using different views of fuzzy concepts. Furthermore, we can analyse topological relations between fuzzy regions based on flou sets. Such models can also be well used in the case of contour maps because many contour lines will form the sharp regions of a fuzzy region and they satisfy the nested relation from smaller to larger as a flou set.

6 CONCLUSIONS AND FUTURE WORK

To solve the problem of modelling fuzzy objects in GIS applications and spatial databases, this paper presents a fuzzy region model based on flou sets. It is different from the traditional fuzzy model that uses membership values and uses the relative relation of sets to represent a fuzzy region as a flou set. The flou-set model uses the calculation properties of flou sets to obtain the properties and relative relations. In this paper, we analyse the possible attributes of a fuzzy region and present two alternatives to calculate relative membership values and absolute membership values. The results of this model in the synthesis of information for minerals prognosis show that it is valuable in fields such as GIS, geography and spatial databases.

Future work includes: modelling high order fuzziness, analysing the topological relations among fuzzy regions based on flou sets and the relations among fuzzy regions based on different dimensions. In addition, the application of this model in spatial information systems and spatial databases is still a long-term goal.

ACKNOWLEDGEMENT

The research is supported by Youth Foundation of Jilin University, NSFC Major Program/China under Grant No.60496321, NSFC/China under Grant No. 60373098, the National High-Tech Research and Development Plan of China under Grant No.2003AA118020, the Major Program of Science and Technology Development Plan of Jilin Province under Grant No. 20020303, the Science and Technology Development Plan of Jilin Province under Grant No. 20030523.

REFERENCES

Altman, D. 1994. Fuzzy set theoretic approaches for handling imprecision in spatial analysis. *International Journal of Geographical Information Systems* 8: 271–289.

Brown, D. 1998. Classification and Boundary Vagueness in Mapping Presettlement Forest Types. *International Journal of Geographical Information Science* 2: 105–129.

Burrough, P. 1996. Natural Objects with Indeterminate Boundaries. In*Geographic Objects with Indeterminate Boundaries*: 3–28. London: Taylor & Francis.

Cohn, A.G. & Gotts, N.M. 1996. The 'Egg-Yolk' Representation of Regions with Indeterminate Boundaries. In P.A Burrough & AU Frank (eds),*Geographic Objects with Indeterminate Boundaries*: 171–187. London: Taylor & Francis.

Gentilhomme, Y. 1968. Les ensembles flous en linguistique. *Cahiers Linguistique Theoretique Appliqee* 5: 47–63.

Guesgen, H. & Albrecht, J. 2000. Imprecise Reasoning in Geographic Information Systems. *Fuzzy Sets and Systems* 1: 121–131.

Kerre, E. 1999. *Fuzzy sets and Approximate Reasoning*: 28–32. Xi'an: Xi'an Jiaotong University Press.

Kulik, L. 2003. Spatial Vagueness and Second-Order Vagueness. *Spatial Cognition & Computation* 2-3: 157–183.

Molenaar, M. & Cheng, T. 2000. Fuzzy spatial objects and their dynamics. *ISPRS Journal of Photogrammetry & Remote Sensing* 55: 164–175.

Schneider, M. 1999. Uncertainty Management for Spatial Data in Databases: Fuzzy Spatial Data Types. In *Advances in Spatial Databases. Lecture Notes in Computer Science, Vol. 1651*: 330–351. Berlin: Springer-Verlag.

Schneider, M. 2003. Design and Implementation of Finite Resolution Crisp and Fuzzy Spatial Objects. *Data and Knowledge Engineering* 1: 81–108.

Tinghua, A. 1998. A topological relation description for spatial objects with uncertainty boundaries. In *Spatial Information Science, Technology and its Applications*: 394–398. Wuhan: Wuhan Technical University of Surveying and Mapping Press.

Wenbao, L. & Min. D. 2002. Analyzing spatial uncertainty of geographical region in GIS. *Journal of Remote Sensing* 6: 45–49.

Zhan, F. 1998. Approximate Analysis of Binary Topological Relations between Geographic Regions With Indeterminate Boundaries. *Soft Computing* 2: 28–34.

Spatio-temporal modelling

Advances in Spatio-Temporal Analysis – Tang et al. (eds)
© 2008 Taylor & Francis Group, London, ISBN 978-0-415-40630-7

A spatial database extension of analytical fields representing dynamic and continuous phenomena*

Bin Chen, Fengru Huang & Yu Fang
Institute of Remote Sensing and Geographic Information System, Peking University, Beijing, China

ABSTRACT: In GIS, geographic information can be represented in object or field views. Field modelling is widely applied in many natural and social sciences. In this paper, we introduce a new method to represent spatial phenomena as dynamic and continuous fields and extend spatial SQL for field querying. The issues of structure, data types, operating predicates and functions of analytical fields are discussed. The spatial database extension of analytical fields can facilitate representation, transaction, analyses and display of those continuously distributing and dynamically changing geographic phenomena.

1 INTRODUCTION

In GIS, object and field models are used to represent geographic information of the real world (Goodchild 1989). Objects describe specific geographic entities that have specific boundaries and can be associated with points, lines and polygons, while fields represent geographic phenomena that usually occupy continuous regions and change over time, such as environmental pollution diffusion, distribution of rainfall, gravitational fields, urban heat islands and distribution of population. Dynamic continuous field modelling is widely applied in many natural and social research fields. In many research domains such as environmental sciences, it is important to study and construct field models, which are more suitable to represent their targets. Most of these models are correlative to space and time and are represented by a series of continuous functions. Integrating the ability of GIS to these scientific models for better analysis and prediction forms important research and application aspects in GIS.

The representation, storage and manipulation of object-based geographic information have been well studied to model and manipulate discrete geographic objects (Egenhofer 1994, Fang et al. 1999, Huang & Lin 1999, OGC 1999). However, little research has been carried out on those of field-based models. Recently, increasing attention has been devoted to represent and manage field-based spatial data due to the increasing description and query requirements of continuous geographic phenomena (Yuan 2001, Yuan & Macintosh 2002, Cova & Goodchild 2002, Laurini 2004a, b). At present, spatial phenomena that have field characteristics are mostly represented by discrete data structures such as grids, raster images and irregular point sets (Goodchild 1997, Guo et al. 2000). These structures sample a field into a static matrix; the continuity information about the field variable is lost and the temporal aspect cannot be recorded. Thus, it is insufficient to use a grid or raster to analyse and predict the distribution and evolving process of spatio-temporal phenomena. Database technology has been used for GIS data management, but most spatial data supported by a database are vector structures based on an object model (Egenhofer 1994, Fang et al. 1999, Huang & Lin 1999) and the specification has been formed for object-based data (OGC 1999). However, no specification or standard on how to represent, store, query and manipulate field-based spatial data in a database has been commonly accepted. Thus it is not possible to take advantage of the mature mechanism of DBMS to manage spatial data represented by a field-based model, including data storage, manipulation, security and query, in a non-procedural manner.

---

* Project 40501052 supported by NSFC

Recently, increasing attention has been devoted to the research of continuous fields. However, when considering how to represent dynamic continuous fields in a database and improve query and process ability on field data types, little research has been carried out. Only in 2004, when Laurini etc. developed the visual query language Phenomena, were continuous fields used to represent, query and manipulate geographic phenomenon through defining an extension of the standard OpenGIS SQL, which allowed users to query experimental data by following a SQL-like SELECT-FROM-WHERE scheme (Laurini et al. 2004a), and a validity time period attribute was added to continuous fields (Laurini et al. 2004b), but they did not deal with dynamic fields, which are changing over time. In order to better represent and formalize spatial phenomena, which are continuously distributing and dynamically changing over time, and support the complicated spatio-temporal query and analysis, this paper introduces a new method to represent dynamic and continuous fields and extends SQL to support spatial-temporal querying on field data. We construct an analytical field data type and a series of continuous spatio-temporal operating functions, and then extend them in an ORDB.

The remainder of this paper is organized as follows. In Section 2, we present the conceptual definitions of dynamic and continuous fields in detail, including the operating functions and predicates. In Section 3, we describe the spatial database extension of analytical dynamic continuous field, (ACDF), and explain the extended ADCF data type and the operating functions in categories. In Section 4, we demonstrate the use of the ADCF type and SQL function extensions by several query examples. Section 5 contains some final remarks and a short discussion on future work.

2 DEFINITIONS OF A DYNAMIC AND CONTINUOUS FIELD (DCF)

A dynamic and continuous field offers a solution for representation and manipulation of geographic phenomena that occupy continuous regions and change over time. In order to support better representation and more complicated spatio-temporal queries on those dynamically and continuously changing spatial phenomena, the following framework is defined.

A dynamic continuous field can be considered as a geographic variable, which is changing over time and distributing continuously within a restrained 2-dimensional region. The distribution of a dynamic continuous field intensity can be represented by the function $f : D \times T \to \Re$, where $D \subseteq \Re^2$ represents a restrained 2-dimensional region, $T \subseteq N$ represents a limited subset of natural numbers indicating the time period, and $f(D \times T) \subseteq \Re$ represents a valid value set of dynamic continuous field intensity. Thus, we define a ternary framework to represent dynamic continuous fields, that is, $\psi = (D, T, F)$, where:

- D is the 2-dimension domain of the dynamic continuous field,
- T is the life period of the dynamic continuous field; it evolves during T,
- F is a ternary function $f(x, y, t)$ whose definition domain is $D \times T$ and value domain is real number; the structure of f is an analytical expression.

There follows some basic but important rules defined on dynamic continuous fields and their operations.

Two dynamic continuous fields $\psi_1 = (D_1, T_1, F_1)$ and $\psi_2 = (D_2, T_2, F_2)$ are called "spatially compatible" if and only if $D_1 = D_2$; they are called "temporally compatible" if and only if $T_1 = T_2$. Two dynamic and continuous fields are "compatible" if they are not only "spatially compatible" but also "temporally compatible".

A series of operations and function are defined to compute dynamic and continuous fields:

- Numeral addition: $Grow(\psi, r) \equiv (D, T, F + r)$;
- Numeral multiplication: $Amplify(\psi, k) \equiv (D, T, k \cdot F)$;
- Addition of two compatible dynamic and continuous fields:

$$Add(\psi_1, \psi_2) \equiv (D_1, T_1, F_1 + F_2);$$

- Multiplication of two compatible dynamic and continuous fields:

$$Multiply(\psi_1, \psi_2) \equiv (D_1, T_1, F_1 \cdot F_2);$$

- Spatial union of two temporally compatible dynamic and continuous fields:

$$SpatialUnion(\psi_1, \psi_2) \equiv (D_1 \cup D_2, T_1, F_1 \cup F_2), \text{ where } D_1 \cap D_2 = \phi;$$

- Temporal union of two spatially compatible dynamic and continuous fields:

$$TemporalUnion(\psi_1, \psi_2) \equiv (D_1, T_1 \cup T_2, F_1 \cup F_2), \text{ where } T_1 \cap T_2 = \phi;$$

The following operating functions and predicates defined on field intensity can provide great support for complex spatio-temporal analysis on dynamic continuous fields.

- Average field intensity at time T and within the range of geometry object G, there are 3 situations:
$Intensity(\psi, t, G) \equiv F(G.x, G.y, t)$, where G represents a geometry object of point data type,
$Intensity(\psi, t, G) \equiv \int_G F(x(u), y(u), t)du / \int_G du$, where G represents a geometry object of line data type,
$Intensity(\psi, t, G) \equiv \int_G \int F(x, y, t)dxdy / \int_G \int dxdy$, where G represents a geometry object of polygon data type.
- Average field intensity during a provided time period T_0:
$Intensity(\psi, T_0, G) \equiv \sum_{t \in T'} Intensity(\psi, t, G)/|T'|$
- Maximum field intensity at time T and within the range of geometry object G,
$IntensityMax(\psi, t, G) \equiv F(o' \cdot x, o' \cdot y, t) : \forall o \in G(F(o \cdot x, o \cdot y, t) \leq F(o' \cdot x, o' \cdot y, t));$
- Minimum field intensity at time T and within the range of geometry object G:
$IntensityMin(\psi, t, G) \equiv F(o' \cdot x, o' \cdot y, t) : \forall o \in G(F(o \cdot x, o \cdot y, t) \geq F(o' \cdot x, o' \cdot y, t)).$
- Value series of average field intensity during a provided time period T_0:
$IntensitySeries(\psi, T_0, G) \equiv \{\langle t, Intensity(\psi, t, G) \rangle | t \in T_0\};$
- Value series of maximum/minimum field intensity at time T and within the range of geometry object G:
$IntensityMaxSeries(\psi, T_0, G) \equiv \{\langle t, IntensityMax(\psi, t, G) \rangle | t \in T_0\}$ and
$IntensityMinSeries(\psi, T_0, G) \equiv \{\langle t, IntensityMin(\psi, t, G) \rangle | t \in T_0\}.$
- The time when all of the field intensity values within the range of geometry object G have exceeded a provided value v:
$WhenAllExceed(\psi, G, v) \equiv \{t | IntensityMin(\psi, t, G) \geq v\};$
- The time when any of the field intensity values within the range of geometry object G exceeds a provided value v:
$WhenAnyExceed(\psi, G, v) \equiv \{t | IntensityMax(\psi, t, G) \geq v\};$
- The time when the average field intensity values within the range of geometry object G exceeds a provided value v:
$WhenAnyExceed(\psi, G, v) \equiv \{t | Intensity(\psi, t, G) \geq v\}.$
- Accordingly, we can define a series of functions returning field intensity lower than a provided value v:
$WhenLower(\psi, G, v).$

On the basis of the dynamic continuous field model and the related definitions we have described above, we implement the SQL extension of dynamic continuous field data type and its operating functions in a spatial database management system.

3 SPATIAL DATABASE EXTENSION OF ANALYTICAL DYNAMIC CONTINUOUS FIELD

We use an analytical structure to represent dynamic continuous fields and call it "Analytical Dynamic Continuous Field (ADCF)". Based on the analytical structure representation of dynamic continuous fields we have defined, we can make use of some models from specific research domains and make a query on a dynamic continuous field more directly. The spatial database extensions of data storage, query and manipulation of ADCF support dynamic analysis on spatial phenomena with continuous space and changing over the time in application systems.

We extend ADCF based on a typical object-oriented relational database management system: PostgreSQL. PostgreSQL was developed from the POSTGRES software package of the University of California, Berkeley. With decades of development, PostgreSQL is now the most advanced open source database management system in the world and provides support on SQL92, SQL99 and SQL3, provides parallel control on multiple versions, and supports almost all SQL components, including sub-query, transaction and user defined types and functions. PostgreSQL can be extended in many ways including data types, functions, operators, aggregate functions, index and procedural languages (PostgreSQL Global Development Group 1996–2003).

Our work of SQL extension of dynamic continuous fields includes two parts: analytical dynamic continuous field data type extension and operation functions extension. ADCF extended data type contains the three components included in the dynamic continuous field framework defined above. Among these three items, D, the 2-dimension domain of the dynamic continuous field, is represented by a rectangular region whose borders are parallel to the coordinate axes; T, the time period of dynamic continuous field evolvement, is represented by an integer set; F, the field intensity function, is represented by a text expression which contains 3 independent variables: x, y, t. By assigning values to each item, we define an analytical dynamic continuous field instance. "ADCF:sin(x*y)*t, x = −1..1, y = −1..1, t = 1..10" is an example of an ADCF instance.

Moreover, we define a series of operating functions for the purpose of management, computation, query and spatio-temporal analysis on extended ADCF-based data. These operators and functions include several categories:

- Functions of field intensity value calculation: *intensity(ADCF, timestamp, geometry)* returns a real value indicating the average field intensity value of "ADCF" within the boundary of "geometry" at the specific time "timestamp", and "geometry" can be all of the geometry data types in PostgreSQL including point, line and polygon data types. IntensityTimeAverage (ADCF, timestamp1, timestamp2) returns a real value indicating the average field intensity value of the whole field "ADCF" at the time period between "timestamp1" and "timestamp2". Functions *intensity_min(ADCF, timestamp, geometry)* and *intensity_max(ADCF, timestamp, geometry)* return a real value indicating the minimum and maximum field intensity value within the boundary of "geometry" at the specific time "timestamp", respectively.
- Functions of the change in field intensity through the life span of the field: intensity_series(ADCF, timespan, geometry) returns a series of average field intensity values within the boundary of "geometry", each value being the field intensity value at each moment during the specific "timespan". Functions intensity_min_series(ADCF, timespan, geometry) and intensity_max_series(ADCF, timespan, geometry) return the minimum and maximum field intensity values within the boundary of "geometry" during the specific "timespan", respectively.
- Analytical Functions on field intensity evolution: *when_all_exceed(ADCF, real, geometry)* returns an integer value indicating the first time when all of the field intensity values of the field "ADCF" within the boundary of "geometry" exceed the specific "real" value; *when_any_exceed(ADCF, real, geometry)* returns an integer value indicating the first time when any field intensity value of the field "ADCF" within the boundary of "geometry" exceeds the specific "real" value; *when_average_exceed(ADCF, real, geometry)* returns an integer value indicating the first time when the average field intensity value of the field "ADCF" within the boundary of "geometry" exceeds the specific "real" value. In addition, we define functions when_all_lower, when_any_lower and when_average_lower to analyse the opposite situations.
- Functions of compatible judgement on 2 fields: *is_compatible(ADCF1, ADCF2)* returns a boolean value indicating if ADCF1 is compatible with ADCF2, and return True only if both *is_spatial_compatible(ADCF1, ADCF2)* and *is_time_compatible(ADCF1, ADCF2)* are true; *is_spatial_compatible(ADCF1, ADCF2)* returns a boolean value indicating if ADCF1 is spatially compatible with ADCF2; *is_time_compatible(ADCF1, ADCF2)* returns a boolean value indicating if ADCF1 is temporally compatible with ADCF2.

- Functions of field operation: *amplify(ADCF, real)* returns a new ADCF, which is "real" times the previous "ADCF". *grow(ADCF, real)* returns a new ADCF, which is the field "ADCF" added to a certain value "real". Functions *add(ADCF1, ADCF2), minus(ADCF1, ADCF2)* and *multiply(ADCF1, ADCF2)* return a new ADCF resulting from the field "ADCF1" plus, minus or multiplied by the other field "ADCF2". Functions *spatial_union(ADCF1, ADCF2)* and *time_union(ADCF1, ADCF2)* accomplish the spatial union of two temporally compatible dynamic fields and the temporal union of two spatially compatible dynamic fields, respectively.
- Functions of field update: *update_function(ADCF, string)* updates the field "ADCF", returns a new ADCF instance, which has a new field intensity function expression, a new spatial domain or a new time period, or combination of any two of these three factors, or even all of them.

These extended operating predicates and functions can fulfil the management, access and spatio-temporal analysis of dynamic continuous field data. Moreover, they integrate geometry data types in the spatial database and the extended ADCF data type; we can make integrative transactions on vector data and field data from the SQL level and have a unified way to represent and handle spatial information based on field models.

4 QUERY EXAMPLES OF ADCF SQL EXTENSIONS

With the ADCF extension in the PostgreSQL database management system, we can use the ADCF extended data type to represent those spatial phenomena based on dynamic continuous fields and carry out spatio-temporal analysis by using the extended operating functions. The following are examples of the use of ADCF SQL extensions.

In an experimental application for pollutant diffusion forecasting, a city area with several environmental stations and residential districts was studied. The city is polluted by different kinds of pollutants. The environmental stations and residential districts are represented by point and polygon objects, respectively. We use dynamic continuous fields to represent the pollutants' spatial distribution and temporal resolution. In PostgreSQL, we can query and forecast the pollutant diffusion using the ADCF extended SQL.

- Get SO2 intensity of all of the monitor stations at 1:00
  ADCF extended SQL: select m.name, intensity(p.extend, m.location, 1)
  from monitor_station m, pollution p
  where p.name = 'SO2'
- Get SO2 intensity of all of the districts at 8:00
  ADCF extended SQL: select d.name, intensity (p.extend, d.location, 8)
  from district d, pollution p
  where p.name = 'SO2'
- When does the SO2 intensity of every district begin to exceed 0.85?
  ADCF extended SQL: select d.name, when_average_exceed(p.extend, 0.85, d.location)
  from district d, pollution p
  where p.name = 'SO2';
- Which district is the most heavily polluted at 9:00?
  ADCF extended SQL: select d.name, p.name, intensity(p.extend, d.location, 9)
  From district d, polution p
  where intensity(p.extend, d.location, 9)>=all(select intensity(p.extend, d.location, 9)
  from district d, polution p2
  where p.name=p2.name)

5 CONCLUSIONS AND FURTHER WORK

In the present paper, we have discussed the feasibility and the implementation of the spatial database extension of Analytical Dynamic Continuous Fields (ADCF). The use of the ADCF SQL extension has also been illustrated. We implement the spatial database extension of Analytical Dynamic Continuous Field through a method of extending data type and functions on an object-oriented relation database management system, based on our definition of a dynamic and continuous field and its operating functions and predicates. This paper provides a feasible and effective way to deal with such issues as how to represent the dynamically and continuously changing spatial phenomena in databases and perform spatio-temporal analysis on them.

This paper demonstrates the SQL extended functions of ADCF through several examples. The results indicate that the ADCF extension in the spatial database helps the use of analytical expressions to represent dynamic and continuous field models. It is an effective way for representation and manipulation on field based model data. Using this method, we can make integrative transactions on vector data and field data and have a unified way to represent and handle spatial information based on the field model. In addition, extended ADCF can be integrated closely with other spatial data types present in current spatial databases. SQL extension of ADCF helps us solve spatial information transactions with question-oriented and non-procedural problems. The implementing of the spatial database extension of ADCF will greatly improve the ability of representation, transaction, analyses and display of those continuously distributing and dynamically changing geographic objects and phenomena.

Several aspects will be focused on in our future work. First, further efforts will be made to enhance the symbolic representation capability of the extended ADCF through integrating the ADCF extended spatial database with other mathematical software tools. Second, more work will be done to simulate the spatial analyses procedures and display of ADCF results, and to enrich the extended functions to improve the analysis ability through augmenting spatial and temporal relationship operations. Finally, we plan to construct a uniform description framework to represent continuous fields and discrete fields of previous research, to further represent field-based and object-based models. Based on such a uniform framework, a spatial-temporal extended database prototype is being developed.

REFERENCES

Cova, T.J. & Goodchild, M.F. 2002. Extending geographical representation to include fields of spatial objects. *Geographical Information Science* Vol. 16, No.6: 509–532.
Egenhofer, M.J. 1994. Spatial SQL: A Query and Presentation Language. *IEEE Transactions on Knowledge and Data Engineering* 6(1): 86–94.
Fang, Y., Chu, F. & Chen, B. 1999. Spatial Structural Query Language—G/SQL. *Journal of Image and Graphics* Vol. 4 (A), No. 11: 901–910.
Goodchild, M.F. 1989. Modelling error in objects and fields. In *Accuracy of Spatial Databases[C]*: 107–113. London: Taylor & Francis.
Goodchild, M.F. 1997. Representing Fields. *Core Curriculum in GIScience*. Santa Barbara: National Center for Geographic Information and Analysis (NCGIA).
Guo, D., Hu, Z. & Chen, Y. 2000. Study on Uncertainties of Spatial Objects in GIS. *Journal of China University of Mining & Technology* Vol. 29, No.1: 20–24.
Huang, B. & Lin, H. 1999. GeoSQL: a Visual Spatial SQL. *Journal of Wuhan Technical University of Surveying and Mapping* Vol. 24, No. 3: 199–203.
OGC. 1999. OpenGIS Simple Feature Specification for SQL 1.1, http://www.opengis.org/docs/99-049.pdf.
PostgreSQL Global Development Group, 1996–2003, PostgreSQL 8.0.0beta1 Document.
Laurini, R., Paolino, L. & Sebillo, M. 2004a. Dealing with Geographic Continuous Fields – the Way to a Visual GIS Environment. *ACM*: 336–343.
Laurini, R., Paolino, L. & Sebillo, M. 2004b. A Spatial SQL Extension for Continuous Field Querying. *Proceedings of the 28th Annual International Computer Software and Applications Conference, IEEE*.
Yuan, M. 2001. Representing Complex Geographic Phenomena in GIS. *Cartography and Geographic Information Science* 28(2): 83–96.
Yuan, M. & McIntosh, J. 2002. A typology of spatiotemporal information queries. In K. Shaw, R. Ladner & M. Abdelguerfi (eds), *Mining Spatiotemporal Information Systems*: 63–82. Dordrecht, Kluwer Academic Publishers.

Advances in Spatio-Temporal Analysis – Tang et al. (eds)
© 2008 Taylor & Francis Group, London, ISBN 978-0-415-40630-7

Design and implementation of a unified spatio-temporal data model

Peiquan Jin, Lihua Yue & Yuchang Gong
Department of Computer Science and Technology, University of Science and Technology of China, Hefei, China

ABSTRACT: A spatio-temporal data model deals with the representation and manipulation of spatio-temporal data. So far, many spatio-temporal data models have been proposed, but most of them are geared towards specific applications and few are for general purposes. Aiming to provide general support for spatio-temporal applications in DBMS, our idea is to design a unified spatio-temporal data model and then to implement a general spatio-temporal DBMS. In this paper, we study several issues concerning the unified spatio-temporal data model, including the data structure, the querying operations and the implementation issues. The spatio-temporal changes of spatio-temporal objects are represented by spatio-temporal data types, and queries on spatio-temporal data and changes are supported by the operations defined on those data types. The unified spatio-temporal data model has been partly implemented on Informix, and a real example on the China Historical Geographic Information System (CHGIS) is discussed. The experimental results show that the unified spatio-temporal data model is able to represent, store and query spatio-temporal changes successfully.

1 INTRODUCTION

Spatio-temporal data models deal with the representation and operation of spatio-temporal data, and they have assumed an important role in the research on spatio-temporal databases and temporal GIS. The key and difficult issue of spatio-temporal data models is the representation and querying of spatio-temporal changes, since spatio-temporal change is the main property that separates spatio-temporal databases from spatial databases.

To represent spatio-temporal changes of spatial objects, many researchers have added timestamps to the spatial and thematic attributes of spatial objects, and many spatio-temporal data models have been developed, e.g. the spatio-temporal snapshot model (Langran 1992), the base state and amendment model (Langran 1992), the spatio-temporal cube model (Langran 1992), the spatio-temporal composite model (Langran 1992) and the spatio-temporal object model (Worboys 1994). In these models, timestamp and version were used to represent the states of spatio-temporal objects, and spatio-temporal changes could be queried through the comparison of versions. However, these models provided weak support for spatio-temporal changes, e.g. they could not represent those changes that involve more than one object (split, mergence, etc.). At the same time, the querying efficiency was very low, since in each spatio-temporal query many versions had to be referred to in order to discover the spatio-temporal changes.

Forlizzi et al. (2000) presented a spatio-temporal data model based on data types. In this model, type constructors are used to represent spatio-temporal changes. A type constructor is a function that defines new types depending upon time. Spatio-temporal changes are queried by the operations defined on type constructors. This model is more practical than other models, because it can be implemented in commercial object-relational database management systems using their extended abilities. But it still cannot model changes related to more than one object.

The event-based spatio-temporal data model (Peuquet & Duan 1995) regards each spatio-temporal change as an event, and represents spatio-temporal changes through an event list. However, it still does not support changes related to many objects. Chen & Jiang (2000) improved this model by introducing changing reasons, but their model can only support land subdivision applications and is not suitable for general purposes.

Other researchers have studied spatio-temporal data modelling technologies from an object-oriented view. Hornsby & Egenhofer (2000) presented an identity-based framework for the representation of spatio-temporal changes, but this approach cannot support changes of thematic attributes and spatial attributes that do not change any object identities. The history-graph model (Renolen 1997) utilized the OMT (Object Modelling Technology) model and presented a semantic data model for spatio-temporal information representation, but it can only represent discrete spatio-temporal changes. The same problems exist in the models developed by Yi et al. (2002) and Zheng et al. (2001).

The common limitation of all the above spatio-temporal data models is that they do not support continuous spatio-temporal changes. In Sistla et al. (1997), a moving-object data model was proposed to model the changes of moving objects, such as cars and ships. But the moving-object data model only suits those applications that need only to trace the continuous changes of locations. Cai et al. (2000) and Wang et al. (1999) researched a spatio-temporal data model based on a constraint database. The parametric rectangle model represented a spatial object as a set of parametric rectangles and the continuous changes of a spatial object were represented by the changes of all parametric rectangles that the object contained. The problem with this model is that the set of parametric rectangles is difficult to obtain in real applications. This problem exists in all data models based on a constraint database.

The problems of previous spatio-temporal data models can be summarized as follows:

(1) They did not present a unified way to represent and query spatio-temporal changes.
(2) They did not support both discrete spatio-temporal changes and continuous spatio-temporal changes.
(3) They were short of effective technologies to implement their data model (Paton et al. 1998).

To solve these problems, a new unified spatio-temporal data model is proposed in this paper. The main advantages of this data model can be described as follows:

(1) The new model is based on the object-relational data model. It uses abstract data types (ADT) to represent and query spatio-temporal changes. Therefore, the model can be transformed easily into object-relational database management systems (ORDBMS) for implementation.
(2) It can represent more types of spatio-temporal change than other models, including discrete changes and continuous changes, spatial changes and thematic changes, as well as an object's life.

The remainder of this paper is structured as follows. Section 2 discusses the data structure of the unified spatio-temporal data model. Section 3 presents the operations for manipulating spatio-temporal data. Section 4 introduces the algorithm for transforming the unified spatio-temporal data model into ORDBMS for implementation. Section 5 discusses the experiment of implementing the unified spatio-temporal data model and conclusions are presented in Section 6.

2 DATA STRUCTURES

In the unified spatio-temporal data model, a spatio-temporal object is represented as a tuple, and we define it as a spatio-temporal tuple. A spatio-temporal relation is a set of spatio-temporal tuples. The key idea of our data model is to design new abstract data types for spatio-temporal changes and new operations to query spatio-temporal changes.

Definition 1: Given n domains $D_1, D_2, \ldots, D_n$, and at least one domain being a spatio-temporal domain, every element in the product $D_1 \times D_2 \times \ldots \times D_n$ is called a spatio-temporal tuple.

Definition 2. A spatio-temporal relation is a set of spatio-temporal tuples.

The basic data structure of spatio-temporal objects is shown in Figure 1, in which the spatio-temporal attributes are extended data types representing spatio-temporal changes.

The type system of the unified spatio-temporal data model is shown in Figure 2 and is defined by the *Second-Order Signature* (Güting 1993), a tool for the formal definition of a data model. It can define not

Primary Key	Thematic Attributes	Spatiotemporal Attributes

Figure 1.   The data structure of the data model.

kinds	IDENT, DATA, SDATA, STDATA, HTSTATE, HT STREL, STTUPLE	
type constructors		
IDENT	*ident*	
DATA	*int, char, bool, real,*	
	point, line, region, composition,	
	instant, period	
SDATA	*point, line, region, composition,*	
HTSTATE	*htstate*	
HT	*mht*	
DATA→STDATA	*std, etc*	
(ident×DATA) + × (*ident×STDATA*) + × [*ident×*HT]	STTUPLE sttuple	
STTUPLE →STREL	*strel*	

Figure 2.   The type system of the data model.

only the relational data model but also other advanced data models, e.g. image data models and spatial data models.

According to the *Second-Order Signature*, a data model is defined by two signatures. The first signature defines the type system of the data model, and the second defines the operations on the type system. In this section, we discuss the type system, which is the first signature. The second signature, defining the operations on the type system, will be discussed in Section 3.

In the type system shown in Figure 2, the *kind* is a set of *types*. The *type constructor* is a function that produces new types from existing types. For example, the type constructor *std* and *stc* produce data types such as *std*(*point*) and *stc*(*region*) to represent discrete spatial changes and continuous spatial changes.

One limitation that researchers of spatio-temporal databases and temporal GIS seem to impose on their models is that objects can only be created, changed and eventually removed (Claramunt & Theriault 1995). However, this is too simplistic a view in a spatial context. Spatial entities may also split into two or more entities, and two or more entities may also be merged into a single one. There are different kinds of changes existing in the real world, and current spatio-temporal data models are usually short in completely supporting different types of changes. This is mainly because of the insufficient cognition to the real world.

According to the object-oriented view, the objects in the real world are identified by identifiers, and the state of an object is represented by its internal attributes, which consist of spatial and non-spatial attributes. The former describe the position and region occupied by an object and the latter are those attributes that are related to the applications an object is involved in, which are called *thematic attributes*. So according to the structure of an object, the spatio-temporal changes can be divided into two categories, which are: (1) *life*: means changes of object identifiers and (2) *processes*: changes of internal attributes. The terminology *life* and *process* come from Sellis (1999) and *processes* can be further classified into *spatial processes* and *thematic processes*, according to what part of a spatio-temporal entity changes.

The classification of changes so far is based on the changing contents. On the other hand, when we consider the changing styles of spatio-temporal objects, we find that it can be *continuous* or *discrete*. This has been discussed in many previous papers (Claramunt & Theriault 1995, Claramunt & Theriault 1996, Chomicki 2001). Thus, a spatio-temporal change of an object could be any one among the following five types:

(1) **TYPE 1** (*continuous spatial processes*): the spatial attributes of an object change continuously with time, such as the spread of fire, flowing of floodwaters and movement of a car. These changes are always related to a period of time.

Table 1. Corresponding data types in the unified spatio-temporal data model for the five types of spatio-temporal change.

Spatio-temporal change	Types in the unified spatio-temporal data model
TYPE 1 and TYPE 3	Types produced by the type constructor *std*
TYPE 2 and TYPE 4	Types produced by the type constructor *stc*
TYPE 5	*mht*

(2) **TYPE 2** (*discrete spatial processes*): the spatial attributes stay static during a period and suddenly change to another value. A typical example is the change in the boundary of a parcel of land. Discrete spatial processes always happen at a specific instant.

(3) **TYPE 3** (*continuous thematic processes*): the thematic attributes of an object change continuously with time, such as changes of soil type.

(4) **TYPE 4** (*discrete thematic processes*): the thematic attributes are basically static, but they may be changed into another value suddenly, e.g. the change of ownership of a parcel of land.

(5) **TYPE 5** (*discrete life*): sudden changes that result in the creation of new objects or deletion of existing objects, such as division of land or mergers of several land parcels.

The reason we do not consider the change of continuous life is that continuous life does not exist in the real world. The creation or deletion of an object always happens right away.

Most of the spatio-temporal changes fall into these five spatio-temporal classes. Since the unified spatio-temporal data model is designed to represent these five types of spatio-temporal change, the model can suit most spatio-temporal applications. Table 1 shows the corresponding types in the unified spatio-temporal data model for these five changes.

The kind *HT* contains only one const type, which is *mht*. This type is also used to represent spatio-temporal changes, but it differs from the type constructors *std* and *stc*. The types produced by the type constructors *std* and *stc* represent spatial processes and thematic processes, which only result in changes in the attributes of an object, e.g. moving of an object (in this case only the position of the object changes with time). The type *mht* represents the life cycle of a spatial object. A spatial object may be created or deleted and we use *mht* to represent such changes. The *mht* is a list of *htstate*, each of which represents one life change of the object.

The kind *IDENT* contains only one const type *ident*, which represents a domain of identifiers (these identifiers can be used as attribute names). For any spatial data type *s* in the kind *SDATA*, the type constructor *std(s)* or *stc(s)* constructs the corresponding spatio-temporal data type, which is in the kind *STDATA*, e.g. *stc(point)* and *stc(region)* are spatio-temporal types representing continuously moving points and discrete moving regions, respectively. For a type *t* in the kind *STTUPLE*, *strel(t)* is a data type in the kind *STREL*. The type *strel(t)* represents a spatio-temporal relational schema, and *t* represents a spatio-temporal tuple. The symbol S^+ means a list of *S*, thus $(ident, DATA)^+$ represents a definition list of attributes. The symbol [*S*] means *S* is optional. The following is an example of the definition of type *sttuple* and type *strel*:

```
sttuple(<(No,int),(owner,string)>,<(boundary,std(region))>,(history,
mht))
strel(sttuple(<(No,int),(owner,string)>,<(boundary,std(region))>,
(history, mht)))
```

3 DATA OPERATIONS

The operations on spatio-temporal relations are similar to relational algebra. There are five basic operations in relational algebra, and other operations can be expressed by these five operations. The following definitions are for the five basic extended relational algebraic operations in the unified spatio-temporal data model. Since the five operations form a complete set of relational algebra, other operations on spatio-temporal relation can also be implemented using these five basic operations.

(1) $R \cup^{st} S$ (*Spatio-temporal Union*): suppose R and S are two spatio-temporal relations, the spatio-temporal union of R and S is the set of R's tuples and S's tuples. Spatio-temporal union $R \cup^{st} S$ is defined as follows:

$R \cup^{st} S \underline{\underline{\Delta}} \{t | t \in R \vee t \in S\}$, where t is a variable representing any spatio-temporal tuple.

(2) $R -^{st} S$ (*Spatio-temporal Difference*): The spatio-temporal difference $R -^{st} S$ is a set containing those tupes in R but not in S.

$$R -^{st} S \underline{\underline{\Delta}} \{t | t \in R \wedge t \notin S\}$$

(3) $R \times^{st} S$ (*Spatio-temporal Product*): suppose a spatio-temporal relation R has r attributes, and S has s attributes. The spatio-temporal product of R and S is a set of spatio-temporal tuples. And each spatio-temporal tuple has $(r+s)$ attributes. The former r attributes of each spatio-temporal tuple come from R, and the latter s attributes come from S.

$$R \times^{st} S \underline{\underline{\Delta}} \{t | t =< t^r, t^s > \wedge t^r \in R \wedge t^s \notin S\}$$

(4) $\sigma_F^{st}(R)$ (*Spatio-temporal Selection*): The spatio-temporal selection makes a horizontal partition on a spatio-temporal relation based on some conditions; that means to select the spatio-temporal tuples that satisfy the given conditions. The conditions are represented by a predication formula F.

$$\sigma_F^{st}(R) \underline{\underline{\Delta}} \{t | t \in R \wedge F(t) = true\}$$

(5) $\pi_{A_{i1}, A_{i2}, \ldots, A_{im}}^{st}(R)$(*Spatio-temporal Projection*): The spatio-temporal projection makes a vertical partition on a spatio-temporal relation. Suppose the attributes of R are denoted as $(A_1, A_2, \ldots, A_r)$, the value of attribute A_i is denoted as t_{Ai}, the spatio-temporal projection $\pi_{A_{i1}, A_{i2}, \ldots, A_{im}}^{st}(R)$ is a set of tuples, each of which has m attributes.

$$\pi_{A_{i1}, A_{i2}, \ldots, A_{im}}^{st}(R)$$
$$\underline{\underline{\Delta}} \{t | t =< t_{A_{i1}}, t_{A_{i2}}, \ldots, t_{A_{im}} > \wedge < t_{A_1}, t_{A_2}, \ldots, t_{A_r} > \in R\}$$

Figure 3 shows the signature of the above operations.

In order to query spatio-temporal changes, we also need to define some other operations on spatio-temporal data types, which include temporal, spatial, life cycle and spatio-temporal operations. Figures 4 to 7 show the signatures of these operations.

Some symbols are used in Figures 4 to 7, and are described as follows:

α: zero dimension spatial data types.

β: one dimension spatial data types.

γ: two dimensions spatial data types.

ε: zero, one or two dimensions spatial data types

$\otimes$: $a \otimes b$ means $a \times b$ or $b \times a$.

types operators	strel, sttuple, bool
$\cup^{st}$	$strel \times strel \rightarrow strel$
$-^{st}$	$strel \times strel \rightarrow strel$
$\times^{st}$	$strel \times strel \rightarrow strel$
σ^{st}	$strel \times (sttuple \rightarrow bool) \rightarrow strel$
π^{st}	$strel \rightarrow strel$

Figure 3. The operations on spatio-temporal relations.

types	period, bool
operators	
coalesce	$period \times period \rightarrow period$
duplicate	$period \times period \rightarrow period$
before	$period \times period \rightarrow bool$
equals	$period \times period \rightarrow bool$
meets	$period \times period \rightarrow bool$
overlaps	$period \times period \rightarrow bool$
during	$period \times period \rightarrow bool$
starts	$period \times period \rightarrow bool$
finishes	$period \times period \rightarrow bool$

Figure 4.   The temporal operations.

types	$real, bool\ \alpha, \beta, \gamma, \varepsilon$
operators	
intersect	$\gamma \times \gamma \rightarrow \gamma$
cross	$\beta \times \beta \rightarrow \beta$
	$\beta \times \gamma \rightarrow \gamma$
union	$v \times v \rightarrow v$
center	$\beta \rightarrow \alpha$
	$\gamma \rightarrow \alpha$
disjoint	$\varepsilon \times \varepsilon \rightarrow bool$
meet	$\varepsilon \times \varepsilon \rightarrow bool$
overlap	$\gamma \times \gamma \rightarrow bool$
	$\beta \times \beta \rightarrow bool$
intersects	$\beta \otimes \gamma \rightarrow bool$
contain	$\gamma \times \alpha \rightarrow bool$
	$\gamma \times \beta \rightarrow bool$
	$\gamma \times \gamma \rightarrow bool$
	$\beta \times \alpha \rightarrow bool$
	$\beta \times \beta \rightarrow bool$
equal	$\varepsilon \times \varepsilon \rightarrow bool$
distance	$\alpha \times \alpha \rightarrow real$
area	$\gamma \rightarrow real$
perimeter	$\gamma \rightarrow real$
length	$\beta \rightarrow real$

Figure 5.   The spatial operations.

types	$instant, period, \alpha^{st}, \beta^{st}, \gamma^{st}, \varepsilon^{st}, bool$
operators	
when	$\varepsilon^{st} \times instant \rightarrow \varepsilon$
history	$\varepsilon^{st} \times period \rightarrow \varepsilon$
stDisjoint	$\varepsilon^{st} \times \varepsilon^{st} \times period \rightarrow bool$
stMeet	$\varepsilon^{st} \times \varepsilon^{st} \times period \rightarrow bool$
stOverlap	$\beta^{st} \times \beta^{st} \times period \rightarrow bool$
	$\gamma^{st} \times \gamma^{st} \times period \rightarrow bool$
stIntersects	$(\beta^{st} \otimes \gamma^{st}) \times period \rightarrow bool$
stContain	$\gamma^{st} \times \alpha^{st} \times period \rightarrow bool$
	$\gamma^{st} \times \alpha^{st} \times period \rightarrow bool$
	$\gamma^{st} \times \alpha^{st} \times period \rightarrow bool$
	$\beta^{st} \times \alpha^{st} \times period \rightarrow bool$
	$\beta^{st} \times \beta^{st} \times period \rightarrow bool$
stEqual	$\varepsilon^{st} \times \varepsilon^{st} \times period \rightarrow bool$

Figure 6.   The spatio-temporal operations.

types	htstate, mht, instant, period
operators	
htWhen	$mht \times instant \rightarrow htstate$
htHistory	$mht \times period \rightarrow mht$

Figure 7.   The life cycle operations.

72

4 TRANSFORMING THE UNIFIED SPATIO-TEMPORAL DATA MODEL INTO ORDBMS

The unified spatio-temporal data model can be transformed into object-relational database management systems (ORDBMS) for implementation. Actually, in recent years, more and more people advocate that the ORDBMS is the appropriate fundamental DBMS to implement a spatio-temporal DBMS (Forlizzi et al. 2000, Yang et al. 2000, Cindy & Zaniolo 2000), because the ORDBMS offers extended facilities for users to add support for complex data management. An ORDBMS provides extensibility of User Defined Types (UDT) and User Defined Routines (UDR) (Stonebraker & Moor 1996). The following algorithm transforms the unified spatio-temporal data model into a set of UDTs and UDRs in an ORDBMS.

Algorithm 1: Transforming the unified spatio-temporal data model into ORDBMS
Input: The type system (as shown in Fig.2) and the operations system (as shown in Figures 3 to 7) of the unified spatio-temporal data model.
Output: Set U and R, where U is the set of UDTs and R is the set of UDRs.
Preliminary: The kind set in the type system is denoted as K, while the operation system is denoted as F. Both U and R are initially empty.

```
1  BEGIN
// The types in kinds IDENT, STTUPLE ands STREL already
// exist in ORDBMS. We needn't transform them.
2  X: = K- IDENT, STTUPLE, STREL, STDATA;
3  For Each S In K Do
4    For Each p In S Do
5         U: = U ∪ {p};
6    End For
7  End For
// Transforming operations
8  For Each q In F Do
9    R: = R ∪ {q};
10 End For
11 END
```

5 EXPERIMENT

In order to demonstrate the implemental feasibility of the unified spatio-temporal data model, we made some experiments on the Informix Dynamic Server 9.21, which is an ORDBMS. We implemented a spatio-temporal data management system on Informix using the extended facilities it provides. The implemented spatio-temporal data management system can represent and store spatio-temporal data. It can also query spatio-temporal changes effectively. We used the data set from the China Historical Geographic Information System (CHGIS) (The CHGIS Project 2003). This data set contains the boundary changes and location changes of each county in Shanghai District since AD 751.

First, for this experimental data set, we created two spatio-temporal relations to store the boundary changes and location changes, as shown below.

```
His_Boundary(no int, name char(50), boundary region, life period)
His_Location(no int, name char(50), location point, life period)
```

The original data set is stored in Mapinfo format, so we first needed to transform the data into the spatio-temporal database in Informix. This was done using the COM interfaces provided by MapInfo. Figure 8 shows the final data transformed into the spatio-temporal database.

Since we build our spatio-temporal database on the ORDBMS Informix, we can directly use standard SQL to access the spatio-temporal database. The underlying interface used was ODBC. By employing spatio-temporal operations defined in Figures 4 to 7, we can query spatio-temporal changes. For instance,

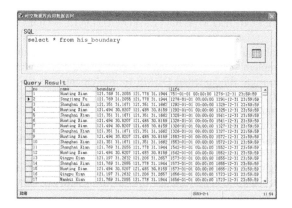

Figure 8.   Spatio-temporal data transformed from CHGIS.

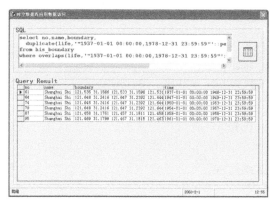

Figure 9.   Find the boundary changes of Shanghai from 1937 to 1978.

if we want to find the boundary changes of Shanghai from 1937 to 1978, we can issue the following SQL statements to obtain the result:

```
SELECT no, name, boundary, duplicate(life,'"1937-01-01
00:00:00,1978-12-31 23:59:59"'::period) as time
```

```
FROM his_boundary
```

```
WHERE overlaps(life,'"1937-01-01 00:00:00,1978-12-31
23:59:59"'::period) and name = 'Shanghai Shi'
```

Figure 9 shows the result of the SQL query.

6   CONCLUSIONS

A unified spatio-temporal data model that can support different types of spatio-temporal changes is the foundation of spatio-temporal data management. In this paper, we present a unified spatio-temporal data model based on the object-relational data model. According to the new model, spatio-temporal objects are represented as spatio-temporal relations and spatio-temporal tuples, and querying operations are realized by extended relational algebra. Spatio-temporal changes are represented by extended spatio-temporal data types and queried through the operations defined on spatio-temporal data types. By defining new spatio-temporal data types, we can enhance the data model with better support for spatio-temporal changes. At the same time, this data model can be easily transformed into a typical ORDBMS for implementation. We have made some experiments on the ORDBMS-based implementation of the unified spatio-temporal data model and the results show it is a practical way to implement a spatio-temporal DBMS on an ORDBMS.

ACKNOWLEDGMENTS

This work was partially supported by the National Science Foundation of China under grant number 60403020 and by the High Technology Development Project of China grant 2002AA783055 from the Ministry of Technology of China.

The authors would like to thank the CHGIS project hosted by Harvard University and Fudan University for providing downloadable spatio-temporal datasets.

REFERENCES

Cai, M., Keshwani, D. & Revesz, P. 2000. Parametric rectangles: A model for querying and animation of spatio-temporal databases. *Proceedings of the 7th International Conference on Extending Database Technology*: 430–444.
Chen, J. & Jiang, J. 2000. An event–based approach to spatio-temporal data modelling in land subdivision systems. *GeoInformatica* 4: 387–402.
Chomicki, J. 2001. Spatio-temporal data models and management. *ICLP 2001 Workshop on Complex Reasoning on Geographical Data, Paphos, Cyprus*.
Cindy, C. & Zaniolo, C. 2000. SQLST: A spatio-temporal data model and query language. *Proceedings of the 19th International Conference on Conceptual Modeling (ER'00), Salt Lake City, USA*: 96–111.
Claramunt, C. & Theriault, M. 1996. Toward semantics for modeling spatio-temporal processes within GIS. In M. Kraak & M. Molenaar (eds), *Proceedings of 7th International Symposium on Spatial Data Handling*: 47–63. Delft, NL: Taylor & Francis Ltd.
Claramunt, C. & Theriault, M. 1995. Managing time in GIS: an event–oriented approach. In J. Clifford & A. Tuzhilin (eds), A. *Recent Advances in Temporal Databases*: 23–42. Berlin: Springer-Verlag.
Forlizzi, L., Güting, R. H., Nardelli, E. & Schneider, M. 2000. A data model and data structures for moving objects databases. *Proceedings of SIGMOD Conference 2000*: 319–330.
Güting, R. 1993. Second-order signature: a tool for specifying data models, query processing, and optimization. *Proceedings of SIGMOD Conference 1993*: 277–286.
Hornsby, K. & Egenhofer, M. 2000. Identity-based change: a foundation for spatio-temporal knowledge representation. *International Journal of Geographical Information Science* 3: 207–224.
Langran, G. 1992. *Time in Geographic Information Systems*. Taylor & Francis Ltd.
Paton, N., Fernandes, A. & Griffiths, T. 1998. Spatio-temporal databases: contentions, components and consolidation. In A.M Tjoa et al. (eds), *Proceedings of 11th DEXA Workshop on Advanced Spatial Databases (ASDM)*, 851–855. IEEE CS Press.
Peuquet, D.J. & Duan, N. 1995. An event–based spatio-temporal data model (ESTDM) for temporal analysis of geographical data. *International Journal of Geographical Information Systems* 1: 7–24.
Renolen, A. 1997. History graphs: Conceptual modeling of spatio-temporal data. *Proceedings of GIS Frontiers in Business and Science*. Brno, Czech Republic: International Cartographic Association.
Sellis, T. 1999. CHOROCHRONOS: research on spatio-temporal database systems. *Proceedings of DEXA Workshop*: 452–456.
Sistla, A., Wolfson, O., Chamberlain, S. & Dao, S. 1997. Modeling and querying moving objects. *Proceedings of ICDE 1997*: 422–432.
Stonebraker, M. & Moor, D. 1996. *Object–Relational DBMS: The Next Great Wave*. Morgan Kaufmann. The CHGIS Project. 2003. *http://www.people.fas.harvard.edu/~chgis/*.
Wang, Y., Wang, W. & Shi, B. 1999. Interval constraint and its algebraic query language. *Chinese Journal of Computer* 22(5): 550–554.
Worboys, M. 1994. A unified model for spatial and temporal information. *The Computer Journal* 37(1): 26–34.
Yang, J., Cheng, H., Ying, C. & Widom, J. 2000. TIP: a temporal extension to Informix. *Proceedings of the ACM SIGMOD, Dallas, Texas*: 213–223.
Yi, S., Zhang, Y. & Zhou, L. 2002. A spatio-temporal data model for plane moving objects. *Journal of Software* 13(8): 1658–1665.
Zheng, K., Tan, S. & Pan, Y. 2001. A unified spatio-temporal data model based on status and change. *Journal of Software* 12(9): 1360–1365.

Advances in Spatio-Temporal Analysis – Tang et al. (eds)
© 2008 Taylor & Francis Group, London, ISBN 978-0-415-40630-7

Spatio-temporal indexing mechanism based on snapshot-increment

Lin Li, Zhangcai Yin & Haihong Zhu

School of Resource and Environment Science, Wuhan University, Wuhan, China

ABSTRACT: Spatio-temporal indexing is one of the key technologies applied to spatio-temporal database engines. Currently, the existing spatio-temporal indexing techniques are extensions of traditional time indexes or spatial indexes, where the first concern is to handle the spatial domain and the temporal feature is considered as the second issue for dealing with queries, such as RT-tree, 3D R-tree and STR-tree. In this paper, we put forward the snapshot-increment spatio-temporal indexing mechanism, focusing on efficiency in both temporal queries and spatial queries by regarding time and space as dimensions of equal importance.

1 INTRODUCTION

Spatio-temporal data management has received a lot of attention recently, and large amounts of spatio-temporal data are produced daily, so the need to efficiently analyse and query these data requires the development of sophisticated techniques (Marios et al. 2005). Spatio-temporal indexing is one of the key technologies applied to spatio-temporal database engines. Currently, research on spatio-temporal indexes is just at its preliminary stage (Huang & Peng 2002), mainly consisting of extensions of traditional time indexes or spatial indexes. In the category of spatio-temporal indexing methods, the main concern is to handle the spatial domain (Mohamed et al. 2003). Dealing with temporal queries is considered as a secondary issue, such as in RT-tree, 3D R-tree and STR-tree (Pfoser et al. 2000). 3D R-tree (Theodoridis et al. 1996), which simply treats the time as another dimension, is not optimal since spatial and temporal features should be considered differently (JeongKyu et al. 2005), and time slice queries are no longer dependent on the live entries at the query time, but on the total number of entries in the history (Mohamed et al. 2003). RT-tree (Xu et al. 1990) has many limitations in that many index items will be created with RT-tree expanding rapidly (Zhang & Yang 2003) if there are large amounts of objects in change. Temporal relationships among objects are not considered thoroughly.

On the other hand, methods of indexing based on valid-time or transaction-time have also been proposed for queries of historical information; at the same time, the efficiency of spatial queries is neglected also. Therefore, methods for spatio-temporal indexes, which consider efficiency of both temporal queries and spatial queries, have become a hot topic in current research (Luo 2002). There are some typical indexes (Liu et al. 2003), MR-tree, HR-tree, HR+-tree and MV3R-tree, which take both temporal and spatial factors into account, and use the R-tree structure to organize spatial information at any time and store spatial distributions of objects at different times. The ultimate goal is to build a separate R-tree for each time instance (Mohamed et al. 2003). However, the biggest weakness of R-tree is that MBR may overlap with each other, which leads to repeated retrieval for the same path and thus affects efficiency in retrieval (Luo 2002), and these indexes can change to tree structures when the number of moving objects becomes larger (Theodoridis et al. 1998). Moreover, they are inefficient for various queries on moving objects since they cannot capture the characteristics of moving objects (JeongKyu et al. 2005). The performance of time window queries using the MR-tree (Xu et al. 1990) is not efficient as many entries can be replicated (Mohamed et al. 2003). The Historical R-tree (HR-tree) [Nascimento & Silva 1998] is very similar to the MR-tree.

In this paper, we propose a new mechanism for snapshot-increment based spatio-temporal indexes after an exploration of the version-increment spatio-temporal data model, which focuses on efficiency in both temporal queries and spatial queries by regarding time and space as dimensions of equal importance in

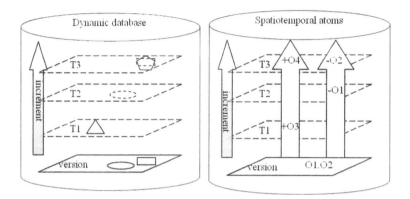

Figure 1.   The spatio-temporal data model and index based on version-increment.

spatio-temporal indexes. We also apply the mechanism for spatio-temporal queries, which are implemented based on a version-increment spatio-temporal model.

2   THE VERSION-INCREMENT BASED SPATIO-TEMPORAL DATA MODEL

2.1   *The spatio-temporal data model based on version-increment*

In the version-increment based spatio-temporal data model—"Version"— is utilized to represent states of geographical phenomena, while "increment" means changes of spatio-temporal phenomena based on version. This model can be regarded as an abstraction of the Basic States Modification Model, the Event-based Spatio-temporal Data Model (Peuquet & Duan 1995), and the Space-time Integration Model. The former two models suit raster data, while the later one splits geometrical space into segments of arcs. The new model chooses spatial features as its object, basing operations on vector data, where "version" is a state of geographical phenomena at a certain time stored in databases; "snapshot" is an instant state of geographical phenomena obtained by operations such as querying, or interpolation among versions. "Version" is the term used in terms of database storage, whereas "snapshot" is used in operations in databases; these two terms have the same meaning in nature. This paper employs the version-increment based spatio-temporal data model and the spatio-temporal indexing mechanism based on snapshot-increment to differentiate concepts of spatio-temporal data models and spatio-temporal indexes.

The geometrical space in the version-increment spatio-temporal data model cannot be further divided as it is the atom of space and time, and instead it can only change in form, such as occurring, disappearing, replacement and so on (Fig. 1). There are two atoms of objects in geometrical space: one an ellipse, the other a rectangle. At the time T1, a triangle appears – drawn in a solid line; at the time T2, the ellipse dies – drawn in a broken line; at the time T3, the rectangle dies – and is drawn in a broken line, and is replaced by a pentagon which appears – drawn in a solid line.

Occurring and disappearing are types of change, and can be integrated into basic types of spatio-temporal changes such as combination, division, replacement and so on. Division means a feature dying with many features then appearing in the same place; combination means many features dying with a single feature appearing in the same place; replacement means many features dying with many features appearing in the same place.

2.2   *The storage structure for spatio-temporal data*

Changes of features include changes in both space geometry and attributes, so three tables, i.e. the table of space geometry, the table of features and the table of attributes, will be created to save geographical information for the convenience of storing spatio-temporal data (Tables 1, 2 and 3). The table of space

Table 1. Space geometry of features.

Space geometry's identifier	Space geometry's shape	Space geometry's lifetime	Features' identifier
Geometry1			Feature1
Geometry2			Feature1
Geometry3			Feature2

Table 2. Features.

Features' identifier	The first space geometry's identifier	Features' lifetime
Feature1	Geometry1	(T11,T12)
Feature2		(T21,T22)

Table 3. Attribute of features.

Features' identifier	Attribute 1	Attributes' lifetime 1	. . . . . .
Feature1	Attr$_1$		

geometry includes space geometry's identifier, space geometry's shape, space geometry's lifetime and features' identifiers. The table of features includes features' identifiers, the first space geometry's identifier, and features' lifetime. The table of attributes includes features' identifiers, attribute name, attribute's lifetime.

When a feature has several geometries with time sequences, only the object which appears first in the geometrical space should be recorded in the table of features, while other objects will be obtained by analysing both the "features' identifier" field and the "space geometry's lifetime" field. The size of the table of features will not change when a feature's geometry changes a great many times. The feature's identifier can be utilized to search for all the objects in the feature's geographical spaces and sort them by the start-time of the lifetime of these geometries. Objects in the geometrical space with the same start-time will appear in the same time; objects in the geometrical space with the same end-time will disappear; objects in the geometrical space with one's start-time the same as the other's end-time will result in a change of replacement. This relating mechanism could also be applied to complex features, features with many objects in the geometrical space. Each feature's geometrical space is stored in disks sorted by time sequence to allow quick access to each feature's objects in the geometrical space.

The version-increment spatio-temporal model shares advantages of both the Sequence Snapshot Model and the Basic States Modification model, as it is an extension of these two models. This can be explained as follows: (1) It absorbs the advantages of the Sequence Snapshot Model, which refer to its intuitional and simple sequences, and so it is just some original representation of changes of geographical phenomena in terms of time, which makes it quite easy to determine characteristics of geographical phenomena. (2) In this model, increments could be related to each other by the "features' identifier" and the "geometrical space's lifetime" with spatial objects' topological characteristics in the time dimension involved, and therefore

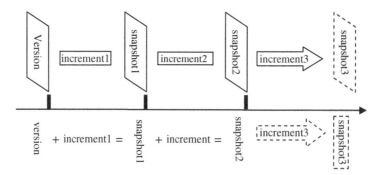

Figure 2.   Relationships among version, increment and snapshot.

it overcomes the shortcoming of no association between objects in the Sequence Snapshot Model. (3) It also has some advantages of the Basic States Modification Model by only storing changes of geographical phenomena, which reduces redundancy of data, makes it simpler and more efficient to perform time-dimension-based queries and analyses, so rules of change for spatial objects can be easily obtained. (4) It overcomes the shortcoming of the Basic States Modification Model by taking vector data as its data source.

3 SNAPSHOT-INCREMENT BASED SPATIO-TEMPORAL INDEXING MECHANISM

3.1 *The basic principles*

One feature may change its geometrical space or attributes many times during its lifetime. The geometrical space is considered as the primary atom of the spatio-temporal index and it only has two changes during its lifetime, with the first being its occurrence or replacing, the second its disappearance or being replaced. The geometrical space can be plus or minus, which is to say that both occurrence and replacing refer to the plus sign (+), while both disappearance and being replaced refer to the minus sign (−). The geometrical space in the initial version should be assumed to be neither plus nor minus (Fig. 1). Therefore, increments of each geometrical space will appear in the form of both a plus atom and a minus atom. The snapshot at any time is equal to the addition of the former snapshot and the increment between them, i.e.

$$\text{Current snapshot} = \text{the former snapshot} + \text{the increment}.$$

Version is the original snapshot (Fig. 2). Scalable Incremental hash-based Algorithm (SINA) stores the positive and negative updates during the course of execution to be sent to the clients for evaluating a set of concurrent continuous spatio-temporal queries (Mohamed et al. 2004).

The snapshot can be seen as the result of logical operations performed on atoms of the geometrical space. In Figure 1, O1 and O2 represent the ellipse and the rectangle, respectively, in the version; + O3 refers to the appearance of the triangle in the increment; −O2 and +O4 show the replacement of the ellipse with the pentagon (Table 4).

The amount of increment will grow larger and larger as time passes, which makes it rather difficult to search for the increment between the version and some snapshot; thus there is a high demand for building a spatio-temporal indexing mechanism based on increment. A spatio-temporal database is considered as a whole data set, and is regarded as a root node. The root node and other nodes include both fields: the set of geometrical spaces and its lifetime (i.e. the time interval), and make up the spatio-temporal indexing tree based on snapshot-increment. The set of geometrical spaces includes the increment data set and the initial snapshot. So, the spatio-temporal indexing mechanism based on snapshot-increment should involve the spatial index by the snapshot and the temporal index by the increment.

The increment data set in a node can be divided into many child increment data sets according to the time interval, and the child sets become the child-nodes of the father-node. Hence, we can discover which

Table 4.   Logical sum between space geometry atoms.

T0	Version	:	{O1, O2}	having O1, O2
T1	Increment	:	+ O3	O3 appears
T2	Increment	:	− O1	O1 disappears
T3	Increment	:	− O2, O4	O4 replaces O2
T3	Snapshot	:	{O3, O4}	The result

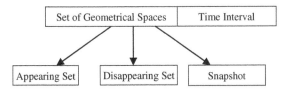

Figure 3.   The node with spatio-temporal double fields.

node a spatio-temporal object is by the time according to the index tree. On the other hand, when a spatio-temporal object is decided in a node of the index tree, we can also query the spatio-temporal object by its coordinates in the snapshot of the node. According to the index tree, we first query the spatio-temporal by one dimension of time, and then by two dimensions of space. At the same time, the temporal index by the increment and the spatial index by the snapshot can be maintained automatically.

3.2   The nodes of the spatio-temporal index

The spatio-temporal double-field node is utilized in the snapshot-increment spatio-temporal index. This node embodies characteristics of space and time and includes two fields: one is the set of geometrical spaces, the other the time interval (Fig. 3). The set of geometrical spaces can be divided into three groups: the set of identifiers of the appearing geometrical space, the set of identifiers of the disappearing geometrical space, and the snapshot at the start-time of the time interval. The former geometrical space and the latter geometrical space, which are involved in the change of replacement, correspond to the disappearing set and the appearing set, respectively. Increments between geometrical spaces in a time interval can be represented by these two sets.

3.3   The structure of the spatio-temporal index

Tree structures can be created based on nodes with double fields (Fig. 4). The snapshot-increment spatio-temporal index can determine the depth of trees by the granularity of time and automatically balance the load of nodes at the same level. Here "balance" means controlling the size of the set of geometrical spaces by adjusting the length of the time interval. The number of nodes depends on the depth of trees and the length of the time interval of nodes.

   The root node is considered as the root of the index tree. The root's time interval $[T_1, T_2]$ is the largest time granularity, and its set of geometrical spaces includes all objects in the database. The deeper the index tree becomes, the smaller the granularity of time will be and also the shorter the length of the time zone will be. However, no matter what level nodes are at, the total length of time zones of nodes at a certain level equals the total length of time zones of nodes at any other level, the total length of time zones of the root nodes. Among those nodes with the same depth, nodes whose time zone has the smallest start-time possess the same snapshot and the same start-time of time zones as its father node.

   The length of the time zone controls the number of appearing or disappearing objects of geometrical spaces. Time zones of nodes with the same depth are sorted by time, so that the time sequence is formed

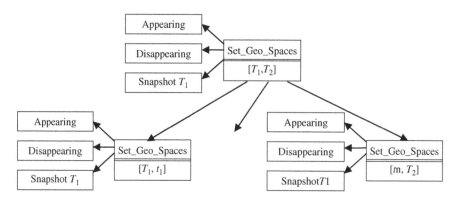

Figure 4.   The spatio-temporal index structure based on snapshot-increment.

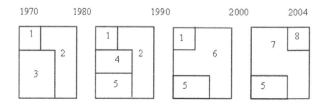

Figure 5.   Process of land subdivision.

with adjacent temporal topological relationships in which the next brother node's snapshot can be obtained logically by summing the last node's snapshot and the increment between them. The length of the time zone can be fixed or variant. If the length is fixed, as for some nodes, the amount of appearing or disappearing objects of geometrical space may become abnormally large. But if the length is variable, it can be adjusted automatically to balance the load of nodes.

4   THE QUERIES BASED ON THE SNAPSHOT-INCREMENT INDEX

4.1   *The preparation for data*

This experiment studies land subdivision (Fig. 5), organizes data using the version-increment spatio-temporal data model and implements queries for snapshots based on the snapshot-increment spatio-temporal indexing mechanism.

The MapInfo table is utilized to manage states of parcels, whose structure includes the parcel's identifier (featureID), the name of the parcel (name), valid start-time (startTime) and end-time (endTime) of the parcel. The MapBasic language is utilized for creating index tables based on the table of states (Fig. 6). For simplification, in a table of states one identifier of features corresponds to one identifier of geometrical spaces and the index table is composed of four fields, namely, operID (Identifier of the operation), featureID (Identifier of the feature), operType (Type of the operation) and operTime (the time when the feature changes) as shown in the following figure. Index tables are sorted by the time when operations occur, whose operType can have the plus sign or the minus sign, meaning appearance or disappearance, respectively.

The method for creating index tables is as follows. Each feature in the table of states has two records sorted by the time when operations are performed in the indexing table. One records the start-time and represents the appearance of the feature with corresponding operation type "+", while the other records the end-time and represents the disappearance of the feature with the corresponding operation type "−". Here, the end-time of active features can be replaced by the current time.

featureID	name	startTime	endTime
F01	1	1970-12-12	2000-12-12
F02	2	1970-12-12	1990-12-12
F03	3	1970-12-12	1980-12-12
F04	4	1980-12-12	1990-12-12
F05	5	1980-12-12	2004-12-12
F06	6	1990-12-12	2000-12-12
F07	7	2000-12-12	2004-12-12
F08	8	2000-12-12	2004-12-12

operID	featureID	operType	operTime
O01	F01	+	1970-12-12
O02	F02	+	1970-12-12
O03	F03	+	1970-12-12
O04	F03	-	1980-12-12
O05	F04	+	1980-12-12
O06	F05	+	1980-12-12
O07	F02	-	1990-12-12
O08	F04	-	1990-12-12
O09	F06	+	1990-12-12
O10	F01	-	2000-12-12
O11	F06	-	2000-12-12
O12	F07	+	2000-12-12
O13	F08	+	2000-12-12
O14	F05	-	2004-12-12
O15	F07	-	2004-12-12
O16	F08	-	2004-12-12

Figure 6.   States and index of spatio-temporal data.

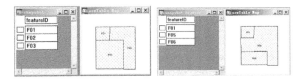

Figure 7.   Original snapshot and successive queried snapshot.

4.2 *The spatio-temporal queries*

The spatio-temporal index is built to allow efficiency in spatio-temporal queries and the convenience of spatio-temporal analyses. Spatio-temporal querying is divided into snapshot querying and evolvement querying. Traditional spatio-temporal models focus on semantic descriptions of space and time, but neglect the efficiency of spatio-temporal queries. In the light of the formal description of the index and characteristics of the indexing table, the snapshot query is made up of two parts: one the last snapshot, the other the increment. When the searched snapshot is the original one, the last snapshot will be empty. The increment query is performed according to the time of the current snapshot t2 (1979-10-10) and the time of the last snapshot t1; if the last snapshot doesn't exist, time t1 is the initial valid time of the database; here t1 is 1970-12-12 (Fig. 7). In the indexing table, records during time t1 and t2 are organized into identifiers of features with the plus or minus sign, and these identifiers and identifiers of features of snapshot t1 are involved in the logical sum operation, which affects efficiency of querying. This experiment shows the "logical sum" operation by operations of the snapshot tables. Firstly, the state of every snapshot is searched by initializing the snapshot table. Then identifiers of features with signs are extracted in the indexing table, and some operations will be performed following the rules given below: records of the identifier of features will be added when the sign is plus, and records will be deleted until all the records in the indexing table are searched for when the sign is minus. Finally, the result of the snapshot table is the feature of the snapshot searched for.

Feature's evolvement querying refers to the search for changes happening in a single feature F during the set period [t1, t2]. Parcels in change have these two characteristics: (1) parcels before changes and after changes have adjacent relationships in time, i.e. the end-time of parcels before changes is just the start-time of parcels after changes. Correspondingly, records of parcels before changes and after changes in the indexing table are also adjacent, which can help find the set of increments ΔF. For example, F3 at the time 1980-12-12 is divided into F4 and F5; operations such as "−F3", "+F4", and "+F5" share the same time, and also are adjacent in the indexing table. (2) As for parcels before and after changes, their spatial positions stay the same, but they have overlapping relationships in space, by which sets of features in change in feature F can be obtained from ΔF. As ΔF decreases the amount of features involved in the overlapping spatial relationship, a great deal of time is saved.

5 CONCLUSION

This paper proposes the version-increment based spatio-temporal data model and its snapshot-increment spatio-temporal indexing mechanism. This mechanism regards space and time as equally important dimensions and can perform quick searches based on time points and intervals, so the efficiency of spatial and temporal queries are both considered. In addition to the above advantages, there are many other merits: (1) The index mechanism is based on operations of changes; (2) It turns the spatio-temporal index into a temporal linear index, which can build multi-level indexes with the help of spatial analysis; (3) The indexing table can be created automatically; (4) The process of indexing is composed of searching and logical operations; (5) Double-direction queries can be performed. Signs of operations in opposite-direction queries will be opposite; (6) Data are separated from indexes.

However, the experiment only shows the case of land subdivision. This snapshot-increment spatio-temporal index needs to be extended into other areas of application.

REFERENCES

Huang, S. & Peng, J. 2002. Research of Index Mechanisms in Discrete Spatio–temporal Environment. *Journal of Soochow University (Engineering Science Edition)* 22(5): 7–12.
JeongKyu, L., JungHwan, O. & Sae, H. 2005. STRG-Index: Spatio-Temporal Region Graph Indexing for Large Video Databases. *SIGMOD Conference*: 718–729.
Liu, J., Yang, X., Yue, L.H. & Zhao, Z.X. 2003. A Spacialtemporal Index Based on R*-tree. *Computer Engineering* 29(14): 60–62.
Luo, X. 2002. *Spatio-temporal Object Model and Application in Cadastral Information Systems*, PhD thesis (Hubei Wuhan University in China).
Marios, H., George, K., Petko, B. & Vassilis, J. 2005. Complex Spatio-Temporal Pattern Queries. *Proceedings of the 31st VLDB Conference, Trondheim, Norway.*
Mohamed, F.M., Xiaopeng, X. & Walid, G.A. 2004. SINA: Scalable Incremental Processing of Continuous Queries in Spatiotemporal Databases. *ACM SIGMOD*: 623–634.
Mohamed, F.M., Thanaa, M.G. & Walid, G.A. 2003. Spatio-temporal Access Methods. *IEEE Data Engineering Bulletin* 26(2): 40–49.
Nascimento, M.A. & Silva, J.R.O. 1998. Towards historical R-trees. *Proc. of the ACM Symp. on Applied Computing, SAC*: 235–240.
Peuquet, D. & Duan, N. 1995. An Event-based Spatiotemporal Data Model (ESTDM) for Temporal Analysis of Geographical Data. *International Journal of Geographical Information System* 9(1): 7–24.
Pfoser, C.D., Jensen, S. & Theodoridis, Y. 2000. Novel Approaches in Query Processing for Moving Object Trajectories. *Proc. of the Intl. Conf. on Very Large Data Bases, VLDB*: 395–406.
Theodoridis, Y., Vazirgiannis, M. & Sellis, T.K. 1996. Spatio-temporal indexing for large multimedia applications. *Proceedings of IEEE ICMCS*: 441–448.
Theodoridis, Y., Sellis, T. & Papadopoulos, A. 1998. Specifications for Efficient Indexing in Spatiotemporal Database. *Capri:Proceedings of SSDBM'98(IEEE)*: 123–132.
Xu, X., Han, J. & Lu, W. 1990. RT-tree: An improved R-tree index structure for spatiotemporal databases. *4th International Symposium on Spatial Data Handling (SDH)*: 1040–1049.
Zhang, S. & Yang, Z. 2003. A Spatio-temporal Indexing Method for GIS. *Geomatics and Information Science of Wuhan University* 28(1): 51–54.

Advances in Spatio-Temporal Analysis – Tang et al. (eds)
© 2008 Taylor & Francis Group, London, ISBN 978-0-415-40630-7

Temporal topological relationships of convex spaces in Space Syntax theory

Hani Rezayan, Farid Karimipour & Mahmoud R. Delavar
Department of Surveying and Geomatic Engineering, Engineering Faculty,
University of Tehran, Tehran, Iran

Andrew U. Frank
Department of Geo-Information, Vienna University of Technology, Vienna, Austria

ABSTRACT: Time lifting hypothesizes the development of a unique and integrated basis for handling static and dynamic GI concepts by the definition of homomorphisms, or more properly functors, between static and dynamic domains. This paper studies time lifting for the convex spaces' topological relationships introduced by Space Syntax theory. This theory illustrates human settlements and societies as a strongly connected space-time relational system between convex spaces. This topological structure is represented as a connectivity graph with some morphologic properties that describe how space and time are overcome. Space Syntax theory introduces more dynamic activities at the local scale. Then the specific problem of the paper is computational modelling of the integrated static and dynamic analyses for an activity based scenario at local scale and the study of how effective these activities overcome space and time. The model is implemented using a functional programming language and validated by being executed for a simple simulated urban bus transportation system. In addition, the paper sets up questions about the computational complexity of mixed usage of static and dynamic data.

1 INTRODUCTION

Time is inherently linked to space (Egenhofer & Mark 1995). So development of an integrated basis for dealing with static and dynamic concepts can be speculated. This paper adopts the time lifting approach, introduced by Frank (2005), to develop such integration for the spatio-temporal concepts introduced by Space Syntax theory.

Time lifting hypothesizes the development of a unique and integrated basis for handling static and dynamic GI concepts by the definition of homomorphisms, or more properly functors, between static and dynamic domains. This approach is based on category theory that abstractly (free from semantics) deals with mathematical structures and the relationships between them. Time lifting follows the general trend of developing GI theory through formalistic usage of mathematics to study structures, changes and spaces.

Space Syntax theory defines human settlements and societies as a strongly connected space-time relational system between convex spaces. This topological structure is represented as a connectivity graph. In addition, the theory introduces some analyses for deriving the graph's morphologic properties, such as connectivity, control, depth and integration. These properties describe how a relational system of convex spaces overcomes space and time.

This paper time lifts the mentioned topological structure of convex spaces by developing a computational model that integrates static and dynamic analyses for an activity-based scenario at local scale, providing more dynamicity among activities rather than the global scale. The model is validated by being executed for a simple simulated urban bus transportation system.

The functional programming paradigms are introduced here as suitable bases for time lifting implementation. They support implementation of algebraic structures, polymorphism, overloading, and definition of structured types. Specifically, this paper uses the Haskell functional programming paradigm[1].

The paper is composed of seven sections. In Section 2, Space Syntax theory is reviewed. Section 3 describes the topological structure of the convex spaces introduced by Space Syntax theory. Then, temporality in Space Syntax theory is illustrated in Section 4. Section 5 describes formalization of time in GI theory and Section 6 illustrates the computational model implementation and execution. Finally, Section 7 provides the conclusions.

2   SPACE SYNTAX THEORY

Space Syntax theory is an urban design theory[2]introduced by Hillier & Hanson (1984). It intends to provide means through which we can explore the structure of human settlements and it enable us to predict and evaluate the likely effects of design alternatives.

Space Syntax theory defines space as a container of relations and interactions (Couclelis 1992, cited in Jiang et al. 2000), i.e. as a configurational entity. The theory uses this definition of space as its foundation for abstraction and integration of general properties, structures and transformations in human settlements and societies (Hillier 1996).

Urban grid is one of the central concepts in Space Syntax theory. Urban grid is the pattern of public space linking the buildings (Hillier 2001). Its is defined as the static core element of an urban system that strongly influences the long term dynamicity of the systems and movement, as the strong force that holds the whole urban system together (Hillier 2001). The theory thus represents a society as a strongly connected space-time relational system (Hillier & Netto 2001).

Space Syntax theory also provides an organic definition for a society: a society is an evolutionary abstraction imposed on space-time. In this society, space is acting as an inverted genotype. It means that the required information to reproduce cultural patterns of space is found in the spatial configurations themselves as relations/interactions. Individuals who make such an organic society (e.g. built areas and activities) are well-defined space-time entities. The spaces between these individuals are filled or overcome by the space-time relational systems, which are convex (Hillier & Netto 2001).

Considering the above-mentioned definitions, Space Syntax theory proposes that the social construction of space in human settlements is mediated by two kinds of spatial laws (Hillier 2001):

1. The local and conservative laws by which different ways of placing buildings gave rise to different spatial configurations.
2. The global and generative laws through which different spatial configurations, by their effect on movement, create different patterns of co-presence amongst people.

This proposition is adapted to the cognitive perception of space, which discriminates space at large scale and space at small scale (Egenhofer & Mark 1995). The large scale space is beyond human perception and cannot be perceived from a single point; the small scale space is presumably larger than the human body, but can be perceived from a single vantage point (Jiang et al. 2000). In Space Syntax theory, residential and cultural factors, which are variants, dominate the local scale and commercial and micro-economic factors, which are invariants, form the global scale.

---

[1] The Haskell expressions represented in this paper are italicized and preceded by " >>" symbol.

[2] Urban design is the process of giving physical design directions to urban growth, conservation and change. It sits at the interface between architecture and planning. While architecture and planning focus on artistic and socio-economic factors, designing emphasizes the physical attributes that usually restrict its scale of operation to arrangements of streets, buildings and landscapes (Batty et al. 1998).

3 TOPOLOGICAL RELATIONSHIPS OF CONVEX SPACES IN SPACE SYNTAX THEORY

Space Syntax theory describes a human settlement as a space-time relational system of convex spaces. The theory represents this topological system as a connectivity graph. A connectivity graph is generated and analyzed as follow:

1. Spatial decomposition of spatial configuration into analysis units (Hillier et al. 1987, Brown 2001):
   - *Bounded spaces* which usually correspond to functional use designations (Fig. 1a)
   - *Convex spaces* which identify the extent of spatial decomposition and usually correspond to localization of space (Fig. 1b)
   - *Axial lines* which are straight lines that identify the extent of spatial continuity and usually correspond to flows and globalization of space. Axial lines connect convex spaces based on their intervisibility (Fig. 1c)
2. Representation of the derived analysis units and their connections as a connectivity graph in which the nodes and links are the analysis units and their connections. These graphs are usually big, shallow, non-dendritic, highly integrated and ringy (with a large number of cycles) rather than tree-like (Hillier & Netto 2001). Considering the degree of linearity in the environment, Jiang et al. (2000) illustrate axial, convex and grid representations for a connectivity graph:
   - Relatively linear/axial representation, where the linear property represents the fact that the built environment is relatively dense, so the free space is stretched in one orientation at most points (e.g. a city, a town, a village or a neighbourhood) (Fig. 2).
   - Non-linear/convex representation, where the free space is partitioned into a finite number of convex spaces (e.g. internal layout of a building) (Fig. 3).

Figure 1.   Extraction of the analysis units for a town in France (Hillier et al. 1987).

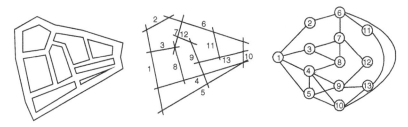

Figure 2.   Axial representation (Jiang et al. 2000).

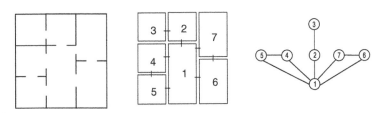

Figure 3.   Convex representation (Jiang et al. 2000).

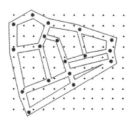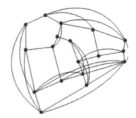

Figure 4. Grid representation (Jiang et al. 2000).

– Non-linear but with more precise spatial presentation / grid representation, where the free space is partitioned into a finite number of points and their visual fields are studied (Fig. 4). A visual field is a space wholly visible from a single vantage point. It is based on the notion of isovist (Benedikt 1979).
3. Evaluation of the graph's morphologic properties (Hillier et al. 1987, Jiang et al. 2000, Brown 2001):
   – *Connectivity value* of a node is the number of its immediate neighbours (1).

$$C_i = k \tag{1}$$

where C_i = Connectivity of ith node
         k = Immediate neighbours

– *Control value* of a node expresses the degree to which the node controls access to its immediate neighbours, taking into account the number of alternative connections of these neighbours (2).

$$ctrl_i = \sum_{j=1-k} 1/C_j \tag{2}$$

where $ctrl_i$ = Control value of ith node
         k = Connected nodes to ith node
         C_j = Connectivity of jth node

– *Depth value* is the smallest number of steps from a node to the other nodes. It is defined as the total depth and mean depth values of a node (3).

$$D_i = \sum_{j=1-n} d_{ij}$$
$$MD_i = D_i/(n-1) \tag{3}$$

where D_i = Total depth value of ith node
         d_{ij} = shortest path between ith and jth node
         n = Number of nodes
         MD_i = Mean depth value of ith node

– *Integration value* is the degree to which a node is integrated or segregated from the system. A node is more integrated if all the other nodes can be reached after traversing a small number of intervening nodes. The integration value of a node is measured as the average depth of the node to other nodes (4).

$$RA_i = 2(MD_i - 1)/(n-2) \tag{4}$$

where RA_i = Relative asymmetry/Integration value of ith node

4 TIME IN SPACE SYNTAX THEORY

Temporality in Space Syntax theory emerges from changes caused under the notion of movement (refer to Section 2). Time is investigated in this theory at two levels of granularity: global and local. During the life of a city, it changes slowly (global scale) while its related activities (interactions and co-presence) change rapidly (local scale).

4.1 *Time at global scale*

"City design is a temporal art, but it can rarely use the controlled and limited sequences of other temporal arts like music" (Lynch 1960, p. 1). The temporal sequence of cities can be reversed, interrupted, abandoned, or cut across (Lynch 1960). It is more like temporality in biology (Frank 2005).

Considering the above-mentioned general temporal characteristic of a city, Space Syntax theory also defines built environments as organic structures and deals with space-time as a genotype (refer to Section 2). Then its main spatio-temporal question at global scale would be how a society is reproduced through time by being realized in space. The answer to this question is based on knowing which structures are reproduced and overcome space, not by dealing with the structures themselves. These are invariant global patterns[3] that control the movement (Hillier 2001). These patterns emerge and change slowly. In this way, a society is defined as a continuous space-time entity (Hillier & Netto 2001).

This notion of temporality and slow change can be represented using each of the three kinds of environments described in Section 3. Considering the large extent of environments at global scale, however, the axial and convex representations can be adopted more efficiently than the grid representation.

4.2 *Time at local scale*

At local scale, we are faced with discrete and freely mobile individuals who carry out activities. No activity generates fast changes and immediate new patterns of space, but they have a certain distribution of demands on co-presence and global movement. For assessment of new activities' impacts on space, we need the range of demands they are likely to make on co-presence rather the contents of these activities (Hiller & Netto 2001).

The main question of space-time at this scale is how independent activities of large numbers of individuals / agents in different locations create the overall pattern of cities. Two main specifications of these activities are (Hiller & Netto 2001):

1. They are movable and do not accumulate into space-time to create larger forms
2. They are governed by social rules and conventions.

At this scale, temporality could be investigated more efficiently using the grid representation of environment, where grid points could represent locations of activities in space and time. Then the concepts of moving objects could be used for modelling the dynamic activities, their interactions and co-presences.

5 TIME IN GI SCIENCE

Space and society change each other. While any changes are due to time, and changes in socio-economic or natural environment attract societies' attention (especially the politicians) (Frank 1998), we should try to effectively formalize and implement time in GIS. Time and space are inherent, fundamental, and different dimensions of reality in which people live (Frank 2005, Egenhofer & Mark 1995).

---

[3] For example, Space Syntax theory introduces a global pattern known as the deformed wheel pattern (Hillier, 2001). It is based on the existence of an integration core in each city. Long roads usually are linked with the core radially and elongated outward to the edge of the city. So a deformed wheel is "a hub, spokes in all main directions, sometimes a partial rim of major lines with less integrated, usually more residential, areas in the interstices forms by the wheel" (Hillier 2001). This pattern is firstly used for explanation of movement. Also the effect of variants on a society are analysed studying the level of deformation having occurred in its deformed wheel pattern.

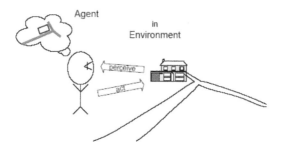

Figure 5.   Closed loop of semantics (Frank 2003).

Much research has been carried out on handling space and time so far. Although space handling is advanced to a point that dominates GIS software is easily understood and used even by novices, time handling is still a controversial topic in GI science and technology. Some reasons are:

- Lack of a comprehensive spatio-temporal ontology (Frank 2003).
- Dealing with time as a discrete or partial continuous property of the world, though our unique reality (space and time) is continuous and governed by differentiable laws (Frank 2003).
- Dominance of analytical approaches suffering from the limitation of computers' numbering systems.
- Underestimation of common behaviours of models that are used in the GI domain (e.g. network, object and field) that resulted in the creation of context-based temporal viewpoints for each of these models (Herring et al. 1990).

A general reason for these problems is lack of effective GI theory. GI theory, like theories in other sciences, should consist of a formal language and rules concerning valid (simple) relationships and facts within the language (Frank 2005). This theory should be broad and comprehensive enough to cover all domains of GI science and technology.

It is speculated that mathematics can provide the required basis for GI theory development. Mathematics is adopted in GI theory as the study of structures, changes and spaces (more than figures and numbers). Besides, GI theory should use computer sciences as the basis for hypotheses implementation and evaluation. This trend is followed in some of the recent advancements such as:

- Development of the basic form ontology for space (SNAP) and time (SPAN) by Grenon and Smith (2004) and their colleagues.
- Development of multi-tier ontology by Frank (2003).
- Development of functional approaches for GI on the basis of category theory by Herring et al. (1990), Frank (2005) and his colleagues.

5.1   *Functional approach to GI theory*

How do we interact simply in the complex real world? Functionalists believe in the existence of abstraction interfaces that absorb complexities of real worlds. We can interact with this abstract world extensionally, free from any computation. The entities of this abstract world are relations, or more properly functions, in the real world.

This functional viewpoint is employed in different domains, such as social sciences, philosophy, and architecture for formalization of concepts. In GI science, Frank (2003) follows this functional viewpoint and introduces the closed loop of semantics. It represents interactions as compositions of a series of functions that are carried out by agents in the real world (Fig. 5). These functions are (5):

$$Reality \rightarrow Observation \rightarrow Modelling \rightarrow Act \rightarrow Reality \tag{5}$$

The commutative version of the closed loop of semantics is shown in (6).

$$
\begin{array}{ccc}
Reality \rightarrow & g \rightarrow Reality \\
\downarrow & \uparrow \\
Observation \rightarrow & g' \rightarrow Act
\end{array}
\tag{6}
$$

where g = Processes/models in real world
   g' = Processes/models in agent's mind

Input to the *observation* function and output from the *act* function are the relations/functions from the real world. The g' function is a functional representation of a GIS. This function is also defined as the composition of a series of functions (7); like what is represented for TIGRIS GIS by Herring et al. (1990).

g' :: *External Modelling* $\rightarrow$ *Conceptual Modelling* $\rightarrow$ *Logical Modelling* $\rightarrow$ *Physical Modelling*   (7)

Herring et al. (1990) introduced category theory as the mathematical basis for functional representation of GI interactions. Category theory is a mathematical theory that abstractly (free from semantics) deals with mathematical structures and the relationships between them. It is an attempt to capture what is commonly found in various classes of related mathematical structures. The fundamental concepts of category theory are *categories* and *functors*.

A category is a collection of primitive element types, a set of operations upon those types, and an operator algebra that is capable of expressing the interaction between operators and elements (Herring et al. 1990). Consider category C then:

- the element types are objects of the category denoted as *obj(C)*;
- the operations are homomorphisms, denoted as *hom(C)*, between the objects. They are morphisms that preserve structures of the objects; and
- the operator algebra is a composition of these homomorphisms.

The associativity of homomorphisms' composition and the existence of identity homomorphism are axioms of a category. For example, Field category contains Fields as its objects and homomorphisms between Fields (8).

$$
\begin{array}{l}
>> m :: Field \rightarrow Field \\
>> e.g.m :: G \rightarrow H \\
>> m(u + v) = m(u) ++ m(v) \\
>> m(u^*v) = m(u)^{**}m(v)
\end{array}
\tag{8}
$$

where G and H = Fields like $(G, +, *)$ and $(H, ++, **)$ which are object of Field category
   m = A homomorphism between Fields
   u and v = Values in Field G

A functor is a homomorphism that associates elements and operations from one category with another and preserves the operator algebra (Herring et al. 1990). Functor F from category C to category D associates to each object x in C an object $F(x)$ in D and to each homomorphism f: $x \rightarrow y$ in C a homomorphism $F(f)$: $F(x) \rightarrow F(y)$. A functor also holds identity and composition association properties.

Categories are types of mathematical structures which preserve their structures during morphisms. Although many mathematical theories attempt to study a particular type of structure just by relating it to another simpler and better understood structure, category theory is used to take ideas to another simpler or even to more complex ones (two-way) by studying the structures and morphisms. For example, our approach in moving from the static to dynamic domain causes more complexity.

By defining all GI concepts as formal mathematical structures and their homomorphisms, GI theory would be formed as a collection of categories. Then any kind of transformation and integration of GI concepts will be possible, using homomorphisms within and among GI categories.

5.2 Functional formalization of time

Consider C_s and C_d as two categories that carry one kind of mathematical structure (like the Space Syntax theory topological structure of convex spaces) in their static and dynamic domains. Then, a functor from the static category to the dynamic category can lift us to the temporal domain. This functor will take a function defined for static data and generate a function that can also be used for dynamic data. This homomorphism from the static to dynamic domain is called time lifting. Although it is possible to define a unique time lift function, for simplicity, separate time lift functions for functions with 0, 1, and 2 inputs are shown in (9). These functions are defined using lambda calculus syntax.

$$
\begin{aligned}
&>> \textit{lift0 } a = \backslash t \to a \\
&>> \textit{lift1 } op\, a = \backslash t \to op\; = (a\; t) \\
&>> \textit{lift2 } op\, a\, b = \backslash t \to op\;(a\; t)\;(b\; t)
\end{aligned}
\tag{9}
$$

where   *lift0*, *lift1*, and *lift2* = Time lift functions
           op = Input function
           a and *b* = Inputs for *op* function
           t = Time parameter
           $\backslash$ = Lambda symbol

For example, time lifting of a function with two inputs, using *lift2* function, is represented in (10).

$$
\begin{aligned}
&>> f :: a \to b \to c \\
&>> g :: (t \to a) \to (t \to b) \to (t \to c) \\
&>> g = \textit{lift2}\, f
\end{aligned}
\tag{10}
$$

where   *a*, *b*, and *c* = Static types
           $(t \to a)$ and $(t \to b)$ = Dynamic input functions
           $(t \to c)$ = Dynamic output function
example: if $f = (+)$ then $f\, 1\, 2 = 1 + 2 = 3$
                                   $g\,(t+1)\,(2t) = 3t + 1$
                                     $g\, 0 = 1;\, g\, 1 = 4;\, g\, 2 = 7$

Then any value type can be structured as a dynamic value type, by being defined as a function from time to that value type (11)

$$
>> \textit{Changing } v = Time \to v
\tag{11}
$$

where v = Any type like Integer, Float, or String
           Time = Time parameter

Polymorphism enables us to use these dynamic types, like their static counterparts, in functions. For this purpose, the dynamic value types must be defined as instances of the classes used by the static types. This process is specifically known as overloading or *ad-hoc* polymorphism. After overloading of dynamic types, functions defined over static types can be used for dynamic types too.

5.3 Analytic issues – floating point numbering system

Dealing with time as a discrete entity and the dominance of analytical approaches are defined as two of the problems that hindered time handling processes in GI science (refer to Section 5). These problems, especially the latter, are due to the boundedness of computers' floating point numbering systems[4]. This causes overflowing and rounding errors that subsequently result in continuity problems.

The computing environments supporting lazy evaluation have dealt with the problem of overflowing by adopting extensional evaluation of expressions. In this approach no expressions are evaluated before they

---

[4] Only a limited number of digits can be used for the integer and floating parts of a number.

are requested. Moreover, the expressions are evaluated until the request is sufficed. Thus the lazy evaluation allows the definition of infinite series. For example in (12) the defined infinite list of natural numbers will remain unevaluated as an extensional expression until the *take* function requests obtaining 5 natural numbers from the list. Then the *naturals* list is evaluated up to number 5 and then the evaluation stops.

$$>> naturals = [1, 2..]$$
$$>> take\ 5\ naturals = [1, 2, 3, 4, 5]$$
(12)

where   *naturals* = Infinite series of natural numbers

One solution to the round off error problem, proposed by Franklin (1984), is substitution of the floating point numbering system with a rational numbering system. The set of all rational numbers[5], denoted as Q, is a linear, dense[6], and totally ordered subset of real numbers. One of the major effects of using rational numbering systems is the elimination of error generation and propagation in GI algorithms due to round off error.

The rational numbering system, however, bring about some consequences. Rational numbers are structured types (like records). Then the computing environment supposed to use rational numbers has to be able to overload numerical operators and functions for structured data types. Besides, overloading is feasible in object oriented computing environments, although many of these environments do not support overloading for structured data types, especially for operators like =, +, and *.Moreover, rational number computation is more expensive (at least three times) than floating point number computation, therefore some efficiency problems are to be expected. Finally, the likely occurrence of large prime numbers in rational numbers must be considered. On the whole, rational numbers are recommended for testing hypotheses in an ideally precise situation.

By defining a static rational type as *Ratio Integer*, its dynamic counterpart is defined in (13).

$$>> Changing\,(Ration\,Integer) = Time \rightarrow (Ratio\,Integer)$$
(13)

The *RI* and *CRI* synonyms (14) are used for static and dynamic rational types:

$$>> type\,RI = Ratio\,Integer$$
$$>> type\,CRI = Changing\,(Ratio\,Integer)$$
(14)

The static and dynamic rational types are overloaded on a *Field* class, which contains basic numeric operators (15 and 16). Then these operators can be used for rational numbers too.

$$>> class\,Field\,a\,where$$
$$>> \quad (+),(-),(*),(/) : a \rightarrow a \rightarrow a.$$
(15)

$$>> instance\,Field(RI)\,where$$
$$\dots$$
$$>> instance\,Field(CRI)\,where$$
$$>> \quad (+) = lift2(+)$$
$$>> \quad (-) = lift2(-)$$
$$>> \quad (*) = lift2(*)$$
$$>> \quad (/) = lift2(/).$$
(16)

Besides, any functions that use these basic operators would be valid for both static and dynamic rational types (17).

$$>> dm :: (Field\,aType) => aType \rightarrow aType \rightarrow aType$$
$$>> dmxy = (x + y)^* (x - y)$$
(17)

---

[5] A rational number is a ratio of two integers, written as a/b where b is not zero.

[6] Being a dense subset means that between any two rationals sits another one (in fact infinitely many other ones).

where   *(Field aType)=> =Means that 'aType' can be any types which are*
                    instance of class Field
<u>*example:*</u>   *dm 1/2 1/3 = 5/36*
                    dmt = dm (t/2) (t/3) = 5t/36
                    dmt 1 = 5/36 and dmt 2 = 10/36

6   COMPUTATIONAL MODEL IMPLEMENTATION AND EXECUTION

Focusing on temporality in Space Syntax theory at the local scale, which provides more rapid dynamicity, movement and changes of activities, the grid representation is selected as the basis of the computational model. The grid representation is simpler than axial and convex representations as it uses predefined and finite vantage points (a set of regular grid points). These vantage points are considered as potential connections of convex spaces.

The linkages of convex spaces are defined by checking their intervisibilities. The resulting connectivity graph will contain positions of static and dynamic activities and their intervisibility relations as its nodes and links.

The intervisibility analysis is done within an underlying static environment. This environment consists of barriers (e.g. buildings) and passages (e.g. streets) that constrain movement and intervisibilities.

6.1   *Model implementation*

The required components for the computational model implementation are:

1. dynamic points,
2. connectivity graph,
3. static environment, and
4. functions.

6.1.1   *Dynamic points*
A point is defined as an algebraic data type (18).

$$>> \ data\ Point\ a = Point\ Id\ a\ a. \tag{18}$$

where   *Point* = Type name and constructor of a point;
              a = Type of the coordinates like *RI* or *CRI*; and
              Id = Unique integer identifier for a point.

Then *Point (RI)* and *Point (CRI)* define static and dynamic *Point* types.

Considering that multiplication and division of points are not required, the basic functions for *Point* data types are defined in *Points* class (19).

$$
\begin{aligned}
&>> \quad class\ Points\ p\ s\ where \\
&>> \quad x,y :: ps \rightarrow s \\
&>> \quad x(Point\_cx\_) = cx \\
&>> \quad y(Point\_\_cy) = cy \\
&>> \quad xy :: s \rightarrow s \rightarrow ps \\
&>> \quad xy\ cx\ cy = Point(-1)cx\ cy \\
&>> \quad (+) :: ps \rightarrow ps \rightarrow ps \\
&>> \quad (-) :: ps \rightarrow ps \rightarrow ps
\end{aligned}
\tag{19}
$$

where   *Points* = Class name and constructor;
        xy = Constructs a point from x and y coordinates;
        x or *y* = Returns x or y coordinate; and
        (+) and (−) = Summation and subtraction of points.

While most of the functions in *Points* class are general and have default definitions, just (+) and (−) functions are overloaded for static (20) and dynamic (21) *Point* data types.

$$
\begin{aligned}
&>> \ \textit{instance Points Point a where} \\
&>> \quad (\,+\,)\,(\textit{Point \_x1 y1})\,(\textit{Point \_x2 y2}) = \textit{Point}\,(\,-\,1)\,(x1 + x2)\,(y1 + y2) \\
&>> \quad (\,-\,)\,(\textit{Point \_x1 y1})\,(\textit{Point \_x2 y2}) = \textit{Point}\,(\,-\,1)\,(x1 - x2)\,(y1 - y2)
\end{aligned}
\tag{20}
$$

$$
\begin{aligned}
&>> \ \textit{instance Points Point (Changing a) where} \\
&>> \quad (\,+\,) = \textit{lift2}(\,+\,) \\
&>> \quad (\,-\,) = \textit{lift2}(\,-\,)
\end{aligned}
\tag{21}
$$

where _ = Any value.
    The other basic operations, like equality and ordering of points, are defined similarly by overloading the types on *Eq* and *Ord* classes (omitted here).

6.1.2   Connectivity graph

The proposed connectivity graph is a set of binary relations between points (22) (Thompson 1998).

$$
\begin{aligned}
&>> \ \textit{data Set a} = \textit{Set } [a] \\
&>> \ \textit{type Relation a} = (a,\ a) \\
&>> \ \textit{type Graph a} = \textit{Set (Relation a)}
\end{aligned}
\tag{22}
$$

where   *Set* = Type constructor of a set;
        Relation = Type constructor of a binary relation; and
        Graph = Type constructor of a graph.
Node is defined as an algebraic data type (23).

$$
>> \ \textit{data Node} = \textit{Node Id }([\textit{Id}],\ C,\ \textit{Ctrl},\ D,\ \textit{MD},\ \textit{RA})
\tag{23}
$$

where   *Node* = Type constructor of a node;
        Id = Unique identifier for a node that is the same as id of its corresponding point;
        [Id] = List of connected nodes' ids; and
        C, Ctrl, D, MD, and *RA* = Static parameters as connectivity (C), control (Ctrl), total depth (D), mean depth (MD), and integration (RA).

The manipulation functions of nodes are defined as some *set* and *get* functions (24).

$$
\begin{aligned}
&>> \ \textit{setNodeId } :: \ \textit{Node} \rightarrow \textit{RI} \rightarrow \textit{Node} \\
&>> \ \textit{setNodeId (Node id param) id}' = \textit{Node id}'\ \textit{param} \\[4pt]
&>> \ \textit{getNodeId } :: \ \textit{Node} \rightarrow \textit{RI} \\
&>> \ \textit{getNodeId (Node id \_)} = \textit{id} \\[4pt]
&>> \ \textit{getConnections } :: \ \textit{Node} \rightarrow [\textit{RI}] \\
&>> \ \textit{setConnections } :: \ \textit{Node} \rightarrow [\textit{RI}] \rightarrow \textit{Node} \\[4pt]
&>> \ \textit{getConnectivity, getControl, getTotalDepth,} \\
&>> \ \textit{getMeanDepth, getIntegrability } :: \ \textit{Node} \rightarrow \textit{RI} \\[4pt]
&>> \ \textit{setConnectivity, setControl, setTotalDepth,} \\
&>> \ \textit{setMeanDepth, setIntegrability } :: \ \textit{Node} \rightarrow \textit{RI} \rightarrow \textit{Node}
\end{aligned}
\tag{24}
$$

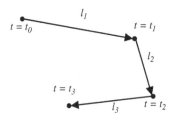

Figure 6.   A path consists of three directional straight lines at different times.

6.1.3   Static environment

The static environment is modelled as a set of planar polygons that do not intersect or include each other (25). The polygons block movement, but the spaces between them form passages that allow movement.

$$>> \ data \ Environment \ a = [Polygon \ a] \qquad (25)$$

where   *Environment* = Type constructor of an environment.

A polygon is defined by its bounding straight line segments. *Line* (26) and a *Polygon* (27) are defined as algebraic data types.

$$>> \ data \ Line \ a = Line \ (Point \ a) \ (Point \ a) \qquad (26)$$

where *Line* = Type constructor for a line.

$$>> \ data \ Polygon \ a = Polygon \ [Line \ a] \qquad (27)$$

where *Polygon* = Type constructor for a polygon.

Passages are modelled as paths that guide movements of the activities. A path consists of a set of directed straight lines meeting each other at different times (Fig. 6).

A path is constructed as a time based conditional statement of multiple directional straight lines (28).

$$>> \ path = cond \ [c1, \ c2, \ldots] \ [l1, \ l2, \ l3, \ldots] \qquad (28)$$

where   *cond* = Checks a series of conditions, finding the first valid condition and selecting its respective value.
$c1 = t0 < t < t1$
$c2 = t1 < t < t2$
$l1, l2$ and $l3$ = Line equations

6.1.4   Functions

Most of the functions implemented in this model are also generalized over lists. The functions that are defined over a list are preceded by '*m*'.

The *areIntervisible* function, as one of the basic functions in this model, checks for the existence of an intervisibility relation between two points in the environment (29).

$$>> \ areIntervisible \ :: \ (Bools \ b, \ Points \ Point \ a) \ => \ Environment \ a \rightarrow Point \ a \rightarrow Point \ a \rightarrow b \quad (29)$$

where   Bools = Type constructor of Boolean class on which static and dynamic Booleans are overloaded.

The *areIntervisible* function defines the intervisibility of two points by counting the intersections of the line that connects them with the environments' polygons (the detailed definitions of this function are omitted).

The *deriveIntervisibles* (30) and the *deriveRelations* (31) functions define all intervisibility relations for a point. Also, *mDeriveRelations* (32) function generalizes *deriveRelations* function over all points in a list.

$$>> \quad deriveIntervisibles :: (Points\ Point\ a) => Environment\ a \rightarrow [Point\ a] \rightarrow Point\ a \rightarrow [Point\ a]$$
$$>> \quad deriveIntervisibles\ env\ ps\ p = filter\ (areIntervisible\ env\ p)\ ps$$
(30)

where *filter aFunction aList* = Applies *aFunction*, a Boolean function, over *aList* and omits the elements which return *False*.

$$>> \quad deriveRelations :: (Points\ Point\ a) => Environment\ a \rightarrow [Point\ a] \rightarrow Point\ a \rightarrow [Relation\ RI]$$
$$>> \quad deriveRelations\ env\ ps\ p = [(getIDp,\ getIDq)|q \leftarrow (deriveIntervisibles\ env\ ps\ p)]$$
(31)

$$>> \quad mDeriveRelations :: (Points\ Point\ a) => Environment\ a \rightarrow [Point\ a] \rightarrow [[Relation\ RI]]$$
$$>> \quad mDeriveRelations\ env\ ps = map\ (deriveRelations\ env\ ps)\ ps$$
(32)

where *map aFunction aList* = applies *aFunction* over *aList*.

Also, some functions are used for flattening the output of the *mDeriveRelations* function into a list of relations (without nested lists) and also for removing duplicate relations. These functions are omitted here.

The *cGraph* function constructs the connectivity graph (33).

$$>> \quad cGraph :: (Points\ Point\ a) => Environment\ a \rightarrow [Point\ a] \rightarrow Graph\ RI$$
$$>> \quad cGraph\ env\ ps = Set\ (mDeriveRelations\ env\ ps)$$
(33)

The makeNodes function generates nodes from a list of points (34).

$$>> \quad makeNodes :: (Points\ Point\ a) => [Point\ a] \rightarrow [Node]$$
$$>> \quad makeNodes\ [] = []$$
$$>> \quad makeNodes\ (p : ps) = Node\ (pID\ p)\ ([], 0, 0, 0, 0, 0)\ :\ makeNodes\ ps$$
(34)

The morphologic analyses of the connectivity graph are implemented as follow:

1. The list of connected nodes (35 and 36):

$$>> \quad deriveConnections :: Graph\ RI \rightarrow Node \rightarrow [Node] \quad (35)$$
$$>> \quad deriveConnections\ g\ n = [b|(a, b) \leftarrow g,\ a == (getNodeID\ n)] + +$$
$$[a|(a, b) \leftarrow g,\ b == (getNodeID\ n)]$$
(35)

$$>> \quad mDeriveConnections :: Graph\ RI \rightarrow [Node] \rightarrow [[Node]]$$
$$>> \quad mDeriveConnections\ g\ ns = map\ (deriveConnections\ g)\ ns$$
(36)

2. The connectivity value (37 and 38):

$$>> \quad connectivity :: Node \rightarrow Node$$
$$>> \quad connectivity = setConnectivity \cdot length \cdot getConnections$$
(37)

where *length aList* = Returns number of elements in a list.
    $\cdot$ = Composition function.

$$>> \quad mConnectivity :: [Node] \rightarrow [Node]$$
$$>> \quad mConnectivity\ ns = map\ connectivity\ ns$$
(38)

3. The control value (39, 40, and 41):

$$>> \quad control :: Graph\ RI \rightarrow Node \rightarrow Node$$
$$>> \quad control\ g\ n = setControl\ (sum\ (map\ reci.getConnectivity\ (findNodes\ g\ (getConnections\ n)))) n$$
(39)

where *findNodes* = Gets ids and returns their relevant nodes.
  reci = Returns a reciprocal of a number.
  sum = Returns summation of a list of numbers.

$$>> \ mControl \ :: \ Graph \ RI \rightarrow [Node] \rightarrow [Node] \quad)$$
$$>> \ mControl \ g \ ns = map \ (control \ g) \ ns \tag{40}$$

4. The depth value (41, 42, 43, 44, and 45):

$$>> \ depth \ :: \ Graph \ RI \rightarrow Node \rightarrow Node \rightarrow RI \tag{41}$$

Readers are referred to Thompson (1998, 332–334) for definition of the *depth* function.

$$>> \ totalDepth \ :: \ Graph \ RI \rightarrow Node \rightarrow RI$$
$$>> \ totalDepth \ g \ p = setTotalDepth \ (sum \ [depth \ g \ p \ q | q \leftarrow graph]) \ p \tag{42}$$

$$>> \ mTotalDepth \ :: \ Graph \ RI \rightarrow [Node] \rightarrow [Node]$$
$$>> \ mTotalDepth \ g \ ns = map \ (totalDepth \ g) \ ns \tag{43}$$

$$>> \ meanDepth \ :: \ Graph \ RI \rightarrow Node \rightarrow Node$$
$$>> \ meanDepth \ g \ n = setMeanDepth \ ((getTotalDepth \ n)/(length \ g - 1)) \tag{44}$$

$$>> \ mMeanDepth \ :: \ Graph \ RI \rightarrow [Node] \rightarrow [Node]$$
$$>> \ mMeanDepth \ g \ ns = map \ (meanDepth \ g) \ ns \tag{45}$$

5. The integration value (46 and 47):

$$>> \ integration \ :: \ Graph \ RI \rightarrow Node \rightarrow Node$$
$$>> \ integration \ g \ n = setIntegration \ y \ (2*(getMeanDepth \ n - 1)/(length \ g - 2)) \tag{46}$$

$$>> \ mIntegration \ :: \ Graph \ RI \rightarrow [Node] \rightarrow [Node]$$
$$>> \ mIntegration \ g \ ns = map \ (integrability \ g) \ ns \tag{47}$$

Then, all the morphologic analyses are composed as the *calcParam* function (48).

$$>> \ calcParam \ :: \ Graph \ RI \rightarrow [Node] \rightarrow [Node]$$
$$>> \ calcParam \ g = ((mIntegrability \ g) \cdot (mMeanDepth \ g) \cdot (mTotalDepth \ g) \cdot (mControl \ g) \cdot \qquad (48)$$
$$mConnectivity) \cdot (mDeriveConnections \ g)$$

Finally, all functions are composed as the *analyseGrid* function (49).

$$>> \ analyseGrid \ :: \ (Points \ Point \ a) => Environment \ a \rightarrow [Point \ a] \rightarrow Changing \ (Graph \ RI)$$
$$>> \ analyseGrid \ env \ ps = ((calcParam \ g) \cdot (makeNodes \ g)) \ where \ g = cGraph \ env \ ps$$
$$\tag{49}$$

example: *f* = *analyseGrid env1 point1*
  f 10 will generate and analyse graphs at time 10.

6.2 Model execution

The model is executed for a simple simulated urban bus transportation system. The results are used to show how the system overcomes space and time.

The environment is implemented as five static polygons. Then, seven buses are implemented as the dynamic activities (Fig. 7). Two sample definitions of these buses are shown in (50).

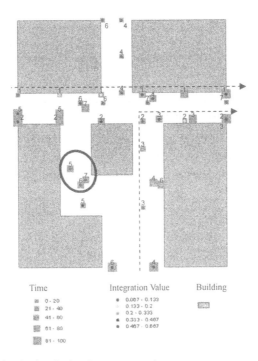

Figure 7. Sample analysis of a simulated urban bus transportation system.

$$>> \ l1\_s1 = Pt \ 1(\backslash t \to 10^{*}t + 100)(\backslash t \to 1440)$$
$$>> \ Bus1 = cond \ [][l1\_s1]$$

$$>> \ l3\_s1 = Pt \ 3(\backslash t \to 700)(\backslash t \to 14^{*}t + 600) \qquad (50)$$
$$>> \ l3\_c1 = \backslash t \to t < 50$$
$$>> \ l3\_s2 = Pt \ 3(\backslash t \to 8^{*}t + 300)(\backslash t \to 1320)$$
$$>> \ Bus3 = cond \ [l3\_c1] \ [l3\_s1, \ l3\_s2]$$

The path of *Bus1* is a straight line and the path of *Bus3* consists of two connected straight lines that are controlled by one temporal condition. These two paths are shown in Figure 7 by dashed arrows.

The *analyseGraph* function (49) is called at times 0, 25, 50, 75 and 100. The results are represented in Figure 7. The integration value for buses is shown over these times.

As a sample analysis of the result, a region of high integration between times 50 and 70 is defined in Figure 7 by a thick ellipse. This region absorbs a high level of travelling demand during the mentioned duration. This finding can be the basis for modification of paths and schedules at the specified region of space and time to obtain more effective integration of the system in space-time. Similar implications about how the system overcomes space and time can be derived using other properties, such as connectivity and control.

7 CONCLUSIONS

The implementation and execution of the computational model validates time lifting for the topological relationships of convex spaces illustrated by Space Syntax theory. This study suggests dividing time lifting of topological structures into two processes: one for deriving analysis units and generating the topological

99

structure and the other for properties evaluation. The former process will be time lifted and the latter will remain unchanged. Then, these two processes will be combined as a unique time lifted process.

In addition, the consistency of Space Syntax theory with GI concepts is implied. The illustrated approach of modelling dynamic activities at local scale and analysing how effective they overcome space and time can be adopted for various real world applications, such as analysing efficiency of urban services.

The Haskell programming language fully supports time lifting. It enables lazy evaluation, definition of algebraic and structured types, polymorphism and overloading.

Finally, more considerations are required about mixed usage of static and dynamic data and its efficiency. Time lifting uses expensive structures and processes like polymorphic algebraic structures and overloading. The models implemented for time lifting should be profiled for finding unnecessary expensive expressions. Then, optimization techniques, like substitution of the expensive expressions with possible cheaper ones, should be followed.

REFERENCES

Batty, M., Dodge, M., Jiang, B. & Smith, A. 1998. GIS and Urban Design. *Center for Advanced Spatial Analyses – CASA*. http://www.casa.ucl.ac.uk/urbandesifinal.pdf (accessed Jan, 2005).

Benedikt, M.L. 1979. To Take Hold of Space: Isovists and Isovist Fields. *Environment and Planning B* (6): 47–65.

Brown, M.G. 2001. Design and value: spatial form and the economic failure of a mall. *3rd International Space Syntax Symposium, Atlanta, USA*.

Egenhofer, M.J. & Mark, D.M. 1995. Naïve Geography. *National Center for Geographic Information and Analysis Publications*.

Frank, A.U. 1998. GIS for Politics. *GIS Planet 1998*. Lisbon, Portugal, IMERSIV.

Frank, A.U. 2003. Ontology for Spatio-Temporal Databases. In M. Koubarakis (ed.), *Spatio-temporal Databases: The Chorochronos Approach. Lecture Notes in Computer Science*: 9–78. Berlin: Springer-Verlag.

Frank, A.U. 2005. Practical Geometry – The Mathematics for Geographic Information Systems. Unpublished Manuscript.

Franklin, W.M. 1984. Cartographic Errors Symptomatic of Underlying Algebra Problems. *International Symposium of on Spatial Data Handling, Zurich, Switzerland*.

Grenon, P. & Smith, B. 2004. SNAP and SPAN: Towards Dynamic Spatial Ontology. *Spatial Cognition and Computation* 4(1).

Herring, J.R., Egenhofer, M.J. & Frank, A.U. 1990. Using Category Theory to Model GIS Applications. In *Proc. 4th International Symposium on Spatial Data Handling, Vol. II, Zurich*.

Hillier B. & Hanson, J. 1984. *The Social Logic of Space*. Cambridge: Cambridge University Press.

Hillier, B., Hanson, J., and Peponis, J. 1987. Syntactic Analysis of Settlements, *Architecture Comport./Architecture Behavior* 3(3): 217-231.

Hillier, B. & Netto, V. 2001. *Society Seen Through the Prism. 3rd International Space Syntax Symposium, Atlanta, USA*.

Hillier, B. 1996. *Space is the Machine*. Cambridge: Cambridge University Press.

Hillier, B. 2001. A Theory of the City as Object: How Spatial Laws Mediate the Social Construction of Urban Space. *3rd International Space Syntax Symposium, Atlanta, USA*.

Jiang B., Claramunt, C. & Klarqvist, B. 2000. An Integration of Space Syntax into GIS for Modelling Urban Spaces. *International Journal of Applied Earth Observation and Geoinformation* (2): 161–171.

Lynch, K. 1960. *The Image of the City*. Massachusetts Institute of Technology, Twenty first printing, 1992.

Thompson, S. 1998. *The Craft of Functional Programming, Second Edition*. Addison – Wesley Press.

Advances in Spatio-Temporal Analysis – Tang et al. (eds)
© 2008 Taylor & Francis Group, London, ISBN 978-0-415-40630-7

Spatio-temporal data model: An approach based on version-difference*

Huibing Wang, Xinming Tang, Bing Lei & Liang Zhai
Key Laboratory of Geo-informatics of State Bureau of Surveying and Mapping, Chinese Academy of Surveying and Mapping, Beijing, China

ABSTRACT: Spatio-temporal data storage and manipulation are two important aspects of spatio-temporal databases. A version-difference spatio-temporal data model is proposed and realized. The terms *version* and *difference* are used to represent spatio-temporal data at a certain time and changes over time. A default version focusing on the current state of the data and always representing the current state is introduced. Data can be reconstructed at any time by version and difference. This model has been applied in the NFGIDD project and achieved good results.

1 INTRODUCTION

Time and space are ubiquitous aspects of the spatio-temporal database, which was developed to monitor and analyse changes of historical information over time. Spatio-temporal applications are increasingly the focus of research activities in the geospatial and database communities.

Generally, the capabilities of any information system largely rely on the design of its data models, which represent the core of an information system. Data models provide the logical framework in which the real world may be described for computing. Consequently, the spatio-temporal data model is the key to handling spatial and temporal data simultaneously. Numerous spatio-temporal data models have been proposed, dealing with data storage and management. The main documented models are the following:

(1) Snapshot model: This model simply gives a new map for each time interval. Each layer is composed of a temporally homogeneous unit. When an event occurs, a new layer is constructed and the occurrence time is stamped onto the layer (all of the information, changed or not changed, is stored in the layer) (Nadi & Delavar 2003).

   The snapshot approach usually results in inconsistency and significant data redundancy.
(2) Space time composite: In the STC model (Langran & Chrisman 1988), the real world is a collection of spatially homogeneous units in a 2D space that changes over time from one unit to another. Each STC has its unique period of change and can be obtained from temporal overlays of snapshot layers.

   The STC can model useful properties of an object such as situation, but it cannot represent changes of attributes such as movement through space. Moreover, the STC approach requires reconstruction of thematic and temporal attribute tables whenever operations involve any changes in spatial objects (shape, size, or configuration) (Yuan 1996).
(3) Spatio-temporal object modelling (Worboys 1994): In this model, the real world is considered as a set of spatio-temporal atoms that are constructed from the integration of a temporal dimension orthogonal to 2D planimetric space. Each of these spatio-temporal atoms is the largest homogeneous unit that can store specific properties in space and time. Thus, this model can store changes in both temporal and spatial dimensions (Nadi & Delavar 2003).

---

* Funded by the Chinese National Fundamental Surveying and Mapping Project (No.1460130524207) and the Open-fund of Key Laboratory of Geo-information of SBSM.

However, as spatio-temporal atoms have a discrete structure, modelling of gradual changes in space or time with this model is impossible. This model is similar to snapshot, and STC models can only represent sudden changes.

Other models, such as event-based models, object-oriented models, etc. have also been well documented by many researchers. However, most of the existing models cover only partly the requirements (they address either spatial or temporal modelling), and most are at the logical level, hence not well suited for database design (Parent & Spaccapietra 1999).

2 THE VERSION-DIFFERENCE MODEL

Building an appropriate spatio-temporal data model concerning both spatial and temporal aspects, which can be used to organize and manipulate spatio-temporal features more effectively and enrich spatio-temporal feature's semantics (attribute, spatial and temporal), is critical to a spatial dynamic system. Therefore, we can use this model to retrospect or track historic data, monitor changes and forecast future transition.

In this paper, we propose a different data model called the version-difference model. In this model, we use the term difference to reflect changes compared with versions that represent a data state at a point in time. The database mainly stores base data and difference data; thus an object representation at any time can be obtained through an aggregation or reconstruction operator of base state and all subsequent changes (differences). Since current data are accessed very frequently, this model takes current data as base data (called the default version). Base data and all historic changes (called version differences) are stored.

This model has the following characteristics compared with the main models introduced above:

(1) Minimize storage size: For the version-difference model, we need not store all information of every state in the area of interest, but to minimize storage only the data at base state and changes between the state considered and the base state. Unchanged features are not duplicated. Each change is labelled with a unique state ID (State ID is incremented) and a changed object is recorded with its predecessors and successors. Datasets may look different at different states.

(2) Access data at any time point easily: This model provides the schema (called version management) to generate data at any time point using reconstruction operators on the base state and all subsequent changes.

(3) Track spatio-temporal object's states easily: Since a changed object is recorded with its predecessors and successors, spatio-temporal object's states can be easily accessed. Thus, this character can be used for spatio-temporal analysis and forecasting.

2.1 *Representation of time attributes*

There are two different notions of time that are relevant for temporal databases. One is called valid time (begin time and end time); the other is transaction time. The former denotes the time period during which an object is true with respect to the real world. The later is the time period during which an object is stored in the database. Note that these two time periods do not have to be the same for a single object.

In this model, the database is bi-temporal; an object's valid time and transaction time are both recorded, to support temporal queries.

2.2 *Event and state*

Event and state are a pair of the most important basic concepts in a temporal database (Tang 1998). One object may have different states during its life span; event is the process from one state to another. In general, we record event by time instance and state by time span in databases. State is an existent form of geographical entity in a given time range, and it is a comparatively stable process. The state of a spatio-temporal object can be divided into attribute state and spatial state, and spatial state can be further divided into spatial topological state and spatial geometry state (Fig. 1).

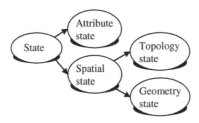

Figure 1.   A spatio-temporal object state.

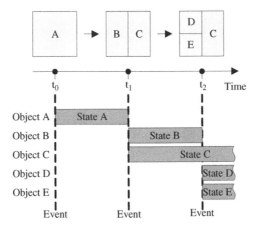

Figure 2.   An example of state changes over time: Object A splits into two objects B and C at time t_1 (an event occurred), and its state began at t_0 and ended at t_1. As for object B, its parent object is A, children objects are D and E; we record this information in the database as a correlation link. Other objects in the figure follow the same rule.

2.3   *Correlation link*

The disappearance of one object is usually coincident with the involvement of a new object's appearance, and vice versa. We record such information in databases as correlation links.

A parent object's death and a child object's birth are triggered by the same event; in other words, parent(s) object and child object(s) have inheritable relationships in their life span. The relationship of an object's variation inheritance is very important in the process of state evolvement and temporal data updates, and it is useful to track an object's various states over time and therefore we need to build spatio-temporal dynamic correlation.

Figure 2 shows the state/event concept by an example.

2.4   *Version and difference*

Version, as the name indicates, explicitly stores states of individual features or objects as they are altered, added and deleted through various events. Each state of a feature or object is stored as a row in a table, along with important transaction information.

In this model, version can be classified into three types:

(1) On-line or On-the-fly Version: This means the version reconstructed by version and difference from the database at a given point of time. This kind of version basically has no redundancy. Users may specify a time point, and the system can generate a dataset belonging to that time point.

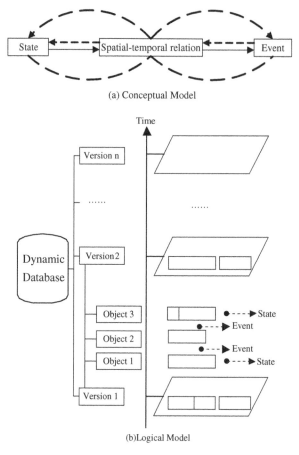

(a) Conceptual Model

(b)Logical Model

Figure 3. "Version-difference" spatio-temporal data model.

(2) Stand-alone or Snapshot Version: Users can create a version that holds all data belonging to a specific state at a given point in time. This version is what is called the stand-alone or snapshot version, and it has redundancy.

The snapshot version has some properties, such as: name, owner, description, create time, last modified time, parent version, privilege, etc. Privilege can be divided into three classes as follows:

Public: Read/write for all users.

Protected: Read/write for owner, read-only for others.

Private: Read/write for owner.

(3) Current Version or Default Version: This is the version comprising the whole current state dataset. Obviously, it is an example of the snapshot version.

Version management should follow these rules:

1. Only the owner can rename, create, modify or delete a version;
2. Creation or modification of a version cannot take effect at once but must wait until the commit operation is complete;
3. Every dataset has a default version;
4. The version's owner can restrict others' privilege for his/her version; and
5. The database should have version verification to identify features' updating.

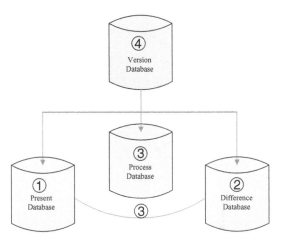

Figure 4. Relationship between the four logical databases: the relation of present database and difference database is recorded in the process database, and the version database is created by the present database, the difference database and with the relation in process.

As for the difference, we must take into consideration several issues as follows. (1) Difference comes from comparing information at two temporal states (say A and B), so it must include information in two parts, one belonging to state A but not B, the other is with state B but not state A. In order to use difference to quickly and continuously fetch data at an instance belonging to a specific state, we must clearly label the two cases in the given difference table. (2) The object relationship between target state and neighbouring states—for example, $0 \rightarrow 1$ (target object emerges.); $1 \rightarrow 0$ (object disappears.); $1 \rightarrow N$ (the number of objects increases from one to many.); $N \rightarrow 1$ (the number of objects reduces to one.)—remains, but properties change. (3) Similar to most spatial database models, the difference table also consists of spatial information, property information and relation information. So as far as each object is concerned, the relationship among the three kinds of information should be recorded in the difference table.

2.5 Storage structure

The "version-difference" model is adopted to support object-oriented design, reduce data redundancy and save storage space. According to the version-difference model, we store spatio-temporal data in four logical databases, each holding different information (Fig. 4):

1. Present database: It stores the whole spatio-temporal dataset of the current state. Because the present dataset will be accessed frequently, we identify it as the base state in order to improve database performance.
2. Difference database: The latest state of an object will be updated to the present database once it has been changed at a valid event; the old state of the object will be moved to the difference database. The events should be arranged by time order and the difference between the current state and old state will be stored when all the conditions or constraints are satisfied. Given a specific time, it is easy to obtain a previous state and rollback the current spatio-temporal relations and implement corresponding calculations.
3. Process database: Transitions are processes representing evolution and therefore subject to constraints, which are preconditions to limit, avoid or force a change. In other words, some change may be uncorrectable and cannot happen in reality; these are named "fake changes". When a "fake change" has occurred, the database should discard the alteration and rollback. Obviously, the process database stores the information of dynamic correlation so as to query and trace history.
4. Version database: Version is a snapshot of the dataset at a given point in time. Users can browse the spatio-temporal data of any time instance and build it as a version in the version database. After this, the

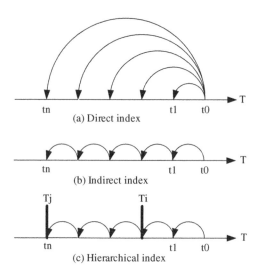

Figure 5.   Spatio-temporal index schema.

version dataset can be browsed and manipulated quickly, and it can be regarded as a base state of other version datasets, whose time stamp is earlier than this one, for the hierarchy index schema (explained in next section).

In short, present database stores the current dataset as the base state; difference database stores the history dataset of changed objects; process database records the process of history changes; version database is used to store snapshot datasets at times specified by users. The relationship of these logical databases is expressed in Figure 4.

2.6   *A hierarchical spatio-temporal index schema*

Program listings or program commands in the text are normally set in typewriter font, e.g. CMTT10 or Courier.

As for the indexing methods of spatio-temporal data models, the large amount of research work that has been done mainly involves two issues: how to build the index and how to design the difference file to precisely measure the changes between two states.

Nowadays, many publications refer to direct index and indirect index. If direct index (Fig. 5a) is used to update a state's objects, all the difference files need to be modified; if an indirect index (Fig. 5b) is implemented, many records in the database have to be searched. However, when the states in the temporal database change frequently over a quite long period of time, the two methods will require considerable amounts of work, either on the calculation of data or in the searching and indexing of large databases.

To improve this situation, one can establish a hierarchical index schema (Fig. 5c), that is, taking the data at time Ti (i = 0, 1, 2 ...) as base state for the data at time tj (j > i) and then comparing different states indirectly.

3   IMPLEMENTATION OF THE NFGIDD PROJECT

We have developed NFGIDD (National Fundamental Geospatial Information Dynamic Database) project's software system (named STDBInfo) based on this model. This project aims to construct a general spatio-temporal data model of national fundamental geospatial information to provide a general framework for monitoring, analysing and forecasting the transition of spatio-temporal objects over time.

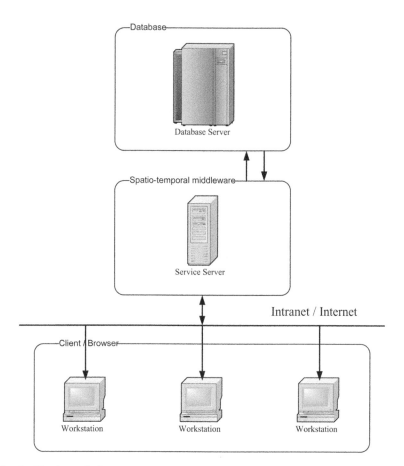

Figure 6.   Sketch of hardware deployment.

3.1   *Hardware deployment*

The system can be deployed as a C/S (Client/Server) or a B/S (Browser/Server) configuration. Its three main components are: client, middle-ware and server (Fig. 6).

(1) Database server: uses a highly efficient database server (e.g. SUN machine) and business database software (e.g. Oracle^{TM}) to store spatio-temporal data and all other data needed by this system.
(2) Spatio-temporal middleware: is a spatio-temporal data engine, which mainly provides the data service interface and acts as the bridge between the user and data. It cannot only respond to user requests and return answers, but also has other capabilities, such as assigning computation tasks automatically. The middleware provides a layer for the query of spatio-temporal information, which delivers more efficient and productive processing of spatio-temporal queries. The middleware is deployed on a separate server.
(3) Client or Browser: provides the user interface, responds to user requests and displays/outputs results.

3.2   *System framework*

The system consists of three main sub-systems: Dynamic data process, Spatio-temporal data engine (middle-ware) and Dynamic data management.

(1) Dynamic data process: Data input, data check, building dynamic correlation;
(2) Spatio-temporal data engine: Mainly provides data input/output interfaces between (1) and (3); and

107

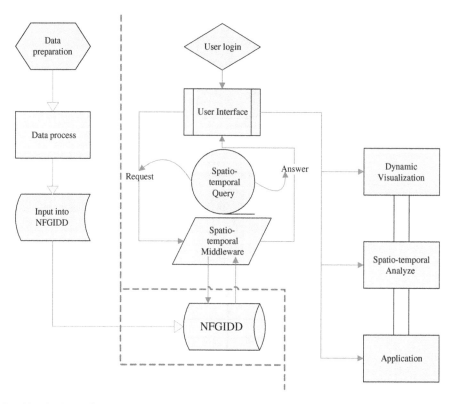

Figure 7.   Sketch of workflow of the NFGIDD system.

(3) Dynamic data management: Spatio-temporal data management, users and security management, dynamic visualization, spatial and temporal query, etc.

Figure 7 explains the system's workflow.

3.3   *Cases studies*

Figure 8 shows the main GUI (Graphic User Interface) of the STDBInfo system, which includes data input, data checking, building of dynamic correlation, users' and security management, dynamic visualization, spatial and temporal query modules, etc.

(1) Example of residential area changes: The system can query and view features' various states over time expediently. Figure 9 shows the transition of residential areas of interest from 1970 to 2000. From it, we can easily find the residential areas that have been extended and where more houses or districts have been built. At the same time, we can add attribute information to these distributions, such as population data.
(2) Example of change statistics: The system provides several methods for change statistics (Fig. 10).

4   CONCLUSION

The "version-difference" spatio-temporal data model presented in this paper organizes and stores the spatio-temporal data through version base state and corresponding differences, and builds the dynamic correlation links and hierarchical index, which can effectively reduce data redundancy and accelerate data

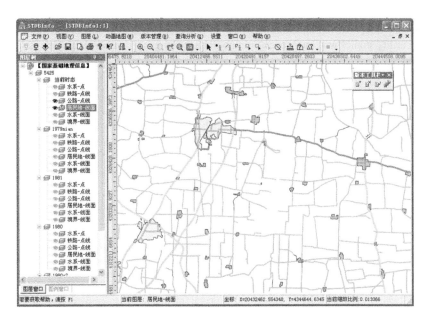

Figure 8.　Main GUI of "Dynamic data management" system.

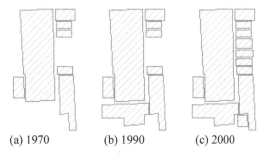

(a) 1970　　　　　(b) 1990　　　　　(c) 2000

Figure 9.　Example of residential area changes.

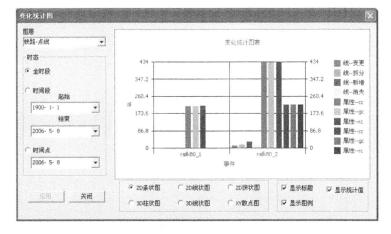

Figure 10.　Example of change statistics.

access or history-trace. An implementation based on this model was also introduced and used to verify this model. Further studies will focus on the temporal query language, spatio-temporal topology and time series analysis.

ACKNOWLEDGMENT

This research was carried out under the support of the National Fundamental Surveying and Mapping Project—National Fundamental Geospatial Information Dynamic Database (NFGIDD).

REFERENCES

Langran G. & Chrisman, N.R. 1988. A Framework for Temporal Geographic Information. *Cartographica* 25(3): 11–14.
Nadi, S. & Delavar, M.R. Spatio-Temporal Modeling of Dynamic Phenomena in GIS. http://www.scangis.org/scangis2003/papers/11.pdf.
Parent, C. & Spaccapietra, S. 1999. Spatio-temporal Conceptual Models: Data Structures + Space + Time. *Proceedings of the 7th international symposium on advances in geographic information systems*: 26–33. ACM Press.
Tang, X.M. 1998. Modeling of Spatiotemporal Data and Identification of Relationships and Change Processes For a Cadastral Information System in China, *ITC MSC thesis*. Enschede, the Netherlands.
Worboys, M.F. 1994. A Unified Model for Spatial and Temporal Information. *The Computer Journal* 37(1): 26–34.
Yuan, M. 1996. Temporal GIS and Spatio-Temporal Modeling. http://ncgica.ucsb.edu/conf/SANTA_FE_CD_ROM/sf_papers/yuan_may/may.html.

Advances in Spatio-Temporal Analysis – Tang et al. (eds)
© 2008 Taylor & Francis Group, London, ISBN 978-0-415-40630-7

An algebra for moving objects

Shengsheng Wang, Dayou Liu, Jie liu & Xinying Wang
College of Computer Science and Technology, Key Laboratory of Symbolic Computation and Knowledge Engineering of Ministry of Education, Jilin University, Changchun, China

ABSTRACT: Moving object databases are becoming more and more popular due to the increasing number of application domains that deal with moving entities. But so far, the research into the qualitative representation and reasoning for road networks of moving objects is quite limited and many key problems still remain unsolved. We propose a new spatial algebra for a road network of moving objects, employing Interval Algebra theory. It is an executable qualitative representation and reasoning method for road networks of moving objects. Renz developed an algebra DIA (Directed Intervals Algebra) to represent one dimensional moving objects based on the spatial interpretation of time intervals. By extending DIA to a road network, we give the formalization of RNDIA, which is compatible with DIA. Then, the reasoning problem is discussed.

1 INTRODUCTION

Spatio-temporal reasoning (STR), the research field aimed at spatial and/or temporal questions, has a wide variety of potential applications in Artificial Intelligence (such as spatial information systems, robot navigation, natural language processing, visual languages, qualitative simulation of physical processes and common-sense reasoning) and other fields (such as GIS and CAD)[Cohn & Hazarika 2001, Teresa Escrig & Toledo 1999].

Existing technology has made it possible to track movements of target objects in the air (e.g. airplanes), on the land (e.g. vehicles, wild animals, people, etc.) and ocean (e.g. ships, animals). The position of these moving objects is usually detected by GPS and presented by a single point. Since a point has no shape or area, its spatio-temporal attribute is the most simple among all shapes.

There is a wide range of research issues concerning moving objects, including modelling and representation of moving objects, query language design, indexing techniques and query optimization, etc. (Su et al. 2001, Vazirgiannis et al. 2001, Porkaew 2001, Cheng et al. 2003).

We believe that moving object research, in common with other spatio-temporal areas, should generally be characterized as: techniques for interpreting the semantic content of spatio-temporal data; extracting and reasoning the spatial knowledge have lagged behind the techniques for storage and query of spatio-temporal data. Furthermore, prior works on spatio-temporal reasoning are relevant but insufficient for moving objects.

There are two types of moving object: those having a free trajectory, such as a bird flying through the sky, and those with a constrained trajectory in which the movement of the object in space is strongly restricted due to physical constraints, e.g. a vehicle driving through a city. Since vehicles equipped with GPS are quite common in modern times, the constrained trajectory, or what we call the road network moving object (RNMO), is more practical.

Little research has been performed on the RNMO reasoning aspect. Van de Weghe & Cohn (2004) built a calculus of relations between disconnected network-based mobile objects in 2004, but the connected topological relations are not considered. Renz (2001) gave an algebra DIA to represent one dimensional moving objects. DIA extends the Interval Algebra, and includes both topology and direction information. He calls his work the first step of a spatial odyssey. But no further research on 2D free or restricted movements is reported as yet.

RNMO is similar to one dimensional moving objects in many aspects. In this paper, we extend the DIA to RNDIA. We give a compatible algebra formalization for RNDIA and the composition reasoning and consistency problem reasoning are discussed.

The rest of the paper is organized as follows. In Section 2, we introduce Renz's Directed Intervals Algebra. In Section 3 and Section 4, Road Network Directed Intervals Algebra is presented. In Section 5, the reasoning method of RNDIA is discussed. Section 6 concludes the paper.

2   DIRECTED INTERVALS ALGEBRA

One particular way of regarding vehicles and their regions of influence (such as safety margin, braking distance, or reaction distance) could be to represent them as intervals on a line, which represents their possibly winding way. Being similar to the well-known Interval Algebra developed for temporal intervals (Allen 1983), it seems useful to develop a spatial interval algebra for RNMO.

There are several differences between spatial and temporal intervals that have to be considered when extending Interval Algebra to deal with spatial applications.

(1) Spatial intervals can have different directions, either the same or the opposite direction as the underlying line.
(2) Roads usually have more than one lane in which vehicles can move, i.e. it should be possible to represent intervals on different lanes and show that one interval is, for example, to the left of, right of, or beside another interval.
(3) It is interesting to represent intervals on road networks instead of considering just isolated roads.
(4) Intervals such as those corresponding to regions of influence often depend on the speed of vehicles, i.e. it should be possible to represent dynamic information.

Renz developed an algebra DIA for qualitative spatial representation and reasoning about directed intervals, identifying tractable subsets, and showed that path-consistency is sufficient for deciding consistency for a particular subset that contains all base relations (Renz 2001).

We say that an interval has a positive direction if it has the same direction as the underlying line and negative direction otherwise. So possible direction constraints could be unary constraints like "x has positive/negative direction" or binary constraints like "x and y have the same/opposite direction".

A base relation consists of two parts: the interval part, which is a spatial interpretation of the Interval Algebra and the direction part d, which gives the mutual direction of both intervals, either $=$ or $\neq$. The 26 base relations of DIA are listed in Table 1.

3   ROAD NETWORK DIRECTED INTERVALS ALGEBRA

The Directed Intervals Algebra surely was a big step towards the real world application of qualitative spatial reasoning, with its large body of work and the large number of results obtained such as algorithms and complex results. Unfortunately, the DIA is only suitable for one dimensional lines; the free or restricted 2D movements are not supported. So there is still a long way to go towards the final goal of putting them into practice. In this paper, we extend DIA to Road Network Directed Intervals Algebra (RNDIA), which could be used to model moving cars in road networks (or other types of restricted 2D movements).

Since a road network is an undirected graph, to apply DIA to it we must define "direction" on the road network. The first step is to set up a frame of reference. The origin is a point in the road network, and may be the centre of the city or the south-west corner of the city or the start/end point of highways. The selection of the origin is critical for the model. For instance, when discussing "the highways near Beijing", choosing Beijing as the origin would be a good choice.

It is more difficult to choose the order, since the road network is an undirected graph. To solve this problem, we use the shortest path. Assume S is the origin of road network $\Re$, SP(S, x) indicates the length of the shortest path in $\Re$ from S to x. Note that both S and x are the points on the edge, not the vertices

Table 1. The 26 base relations of the directed intervals algebra.

Directed Intervals Base Relation	Symbol	Pictorial Example
x behind$_=$ y	b$_=$	-x->
y in-front-of$_=$ x	f$_=$	-y->
x behind$_{\neq}$ y	b$_{\neq}$	<-x-
		-y->
x in-front-of$_{\neq}$ y	f$_{\neq}$	-x->
		<-y-
x meets-from-behind$_=$ y	mb$_=$	-x->
y meets-in-the-front$_=$ y	mf$_=$	-y->
x meets-from-behind$_{\neq}$ y	mb$_{\neq}$	<-x-
		-y->
x meets-in-the-front$_{\neq}$ y	mf$_{\neq}$	-x->
		<-y-
x overlaps-from-behind$_=$ y	ob$_=$	-x->
y overlaps-in-the-front$_=$ x	of$_=$	-y->
x overlaps-from-behind$_{\neq}$ y	ob$_{\neq}$	<-x-
		-y->
x overlaps-in-the-front$_{\neq}$ y	of$_{\neq}$	-x->
		<-y-
x contained-in$_=$ y	c$_=$	-x->
y extends$_=$ x	e$_=$	-y->
x contained-in$_{\neq}$ y	c$_{\neq}$	<-x-
y extends$_{\neq}$ x	e$_{\neq}$	-y->
x contained-in-the-back-of$_=$ y	cb$_=$	-x->
y extends-the-front-of$_=$ x	ef$_=$	-y->
x contined-in-the-back-of$_{\neq}$ y	cb$_{\neq}$	<-x-
y extends-the-back-of$_{\neq}$ x	eb$_{\neq}$	-y->
x contained-in-the-front-of$_=$ y	cf$_=$	-x->
y extends-the-back-of$_=$ x	eb$_=$	-y->
x contained-in-the-front-of$_{\neq}$ y	cf$_{\neq}$	<-x-
y extends-the-front-of$_{\neq}$ x	ef$_{\neq}$	-y->
x equals$_=$ y	eq$_=$	-x->
		-y->
x equals$_{\neq}$ y	eq$_{\neq}$	-x->
		<-y-

of the graph. If S and x are not on the vertices of $\mathfrak{R}$, SP(S, x) can be calculated by adding S and x as new vertices to $\mathfrak{R}$ and performing the classical shortest path algorithm.

For two points x and y in $\mathfrak{R}$, if SP$(S, x) <$ SP(S, y) then $x < y$, If SP$(S, x) \leq$ SP(S, y) then $x \leq y$. (Fig. 1)

Another type of order is defined by two points, such as "the highways from Beijing to Shanghai". The metric value of x from S to T is defined by SP(S,T,x) (Fig. 2):

$$SP(S, T, x) = \frac{SP(S, x)}{SP(S, x) + SP(T, x)} SP(S, T) \qquad (1)$$

Without loss of generality, we only use the first definition of SP in the latter part of this paper.

In DIA, the car movement is represented by a straight line, which is interpreted as a vehicle and its region of influence (such as safety margin, braking distance, or reaction distance). It can also be interpreted as

Figure 1. Road network directed intervals algebra.

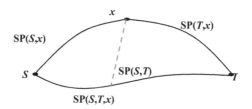

Figure 2. Road network directed intervals algebra.

Table 2. The reversed direction operator.

	+	−	*
~	−	+	*

the car's movement track during a period of past time or in the near future. Future track is estimated by current speed and direction. Whatever the case, the DIA cannot tell us when the car is turning around or when two cars are moving on parallel/crossed streets. In our work, car movement track is represented by a directed curve embedded in the road network.

Track X = <x1, x2> is the movement track from x1 to x2. The direction of track is defined:

$$
D(<x_1,x_2>) = \begin{cases} + & \text{if } \forall p \in < x_1,x_2 > [x_1 \leq p \leq x_2] \\ - & \text{if } \forall p \in < x_1,x_2 > [x_2 \leq p \leq x_1] \\ * & \text{otherwise} \end{cases} \tag{2}
$$

The reversed direction operator is ~ (Table 2).

X is "simplex directed" iff D(X) {*} and A X [D(A) = D(X)] (Figure 3).

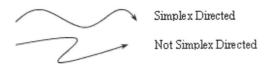

Simplex Directed

Not Simplex Directed

Figure 3. "Simplex Directed" means it never makes a detour. "Not Simplex Directed" means it makes a detour.

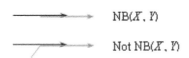

NB(X, Y)

Not NB(X, Y)

Figure 4. No branch.

In this paper, we assume that all tracks are simplex directed. This assumption is reasonable if we only investigate the short tracks (the movement within a small period of past/future time).

The order of two tracks $X = <x_1, x_2>$, $Y = <y_1, y_2>$ is defined as follows:

If max $\{x_1, x_2\} < \min\{y_1, y_2\}$ then $X < Y$.

If max $\{x_1, x_2\} \leq \min\{y_1, y_2\}$ then $X \leq Y$.

The union of X and Y has no branch if

$$\forall A \subset X, \forall B \subset Y[A \cap B = \emptyset \rightarrow A < B \vee B < A],$$

denoted by NB(X, Y) (Figure 4).

NB(X, Y) means the X and Y can be treated as one dimensional directed intervals since $X \cup Y$ has no branch, otherwise it is not the case.

Proposition 1: X is simplex directed $\rightarrow$ NB(X, X)

Proof. We assume that $\forall <a_1, a_2> <b_1, b_2> \subset X$. Without loss of generality, we assume that D(X) = $\{+\}$ and $a_1 < b_1$. So $a_1 < a_2$ and $b_1 < b_2\{<a_1, a_2>\} \cap \{<b_1, b_2>\} = \emptyset \rightarrow a_2 < b_1 \rightarrow <a_1, a_2> < <b_1, b_2>$.

Proposition 2: X, Y are simplex directed, $X \subseteq Y$ or $Y \subseteq X \rightarrow$ NB(X, Y)
Proof. We assume that $X \subseteq Y$, NB(X, Y) = NB(X, X).

Proposition 3: X, Y are simplex directed, $X \leq Y$ or $Y \leq X \rightarrow$ NB(X, Y)

Proof. Given two simplex directed tracks $X = <x_1, x_2>$, $Y = <y_1, y_2>$. We assume that $X \leq Y$. $\forall A = <a_1, a_2> \subset X$, $\forall B = <b_1, b_2> \subset Y$, max$\{a_1, a_2\} \leq$ max$\{x_1, x_2\} \leq \min\{y_1, y_2\} \leq \min\{b_1, b_2\}$. So $A \leq B$.

Proposition 4: If NB(X, Y), the relation between X and Y is equal to the relation between two intervals in one dimension line.

Proof. We build a mapping from X, Y to two intervals in one dimensional line. $\delta(p): p \in X, Y \rightarrow$ SP(S, p). Every $p \in X, Y$ have one and only one $\delta(p)$. No two points p, q of X, Y map to same value ($\delta(p) = \delta(q)$), otherwise the small subtracks $<p> <q>$ contained p and q, have $\neg$NB($<p>$, $<q>$), it will lead to $\neg$NB(X, Y). So δ is one to one mapping. And the relation of X and Y can be exclusively represented by the two intervals.

Next we will give the base RNDIA relations of two simplex directed tracks by the predications defined above.

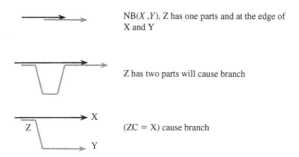

NB(X,Y), Z has one parts and at the edge of X and Y

Z has two parts will cause branch

($ZC = X$) cause branch

Figure 5. Overlay Relations.

4 BASE RNDIA RELATIONS

Firstly, the RNDIA base relations are divided into two parts: NB(X,Y) and not NB(X,Y). If NB(X,Y) is true, X,Y can be simply treated as one dimensional directed intervals, and RNDIA is equivalent to DIA. In other cases, we have to define some new relations.

The points set of track X is $\{X\}$. $\{X\}$ is an infinite set. Unlike X, $\{X\}$ has no direction.

Given two simplex directed tracks $X = <x_1,x_2>$, $Y = <y_1,y_2>$.

By Proposition 2, we know that when one track is totally contained in the other there is surely no branch, there is no need to define new relations under the conditions (1) to (3).

(1) $\{X\} = \{Y\}$
  If $D(X) = D(Y)$ then X eq$_=$ Y
      Else X eq$_{\neq}$ Y
(2) $\{X\} \subset \{Y\}$
  If $D(X) = D(Y)$
          If $x_2=y_2$ then X cf$_=$ Y
    Else If $x_1=y_1$ then X cb$_=$ Y
    Else X c$_=$ Y;
    Else
          If $x_1=y_2$ then X cf$_{\neq}$ Y
    Else If $x_2=y_1$ then X cb$_{\neq}$ Y
    Else X c$_{\neq}$ Y;
(3) $\{Y\} \subset \{X\}$
  If $D(X) = D(Y)$
          If $x_2=y_2$ then X eb$_=$ Y
    Else If $x_1=y_1$ then X ef$_=$ Y
    Else X e$_=$ Y;
    Else
          If $x_2=y_1$ then X ef$_{\neq}$ Y
    Else If $x_1=y_2$ then X eb$_{\neq}$ Y
    Else X e$_{\neq}$ Y;
(4) $\{X\} \cap \{Y\}$ is line
  (4.1) NB(X, Y)

$Z = X \cap Y$. Z is connected (has no detached parts) and Z is not in the middle of X or Y, otherwise there will be branches (Fig. 5).

So we have 4 relations equivalent to DIA relations:

If $D(X) = D(Y)$
          If ($D(X) = \{+\}$ and $x_2 < y_2$) or ($D(X) = \{-\}$ and $y_2 < x_2$) then X ob$_=$ Y

116

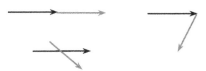

Figure 6.   Intersection is point.

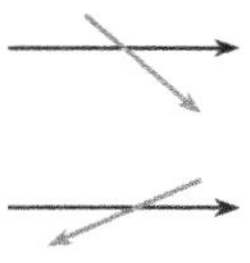

$X\ cr = Y$

$X\ cr \neq Y$

Figure 7.   Cross Relations.

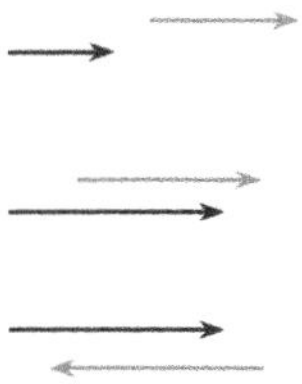

$Xb = Y$

$Xpr = Y$

$Xpr \neq Y$

Figure 8.   Disconnected Relations.

        Else X of$_=$ Y;

Else

        If $(D(X) = \{+\}$ and $x_2 < y_1)$ or $(D(X) = \{-\}$ and $y_1 < x_2)$ then X ob$_{\neq}Y$

        Else X of$_{\neq}$ Y;

(4.2) not NB(X, Y)

Two new relations are proposed:

If $D(X) = D(Y)$

        X po$_=$ Y;

Else

        X po$_{\neq}$ Y;

The latter two cases in Figure 5 are all po$_{\neq}$.

(5) $\{X\} \cap \{Y\}$ is one or more points

In this case, NB(X, Y) if $X \leq Y$ or $Y \leq X$ (Fig. 6). And two new relations are defined when NB(X, Y) is false (Fig. 7). So we define relations as:

If $D(X) = D(Y)$

        If $(D(X) = \{+\}$ and $Y \leq X)$ or $(D(X) = \{-\}$ and $X \leq Y)$ then X mf$_=$ Y;

        ElseIf $(D(X) = \{+\}$ and $X \leq Y)$ or $(D(X) = \{-\}$ and $Y \leq X)$ then X mb$_=$ Y;

        Else X cr$_=$ Y;

Else

        If $(D(X) = \{+\}$ and $X < Y)$ or $(D(X) = \{-\}$ and $Y < X)$ then X mf$_{\neq}Y$;

        ElseIf $(D(X) = \{+\}$ and $Y < X)$ or $(D(X) = \{-\}$ and $X < Y)$ then X mb$_{\neq}Y$;

        Else X cr$_{\neq}Y$;

(6) $\{X\} \cap \{Y\} = \emptyset$

Similar to (5) NB(X, Y) if $X < Y$ or $Y < X$.

If $D(X) = D(Y)$

        IF $(D(X) = \{+\}$ and $Y < X)$ or $(D(X) = \{-\}$ and $X < Y)$ then X f$_=Y$;

117

Table 3. Composition Table of RNDIA (I).

○	b	f	mb	mf	ob	of	c	e	cb	ef	cf	eb	eq	pr	cr	po
b	b	* pr cr po	b mb ob c cb pr cr po	b	b mb ob c cb pr cr po	b mb ob c cb pr cr po	b	b	b	b mb ob c cb pr cr po	b	b	b	b mb ob c cb pr cr po	b mb ob c cb pr cr po	b mb ob c cb pr cr po
f	* pr cr po	f	f mf of c cf pr cr po	f	f mf of c cf pr cr po	f	f mf of c cf pr cr po	f	f mf of c cf pr cr po	f	f	f	f	f mf of c cf pr cr po	f mf of c cf pr cr po	f mf of c cf pr cr po
mb	b	f mf of e ef pr cr po	b	cf eb eq pr cr po	b	ob c cb pr cr po	ob c cb pr cr po	b	mb	mb	ob c cb pr cr po	b	mb	b mb ob c cb pr cr po	b mb ob c cb pr cr po	b mb ob c cb pr cr po
mf	b mb ob e eb pr cr po	f	cb ef eq pr cr po	f	of c cf pr cr po	f	of c cf pr cr po	f	of c cf pr cr po	f	mf	mf	mf	f mf of c cf pr cr po	f mf of c cf pr cr po	f mf of c cf pr cr po
ob	b	f mf of e ef pr cr po	b	of e ef pr cr po	b mb ob	ob of c e cb ef cf eb eq po	ob c cb po	b mb ob e eb	ob	ob e eb	ob c cb po	b mb ob	ob	b mb pr cr po	b mb pr cr po	b mb ob c cb pr cr po
of	b mb ob e eb pr cr po	f	ob e eb cr po pr	f	ob of c e cb ef cf eb eq po	f mf of	of c cf po	f mf of e ef	of c cf po	f mf of	of	of e ef	of	f mf pr cr po	f mf pr cr po	f mf of c cf pr cr po

(Continued)

Table 3. Continued.

O	b	f	mb	mf	ob	of	c	e	cb	ef	cf	eb	eq	pr	cr	po
c	b	f	b	f	b	f	c	*	c	f	c	b	c	b	b	b
					mb	mf				mf		mb		f	f	f
					ob	of				of		ob		mb	mb	mb
					c	c				c		c		mf	mf	mf
					cb	cf				cf		cb		pr	pr	ob
															cr	of
																c
																cb
																cf
																pr
																cr
																po

IF $(D(X) = \{+\}$ and $X < Y)$ or $(D(X) = \{-\}$ and $Y < X)$ then $X \, b_= Y$;

Else $X \, pr_= Y$

Else

IF $(D(X) = \{+\}$ and $X < Y)$ or $(D(X) = \{-\}$ and $Y < X)$ then $X \, f_{\neq} Y$;

IF $(D(X) = \{+\}$ and $Y < X)$ or $(D(X) = \{-\}$ and $X < Y)$ then $X \, b_{\neq} Y$;

Else $X \, pr_{\neq} Y$

{eq=, eq, cf=, cb=, c=, cf, cb, c, ef=, eb=, e=, ef, eb, e, ob=, of=, ob, of, po=, po, mf=, mb=, mf, mb, cr=, cr, b=, f=, b, f, pr=, pr } are 32 base relations of RNDIA. From the above procedure, we know they are JEPD (jointly exhaustive and pair-wise disjoint) relations. According to Proposition 4, { eq=, eq, cf=, cb=, c=, cf, cb, c, ef=, eb=, e=, ef, eb, e, ob=, of=, ob, of, mf=, mb=, mf , mb , b= , f= , b , f } are equivalent to DIA, since they have no branch (NB(X , Y)).

When the road network has no branch, such as a single highway without exit, and S is the start point of the highway, NB(X,Y) is always true. In this case, from the definition we can deduce that NRDIA is equal to DIA. So DIA is a special case of RNDIA (DIA RNDIA).

5 SPATIO-TEMPORAL REASONING OF RNDIA

Two main topics of STR are representation and reasoning. Composition reasoning is the basic operation for further reasoning works such as CSP. The composition is to determine xTz by xRy and ySz. In most cases, T cannot be unique. And {*} means the union of all base relations. The composition of relation S,T is represented by S $\bigcirc$ T. The composition of DIA has been specified by Renz (Renz 2001).

The composition of RNDIA cannot be automatically calculated since RNDIA is more complex than DIA. For instance, b = $\bigcirc$ f= = {*} = (If x behind y and y in front of z then x and z can be any relation without considering directions.)

Unlike DIA, we have no idea about automatically calculating the composition of RNDIA. By enumerating all possible compositions, we give the composition table of RNDIA (Tables 3, 4, 5).

Directions {=, $\neq$} are not considered in the composition table of RNDIA; they are determined by the following rule:

$$R_p, S_q \in \text{RNDIA}$$

$$R_p \circ S_q \begin{cases} (R \circ S)_p & \text{if } q = \{=\} \\ (R \circ S)_{\neg q} & \text{if } q = \{\neq\} \end{cases}$$

Table 4. Composition Table of RNDIA (II).

○	b	f	mb	mf	ob	of	c	e	cb	ef	cf	eb	eq	pr	cr	po
e	b mb ob e eb pr cr po	f mf of e ef pr cr po	ob e eb pr cr po	of e ef pr cr po	ob e eb po	of e ef po	ob of c e cb ef cf eb eq po	e	ob e eb po	e	of e ef po	e	e	pr cr po	cr po	po
cb	b	f	b	mf	b mb ob	of c cf	c	b mb ob e Eb	cb	cb ef eq	c	b mb ob	cb	b mb pr	b mb pr cr	b mb ob c cb pr cr po
ef	b mb ob e eb pr cr po	f	ob e eb pr cr po	mf	ob e eb po	of	of c cf po	E	cb ef eq po	ef	of	e	ef	pr cr po	cr po	po
cf	b	f	mb	f	ob c cb	f mf of	c	f mf of e ef	c	f mf of	cf	cf eb eq	cf	f mf pr	f mf pr cr	f mf of c cf pr cr po
eb	b	f mf of e ef pr cr po	mb	of e ef pr cr po	ob	of e ef po	ob c cb po	e	ob	e	cf eb eq po	eb	eb	pr cr po	cr po	po
eq pr	b b mb ob e eb pr cr po	f f mf of e ef pr cr po	mb b mb ob e eb pr cr po	mf f mf of e ef pr cr po	ob b mb pr cr po	of b mb pr cr po	c pr cr po	e b f mb mf pr	cb cb pr cr po	ef pr	cf pr cr po	eb pr	eq pr	pr *	cr b f mb mf ob of c cb	po b f mb mf ob of c cb

Table 5. Composition Table of RNDIA (III).

○	b	f	mb	mf	ob	of	c	e	cb	ef	cf	eb	eq	pr	cr	po
															cf pr cr po	cf pr cr po
po	b mb ob e eb pr cr po	f mf of e ef pr cr po	b mb ob e eb pr cr po	f mf of e ef pr cr po	b mb ob e eb pr cr po	f mf of e ef pr cr po	po	b f mb mf ob of e ef eb pr cr po	po	f mf of e ef pr cr po	po	b mb ob e eb pr cr po	po	b f mb mf ob of c cb cf pr cr po	b f mb mf ob of c cb cf pr cr po	*

The main reasoning problem in spatial and temporal reasoning is the consistency problem (RSAT), which is formally defined as follows.

Given a set Θ of constraints of the form xRy, where x,y are variables and R is an spatio-temporal relation such as DIA or RNDIA. Deciding the consistency of Θ is called RSAT. An assignment of spatio-temporal regions such that all the constraints are satisfied is a model of Θ.

Every pair of variables has constraints in Θ. If no information is given about the relation holding between two variables x and y, then the universal constraint x {*} y is contained in Θ.

We say that a set of constraints Θ' is a refinement of Θ if and only if the same variables are involved in both sets, and for every pair of variables x,y, if $xR'y \in \Theta'$ and $xRy \in \Theta$ then $R' \subseteq R$. Θ' is a consistent refinement of Θ if and only if Θ' is a refinement of Θ and both Θ' and Θ are consistent. A consistent scenario Θ_s of a set of constraints Θ is a consistent refinement of Θ where the relation of every constraint in Θ_s is a basic relation. The model of consistent constraints Θ could be described by Θ_s if all the basic relations between the variables are determined.

There are several ways of deciding the consistency of a given set of constraints over a set of relations. The most common way is to use backtracking over Θ. This is done by applying for each triple of constraints xTz , xRy , ySz the operation $T = T \cap (R \bigcirc S)$.

If the empty relation is not contained, the resulting set is path consistent. But the RSAT of RNDIA is at least NP-hard since the consistency problem of the DIA and Interval Algebra are NP-hard. Obtaining the tractable subset of RNDIA is our next work.

6 CONCLUSION

With the development of wireless communications and positioning technologies, the concept of moving object databases (MOD) has become increasingly important, and has posed a great challenge to spatio-temporal databases and spatio-temporal reasoning. Existing methods are not well equipped to handle qualitative representation and reasoning of a road network of moving objects. So we develop RNDIA in this paper.

The main contribution of this paper is a methodology for representing and reasoning a road network of moving objects. We extended the Directed Interval Algebra to Road Network Directed Intervals Algebra for dealing with road networks of moving objects. By applying a shortest path based order system, we established a mapping from the road network to a one dimensional line. RNDIA base relations are defined to be compatible with DIA relations. DIA is a special case of RNDIA. Then, the reasoning problems such as RSAT are discussed. Our work improves the application of Interval Algebra theoretical results and proposes an executable qualitative representation and reasoning method for road networks of moving objects.

ACKNOWLEDGEMENTS

Supported by NSFC Major Research Program 60496321, Basic Theory and Core Techniques of Non Canonical Knowledge; National Natural Science Foundation of China under Grant No. 60603030, 60373098, 60573073, the National High-Tech Research and Development Plan of China under Grant No.2003AA118020, the Major Program of Science and Technology Development Plan of Jilin Province under Grant No. 20020303, the Science and Technology Development Plan of Jilin Province under Grant No. 20030523, Youth Foundation of Jilin University (419070100102).

REFERENCES

Allen, J.F. 1983. Maintaining knowledge about temporal intervals. *Comm. ACM,* 26(11): 832–843.
Cheng, R., Prabhakar, S. & Kalashnikov, D.V. 2003. Querying imprecise data in moving object environments. *ICDE 2003. Proc. of the 19th IEEE International Conference on Data Engineering, Bangalore, India.*
Cohn, A.G. and Hazarika, S.M. 2001. Qualitative Spatial Representation and Reasoning: An Overview. *Fundamental Informatics* 46(1–2): 1–29.
Porkaew, K., Lazaridis I. & Mehrotra, S. 2001. Querying Mobil Objects in Spatio–Temporal Databases. In C.S. Jensen, M. Schneider, B. Seeger, V.J. Tsotras (eds), *Advances in Spatial and Temporal Databases*: 59–78. Berlin: Springer.
Renz, J. 2001. A spatial odyssey of the interval algebra: Directed intervals. In B. Nebel (ed.), *Proc. of the 17th Int'l Joint Conf. on AI*: 51–56. Seattle: Morgan Kaufmann.
Su, J., Xu, H. & Ibarra, O.H. 2001. Moving Objects: Logical Relationships and Queries. In C.S. Jensen, M. Schneider, B. Seeger, V.J. Tsotras (eds), *Advances in Spatial and Temporal Databases*: 3–19. Berlin: Springer.
Teresa Escrig, M. & Toledo, F. 1999. *Qualitative Spatial Reasoning: Theory and Practice.* Amsterdam: Ohmsha.
Van de Weghe, N., Cohn, A.G., Bogaert, P. & De Maeyer, P. 2004. Representation of moving objects along a road network. In A. Brandt (ed.), *Proc. of the 12th Int. Conf. on Geoinformatics*: 187–197. Gävle, Sweden
Vazirgiannis, M. & Wolfson, O. 2001. A Spatiotemporal Model and Language for Moving Objects on Road Networks. In C.S. Jensen, M. Schneider, B. Seeger, V.J. Tsotras (eds), *Advances in Spatial and Temporal Databases*: 20–35. Berlin: Springer.

Spatio-temporal analysis

Advances in Spatio-Temporal Analysis – Tang et al. (eds)
© 2008 Taylor & Francis Group, London, ISBN 978-0-415-40630-7

Interfacing GIS with a process-based agro-ecosystem model – Case study North China Plain

Georg Bareth
Department of Geography, GIS & RS, University of Cologne, Cologne, Germany

Zhenrong Yu
College of Resources and Environment, Land Use and Geoinformation, China Agricultural University, Beijing, China

ABSTRACT: The Sino-German Project between the China Agricultural University and the University of Hohenheim, Germany, focused on sustainable agriculture in the North China Plain. One major aim of the project was to set up a GIS-based Agricultural Environmental Information System (AEIS) for the North China Plain. In this contribution, the importance of the linkage of process-based agro-ecosystem models – in this case, the Denitrification and Decomposition Model—with GIS is emphasized. The purpose of the linkage is the modelling of agro-environmental impacts on a regional level. A key issue in the GIS-based regional modelling is the establishment of an adequate geodatabase, which is defined here as the AEIS. By using the AEIS, it is possible to model the emission of N_2O, the volatilization of NH_3, and the leaching of NO_3^- in the winter wheat/summer maize area for the entire North China Plain by employing the DNDC model.

1 INTRODUCTION

The Sino-German Project between the China Agricultural University in Beijing and the University of Hohenheim, Germany, started in November 1998, lasted for 4½ years and was located in Beijing. The focus of the project was on sustainable agriculture in the North China Plain (NCP) (http://www.uni-hohenheim.de/chinaproject). The geographic extent of the NCP is shown in Figure 1. One major aim of the project was the establishment of an experimental field near Beijing to investigate different agricultural practices and their impact on yield and the environment. The crops and vegetables investigated were winter wheat, summer maize, spinach and cauliflower. Intensive measurement equipment was installed in the experiment fields to collect long term data on soil water fluxes, volatilization of ammonia and greenhouse gas emissions. Plant and root data were sampled as well to provide all necessary data for intensive modelling, evaluation and calibration. Additional investigations into pesticide, herbicide and fungicide applications were carried out. Questionnaires were used to collect information concerning agricultural practices and the economical situation of the farmers. Finally, plant and animal monitoring and mapping were completed to investigate the biodiversity in the study region.

Apart from the field experiment, the second task was to set up an Agricultural Environmental Information System (AEIS) for the NCP (Bareth and Yu 2002), which almost exceeds the size of Germany. According to Bill (1999), an Environmental Information System (EIS) is an extended GIS for the description of the state of the environment, referring to critical impacts and loads. An EIS serves for the capture, storage, analysis and presentation of spatial, temporal and attribute data and provides the basis for measures for environmental protection. Consequently, the establishment of an AEIS for the simulation of sustainable scenarios means the setting up of an extensive geo- and attribute database. The modelling of the C- and N-cycles in agro-ecosystems especially require numerous input parameters such as pH, soil texture, fertilizer N-Input, animal waste input, use of irrigation water, dates of sowing and harvest, yield, etc. (Li et al. 2001).

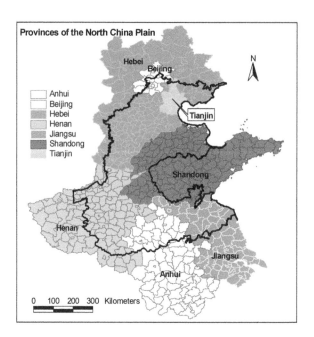

Figure 1.   Provinces and geographical boundary of the NCP.

According to Bareth and Yu (2002), an AEIS for sustainable agriculture includes five different information systems, namely:

- Base Geo Data Information System (BGDIS)
- Soil Information System (SIS)
- Climate Information System (CIS)
- Land Use Information System (LUIS)
- Agricultural Management Information System (AMIS)

Most important for the spatial matching of all data in the AEIS is the integration of a BGDIS. The BGDIS should provide topographical data, elevation lines or a Digital Elevation Model (DEM), and an administrative boundary data set. The SIS is essential for providing soil parameters for the agro-ecosystem modelling. Therefore, the SIS has to include (i) spatial soil information in the form of maps and (ii) a detailed description of the soil types, including soil genesis, physical and chemical soil properties. The CIS provides the necessary climate/weather information. Climate/weather maps can be generated from point data using GIS interpolation methods. Land use data should be provided by a LUIS. Detailed land use information on crops should be stored in the LUIS. For detailed agro-ecosystem modelling, the information level in available land use maps is rather poor. The analysis of multi-spectral, hyper-spectral and/or radar data from satellite or airborne sensors is a standard method of acquiring such information. Finally, the AMIS is a crucial component. For agro-ecosystem modelling, farm management data like fertilizer N-Input, animal waste input, use of irrigation water, dates of sowing and harvest, yield, etc. are a must. Such spatial information is usually not available and has to be established. In the end, the AEIS is the sum of the described information systems. They are linked to each other using GIS technologies. Additionally, models and methods for data analysis have to be integrated in the AEIS. It is possible to link or even integrate complex agro-ecosystem models into GIS (Hartkamp et al. 1999) and consequently into the AEIS. Therefore, the AEIS for the NCP provides information for decision support and could also be regarded as a spatial decision support system (SDSS). The major objectives of the AEIS for the NCP are to provide information (i) about agriculture in the region, (ii) about the impact of agricultural practices on the environment, and (iii) of simulation scenarios for sustainable strategies.

2 INTEGRATION OF AGRO-ECOSYSTEM MODELS INTO GIS

Traditionally, process based agro-ecosystem models or agronomic models in general are developed and used for site or field scales. A major problem for regional applications of such kinds of models is the limited availability of regional data. The scientific and technical progress in spatial sciences, especially in GIS and remote sensing (RS) technologies, has enabled the setting up of soil, land use, climate and agricultural management information systems. Unfortunately, complete data sets for regions are hardly available and therefore the regional application of agro-ecosystem models is still difficult. The lack of data availability can nowadays be solved by using GIS and RS technologies. Consequently, interfacing GIS with (agro-)ecosystem models is becoming more important.

Seppelt (2003) describes the coupling of GIS and models in a theoretical and technical context. Hartkamp et al. (1999) introduce four different ways to interface GIS with agronomic models. They also introduce definitions for the different ways of interfacing. This is an essential point because the use of terminology that has not been defined causes confusion. Hartkamp et al. (1999) define four terms that are frequently used as follows:

Interface: The place at which diverse (independent) systems meet and act on or communicate with each other.
Link: To connect.
Combine: To unite, to merge.
Integrate: To unite, to combine, or incorporate into a larger unit; to end segregation.

The three different methods of interfacing GIS with models are described as follows. Linking of GIS with models is basically just an exchange of files or data. In this case, the model is independent from the GIS and vice versa. Only the results of each system are exchanged. Combining GIS with models involves the processing of data and the automatic exchange of data. Finally, integrating GIS and models describes the real incorporation of one system into the other.

Hartkamp et al. (1999) give a very good overview of examples of interfacing GIS and models. They describe 46 publications: 22 discuss linking; 23 give examples of combining; but only 6 are of the integrating type. None of the integrating GIS-model interfaces described are for agro-ecosystem models focusing on C- and N- dynamics. Most of them deal with topography, soil erosion and groundwater flow.

For the regional modelling of C- and N-dynamics in (agro-)ecosystems on a regional scale, the integration of such models is especially important and can be regarded as a key issue. The available approaches to GIS-model interfacing for agro-ecosystem models hardly use the GIS analysis capabilities. For example, the Denitrification and Decomposition Model (DNDC) (Li et al. 2001) has, in its latest versions, a regional model part included. Spatial parameters such as land use and soil are not considered in their spatial relation. GIS is just used for the display of results on county level using a county map. Plant (1998) describes a GIS-DNDC-interface of the linking type for the Atlantic zone of Costa Rica. This study shows clearly the importance of interfacing agro-ecosystem models with GIS. This approach is not automated and consequently very difficult to use for other regions. Therefore, a method for integrating agro-ecosystem models into GIS is introduced that enables fully automated spatial model simulations.

In the approach, the DNDC model is used for the integration into a GIS. The DNDC Model was developed at the University of New Hampshire in the early 1990s (Li 2000). Its original purpose was the simulation of N_2O and NO gas fluxes from agricultural fields on a daily basis, but it also has a very detailed soil C sub-model. It is currently the most used simulation tool for N related gas fluxes and has been employed for county based national inventories of N_2O for China (Li et al. 2001), the USA (Li et al. 1996), the UK (Brown et al. 2002), Germany (Werner 2003) and several other countries. The soil organic matter sub-model is made up of three different soil C pools, each being divided into a labile and resistant fraction. The plant growth sub-model rather simply models biomass at a potential growth rate that is modified by environmental factors. The key driving variables are climate, soil properties and agricultural management. Even though it has a very simple plant growth sub-model, it is capable of simulating trends in soil C quite accurately (Li et al. 1997). In another simulation study of an arid farmland ecosystem in China, the DNDC model was also able to model the influence of different management practices on soil C content (Liang & Xu 2000).

The method for the integration of agro-ecosystem models is based on the soil-land-use-system approach (SLUSA) (Bareth et al. 2001). The SLUSA is based on the ecosystem approach described by Matson and Vitousek (1990). This approach was furthermore disaggregated in order to estimate and visualize the greenhouse gas emissions (CH_4, CO_2, N_2O) from agricultural soils for a distinct region (Bareth 2000). The disaggregation of the ecosystem approach was undertaken using a GIS and available digital spatial data (Bareth 2000). The SLUSA is a GIS and knowledge based approach for environmental modelling. Its first step sets up a GIS that contains relevant data. In the second step, GIS tools are used to overlay climate, soil, land use, topography, farm management and other data, for example preserved areas or biotope (if available). This procedure is the basis for the spatial related identification of different soil-land-use-systems. In the third step, emission data and process knowledge of the region of interest, as well as from literature, are linked to these systems. For the linkage, potential classes of N_2O emissions are created and knowledge based production rules are programmed. The latter are commonly used to generate new knowledge from expertise in knowledge based systems, such as expert systems (Wright 1993). Available data from the region of interest are considered with higher priority. Finally, GIS software tools are used in the fourth step to visualize and to quantify the generated emission data. The quantitative emission estimation is achieved by multiplying the areas of each N_2O emission potential class with the representative value of each class. The new spatial emission data can be integrated into other regional models, such as the regional farm models (Bareth and Angenendt 2003).

Based on the SLUSA, Figure 2 describes the new and real integration of agro-ecosystem models into GIS on the database level. The agro-ecosystem model now operates like an algorithm in the database. While the steps 1, 2 and 4 remain in the method, the third step of the SLUSA is changed (Fig. 2). The implementation of the knowledge based production rules of the SLUSA is replaced by the integration of the agro-ecosystem model. All input parameters for the model derive from the GIS database. The advantage of this method is the automatic spatial modelling. The GIS integrated DNDC model uses the input parameters from the

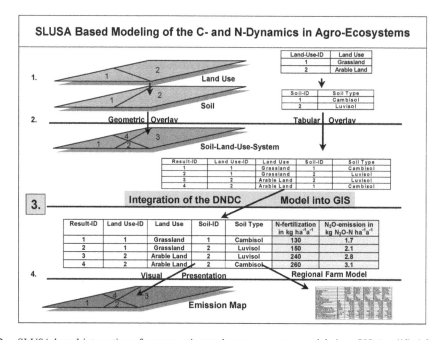

Figure 2.   SLUSA-based integration of process orientated agro-ecosystem models into GIS (modified from Bareth et al. 2001).

fields of the GIS database. Therefore, the DNDC model program code was changed for that purpose and necessary interfaces were programmed (Huber et al. 2002). For the input, fixed field names were created in the DNDC model and the GIS database must use these definitions. The results of the model runs are automatically written into predefined fields of the GIS database. Consequently, the results of the modelling can immediately be used for regional calculations and visualizations using GIS software tools. Finally, a new menu button is integrated in ArcGIS that allows users easy access to DNDC modelling (Huber et al. 2002).

The evaluation of models is a very important step. For this procedure, measurement data are necessary. For Dongbeiwang Township (the measurement site), the evaluation of the DNDC model can be achieved by using measurement data from the experiment field. For example, long-term measurements of gas fluxes with the automatic closed-chambers-technique are available from the research project. The chambers and the automatic measurement system were designed mainly by Martin Kogge of sub-project B2 of the Sino-German Project (http://www.uni-hohenheim.de/chinaproject/martin.htm). The chambers are designed to measure CO_2, CH_4 and N_2O. For the three different treatments in the field experiment, measurements were carried out with three repetitions per plot. The treatments are conventional fertilization and irrigation, optimized fertilization and irrigation, and conventional fertilization and optimized irrigation. N_2O measurements are available for summer maize in 2001. For the traditional treatments, the total emission for the period of 21 June 2001 to 9 October 2001 for summer maize was 2.35 (± 0.48) kg N_2O-N ha^{-1}.

4   RESULTS

In Figure 3, the integrated (geo)data of the AEIS for the North China Plain is displayed. The official digital topographic database (1:250,000) was purchased as the Base Geodata Information System (BGDIS). On the national level, the Institute of Geomatics in Beijing is responsible for topographical data. In 1999, they finished the digital topographical database at the scale of 1:250,000 (DTD250) in vector format. They provide the data as ArcInfo coverage and in many other GIS formats. The data set includes separate files for rivers, lakes, roads, railways, cities, one degree net, 50 m contour lines and administrative boundaries

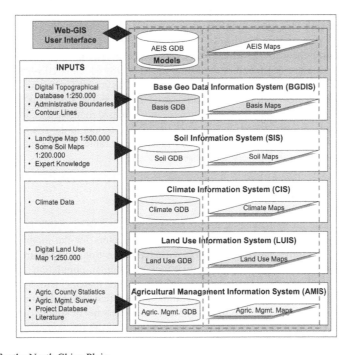

Figure 3.   AEIS for the North China Plain.

Table 1. Annual N_2O-emission, NH_3-volatilization and NO_3^--leaching for the summer maize/winter wheat area of the North China Plain (DNDC model).

Year	N_2O	NH_3	NO_3^-	Total
	Gg N yr^{-1}			
1995	52	1358	426	1837
2000	53	1373	429	1855

(counties, districts, provinces) (Bareth and Yu 2004). The layers were digitized from the topographical maps at the same scale. The dates of the maps range from the early to the late eighties. A DEM, generated from the contour lines, is available too.

Compared to the availability and the access of topographical and land use data, soil data are much more difficult to obtain in China. Several soil maps at a scale of 1:200,000 were digitized. The problem here was the lack of information about the soil properties. Therefore, the 1:500,000 land type map for the North China Plain was digitized and soil properties for the soil units were derived from a Chinese textbook on soil (Hseung and Li 1990).

The Climate Information System (CIS) consists of daily weather data for the years 1999–2002. The weather data were purchased through the German Weather Service (DWD) for nine weather stations in the North China Plain. Thiessen polygons were then created to provide spatial weather data on a daily base. Similar approaches were used by Mathews and Knox (1999).

Digital spatial land use data are available in China at the scales of 1:250,000 and 1:100,000 in ArcInfo vector format. The land use data is derived from LandsatTM land use classifications. The Institute of Remote Sensing Applications (IRSA) of the China Academy of Sciences in Beijing is responsible for the land use classifications. For the North China Plain, the data were obtained at a scale of 1:250,000. Additionally, statistical county data for the same years were obtained.

Finally, the Agricultural Management Information System (AMIS) was created on the basis of statistical county data, available survey data, the data of the research project and from literature. For each weather region, a specific management for the wheat/maize rotation is specified.

All the mentioned data are organized in the frame of the AEIS, which is GIS-based. Consequently, it is possible to identify unique weather-soil-land-use-management systems (agro-ecotopes). Approximately 240,000 polygons were created overlaying the spatial soil, land use and weather data. By using the database functions within the GIS, 285 unique agro-ecotopes for wheat/maize rotation were identified. The parameters of these 285 agro-ecotopes are used as input to the DNDC model in the site mode. The final result of this modelling process for the North China Plain is shown in Table 1. The spatial distribution of the emission of N_2O is presented in Figure 4, as an example.

5 DISCUSSION AND CONCLUSIONS

Li et al. (2001) modelled a national inventory of N_2O emissions from arable lands for the whole of China with the process-based DNDC model. The input parameters for the model are daily minimum and maximum air temperature, daily precipitation, nitrogen deposition, soil texture, pH, organic matter content, individual crop areas, N-fertilizer use, livestock and human populations, tillage management, irrigation management, planting and harvest dates, and crop residue management. Most of the agricultural parameters derive from official census data. The authors modelled low N_2O emissions for the North China Plain. These did not correspond to the daily field measurements of summer maize in 2001, which are presented in this contribution. One explanation could be that the measurements of 2001 cannot be compared to the model results of data of 1990. But from process knowledge, high emissions from summer maize are expected due to high precipitation, high temperatures and high N-input during the growing

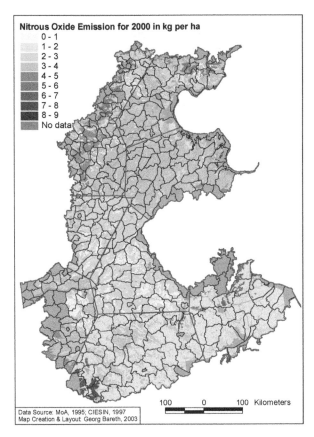

Figure 4. N$_2$O emission from summer maize/winter wheat areas in the North China Plain.

season. Another explanation for the low regional model results could be that the agricultural census data used do not reflect the agricultural practices of distinct crop management. Only average N-fertilizer data per county are available and applied. The use of averaged fertilizer data is a standard method in regional modelling (Brown et al. 2002, Li et al. 2001, Werner 2003) and can be regarded as a key problem in regional modelling. It does not reflect the philosophy of process-based models that are designed to model site specific differences. The modelling of N fluxes especially requires regionally differentiated N input data. In this study, it was generated on the basis of the land use map and on linked knowledge of regional N-input for the specific crops. Additionally, the quality of the census data has to be discussed in general. Tso et al. (1998) mention that the census data are not reliable and there are wide discrepancies, e.g. of about 25–40 per cent in the amount of agricultural land actually under cultivation. The key question therefore is whether regional process-based models should be applied on a poor spatial database.

Additional results of regional modelling of annual N2O-emission, NH3-volatilization and NO3–leaching are presented in Tables 2 and 3. Different modelling approaches were used. In Table 2, the results of the empirical method for national greenhouse gas inventories of the International Panel on Climate Change (IPCC) are presented. In this empirical method, the given emission and leaching factors are multiplied by numbers derived from county statistics, e.g. local input of mineral N-fertilizer. While the total amount of N losses are very similar to the modelling results of the DNDC (Table 1), the fluxes of NH3- and NO3- are almost the opposite.

In Table 3, the results of the GIS- and knowledge-based approach (SLUSA) are listed. The overall N-losses are again in the same range as the previously discussed results. But the losses from N$_2$O-emission, NH$_3$-volatilization and NO$_3^-$-leaching vary significantly from the results of the DNDC model and the IPCC

Table 2. Annual N_2O-emission, NH_3-volatilization and NO_3^--leaching for the summer maize/winter wheat area of the North China Plain (IPCC).

Year	N_2O	NH_3	NO_3^-	Total
	Gg N yr^{-1}			
1995	61	433	1300	1794
2000	62	438	1313	1813

Table 3. Annual N_2O-emission, NH_3-volatilization and NO_3^--leaching for the summer maize/winter wheat area of the North China Plain (SLUSA).

Year	N_2O	NH_3	NO_3^-	Total
	Gg N yr^{-1}			
1995	27	975	960	1962
2000	27	985	970	1982

method. Very interesting is the small amount of simulated N_2O-fluxes. Overall, the results of the SLUSA are considered to be the most realistic because regional measurements are integrated in this approach. Therefore, evaluation and calibration of the DNDC model and the IPCC method have to be carried out on a regional level.

In this contribution, it is clearly shown that there is great potential in linking GIS and agro-ecosystem models. Key problems in this regional modelling approach are still data availability and evaluation and calibration of the agro-ecosystem models. Regional AEIS has to be established for general regional modelling and more regional measurement data have to be available for these purposes, as well as for the assessment of the simulation results. There should be discussion on whether different methods for regional modelling of agro-ecosystems are more appropriate for the available spatial databases. This could mean that approaches like the IPCC method or the SLUSA are real alternatives to process-based modelling on a regional level when the quality of spatial databases is poor.

For the regional use of process-based models, the enhancement of spatial data quality and resolution is a must. Consequently, remote sensing and GIS should become the key tools to generate better land use, soil and weather information systems to derive the input parameters for regional agro-ecosystem modelling. Finally, the impact of using averaged census data as input parameters for regional modelling should be discussed more intensively.

ACKNOWLEDGEMENT

The project was funded by the German Ministry of Education and Research (http://www.bmbf.de).

REFERENCES

Bareth, G. 2000. *Emissions of greenhouse gases from agriculture – regional presentation and estimation for a dairy farm region by using GIS.* Hohenheimer Bodenkundliche Studien. (In German with English summary).
Bareth, G. & Angenendt, A. 2003. Ökonomisch–ökologische Modellierung von klimarelevanten Emissionen aus der Landwirtschaft auf regionaler Ebene. *Ber. über Landwirtschaft* 81: 29–56.

Bareth, G., Heincke, M. & Glatzel, S. 2001. Soil–land–use–system approach to estimate nitrous oxide emissions from agricultural soils. *Nutr. Cycl. Agroecosys* 60: 219–234.

Bareth, G. & Yu, Z. 2004. Availability of digital geodata in China. *Peterm. Geogr. Mitt.* 148: 78–85. (In German with English summary).

Bareth, G. & Yu, Z. 2002. Agri–Enviro–Information–System for the North China Plain. *Proceedings of the International Society for Photogrammetry and Remote Sensing (ISPRS) Commission II Symposium of Integrated Systems for Spatial Data Production, Custodian and Decision Support, 20–23 August 2002, Xian*: 23–29.

Bareth, G., Yu, Z. & Doluschitz, R. 2002. Set–up of a Regional Agricultural Environmental Information System (RAEIS) in China. *Proceedings of the EnviroInfo'2002, 25–27 September 2002, Vienna*: 233–240.

Bill, R. 1999. *Grundlagen in GIS, Band 1*. Heidelberg: Germany Wichmann. (In German).

Brown, L., Syed, B., Jarvis, S.C., Sneath, R.W., Phillips, V.R., Goulding, K.W.T. & Li, C. 2002. Development and application of a mechanistic model to estimate emission of nitrous oxide from UK agriculture. *Atmospheric Environment* 36: 917–928.

Hartkamp, A.D., White, J.W. & Hoogenboom, G. 1999. Interfacing GIS with agronomic modeling: A Review. *Agronomy Journal* 91: 761–772.

Hseung, Y. & Li, C. 1990. *Soils of China*. Beijing, China: Science Press.

Huber, S., Bareth, G. & Doluschitz, R. 2002. Integrating the process–based simulation model DNDC into GIS. *Proceedings of the EnviroInfo'2002, 25–27 September 2002, Vienna*: 649–654.

Li, C. 2000. Modeling trace gas emissions from agricultural ecosystems. *Nutr. Cycl. Agroecosys* 58: 259–276.

Li, C., Zhuang, Y., Cao, M., Crill, P., Dai, Z., Frolking, S., Moore, B., Salas, W., Song, W. & Wang, X. 2001. Comparing a process–based agro–ecosystem model to the IPCC methodology for developing a national inventory of N_2O emissions from arable lands in China. *Nutr. Cycl. Agroecosys* 60: 159–175.

Li, C., Frolking, S., Crocker, G.J., Grace, P.R., Klir, J., Korchens, M. & Poulton, P.R. 1997. Simulating trends in soil organic carbon in long–term experiments using the DNDC model. *Geoderma* 81: 45–60.

Li, C., Narayanan, V. & Harriss, R. 1996. Model estimates of nitrous oxide emissions from agricultural lands in the Unites States. *Global Biogeochemical Cycles* 10: 297–306.

Liang, Y.L. & Xu, B.C. 2000. Simulated carbon and nitrogen contents in arid farmland ecosystem in China using denitrification–decomposition model. *Communications in Soil Science and Plant Analysis* 31: 2445–2456.

Matthews, R. & Knox, J. 1999. Up-scaling of methane emission – experimental results: final report. In *Dept. Natural Resources Management*. Silsoe, UK: Cranfield University.

Matson, P.A. and Vitousek, P.M. 1990. Ecosystem approach to a global nitrous oxide budget. *Bio Science* 9: 667–672.

Plant, R.A.J. 1998. GIS–based extrapolation of land use–related nitrous oxide flux in the Atlantic zone of Costa Rica. *Water, Air, and Soil Pollution* 105: 131–141.

Seppelt, K. 2003. *Computer–based environmental management*. Weinheim, Germany: Wiley–VCH.

Werner, C. 2003. *Erstellung eines N–Spurengas–Emissionskatasters für land– und forstwirtschaftlich genutzte Böden der Bundesrepublik Deutschland*. Master Thesis. Inst. f. Geographie, Bayerische Julius–Maximilians–Universität Würzburg.

Wright, J.R. 1993. GIS and spatial modeling. In J.R. Wright, L.L. Wiggins, R.K. Jain & T.J. Kim (eds), *Expert systems in environmental planning*: 83–84. Berlin: Springer.

133

Advances in Spatio-Temporal Analysis – Tang et al. (eds)
© *2008 Taylor & Francis Group, London, ISBN 978-0-415-40630-7*

An activity-based spatio-temporal data model for epidemic transmission analysis

Tao Cheng
Department of Geomatic Engineering, University College London, United Kingdom
School of Geography and Planning, Sun Yat-sen University, Guang Zhou, China

Zhilin Li
Department of Land Surveying and GeoInformatics, The Hong Kong Polytechnic University,
Hong Kong, China

Jianhua Gong
State Key Laboratory of Remote Sensing Science, Institute of Remote Sensing Applications, Chinese
Academy of Sciences, Beijing, China

ABSTRACT: Recently GIS have been used in the surveillance and monitoring of diseases and the control of epidemics. Most of those mapping systems use aggregated datasets. These aggregated datasets are not sufficient to support the analysis of epidemiological transmission if the disease is spread mostly by person-to-person and from region to region such as SARS—Severe Acute Respiratory Syndrome. This paper develops an activity-based spatio-temporal data model to support epidemic transmission analysis in a GIS environment by identifying spatial and temporal opportunities for activity participation. The model can support tracing and prediction of spatially varying, temporally dynamic and individually based epidemiological phenomena. A prototype system based on the data model is implemented using a case study from Hong Kong.

1 INTRODUCTION

Severe acute respiratory syndrome (SARS) is a highly infectious and potentially lethal atypical form of pneumonia that begins with deceiving, common flu-like symptoms. All around the world during the peak of the outbreak in the first half of 2003, SARS was negatively affecting every aspect of daily life: economic, social, travel, work, at school and home. Due to its spatial analysis and display capability, GIS is well suited for studying associations between location, environment and disease. Recently, GIS have been used in the surveillance and monitoring of diseases and control of epidemics. Many universities and research institutes, and some GIS companies, provided Internet SARS mapping services, which allowed public health decision makers, travellers and local populations at risk to visually monitor and appreciate at a glance changes, trends and patterns buried in different online SARS datasets that were continuously varying with time during the 2003 outbreak (Boulos 2004).

However, most of those mapping systems use aggregated datasets. These are not sufficient to support the analysis of SARS transmission since the disease is spread mostly by person-to-person and from region to region following the network of contacts between them. Through this network, the SARS disease spreads through space and time, just like most infectious diseases. Since the contact between persons is realized through their common activities, this paper develops an activity-based spatio-temporal data model to represent such spatially varying, temporally dynamic, and individually based epidemiological phenomena.

We will first review the transmission process of epidemic disease in order to understand the relationship between individuals and the relationship between individuals and the environment. This will set forth the requirement of the spatio-temporal data model. Then we will present the design of the data model based upon a mobility-oriented view by incorporating the transmission process. Finally, a prototype system will be implemented using a case study based in Hong Kong.

2   CONCEPTUAL FRAMEWORK OF EPIDEMIC DISEASE TRANSMISSION

Individual-based epidemiological modelling requires the data model to consider discrete individuals as the modelling unit. The model should also represent the characteristics and behaviours of the individuals, the relationships between them and the environment, and changes in these characteristics and their interaction through time and space (Bian 2004).

(1) Unique individuals
Individual-based modelling requires that a conceptual model be based on the following assumptions: (1) individuals are different; (2) individuals interact with each other locally; (3) individuals are mobile; and (4) the environment for individuals is heterogeneous. These considerations are widely supported in several disciplines that have experienced a major shift away from population-based and towards individual-based approaches (Adams et al. 1998, Bian 2000, Bousquet et al. 2001).

(2) Disease development
An infection history can be considered as a sequence of distinct periods, each of which begins and ends with a discrete event. The critical periods include the latent, the infectious, and the incubation periods. The critical events include the receipt of infection, the emission of infectious material, and the appearance of symptoms (Baily 1975).

(3) Interaction/contact between individuals
An individual participates in a sequence of activities on a daily basis. Some of the activities are stationary and some are mobile. Stationary activities occur at a physically fixed location, such as a home or a workplace. At these locations, the individual may interact with other individuals in a group activity. When a group dissolves, an individual travels through space and time to another location, often joining another group. Local infections occur at a stationary location and the long-distance dispersions occur through travel (Miller 2004).

(4) Infectious factors
The probability of infection depends on the attributes of the individual, such as age, the infection status of other individuals in the group, and the contact structure within the group (Cliff & Haggett 1990).

3   AN ACTIVITY-BASED SPATIO-TEMPORAL DATA MODEL

3.1   *Activity-based approach*

Recently, activity theory has experienced a renaissance, as researchers have expanded its power and scope. Moreover, there are increasingly applications in the GIS and transport communities through the use of an activity-based approach (Wang & Cheng 2001, Miller 2005). Activity theory focuses on people rather than places as the source of travel and location demands. Individuals participate in periodic activities that have varying levels of necessity and urgency. The resources that satisfy these activities are sparsely distributed in space and time, i.e. at few locations and for limited time intervals. This can include requirements that other individuals be contemporaneous in space, time or both (such as work or socializing). Individuals must distribute their limited time among these activities, using transportation to trade time for space when travelling to activity locations. A major application of activity theory is empirical measurement and analysis of space-time activity (STA) data, or records of where and when individuals conducted activities over a daily, weekly or monthly cycle (Miller 2005).

Activity-based approaches received some early attention in the late 1980s and became a dominant modelling approach (Wang & Cheng 2001). Although most authors believe that GIS, through the capabilities offered for the collection, storage, manipulation, analysis and display of spatially referenced data, could prove critical to the ultimate success of the activity-based approach, all agree that existing GIS techniques need to be further developed to meet the requirements of activity-based modelling. This is partly because, compared with traditional trip-based models, activity-based models are dependent on detailed individual-level data, spatially more accurate representations of the physical environment, and the development of more efficient information processing technologies. More importantly, most models available in GIS are static descriptions of reality. They are not dynamic and do not describe spatial changes and/or change with time (Burrough & Frank 1995), but in activity-based modelling, the activity patterns of individuals are dynamic, and change in space and time.

Due to the increasing prominence of activity-based approaches to transportation and urban analysis, as well as increasing abilities to collect STA data, there have been several recent attempts to develop conceptual and logical data models to support activity analysis. Despite the complex, many-to-many relationships inherent in these data, including the relationships between individuals, households, travel, activities and their spatial and temporal dimensions, one of the major challenges in developing activity-based data models is to eliminate redundancy as much as possible (Shaw & Wang 2000). Another consideration is representing complex temporal dynamics where an individual is sometimes moving along a continuous space-time trajectory but at other times is stationary in space (Wang & Cheng 2001). Shaw & Wang (2001) developed a relational data model for handling disaggregate STA data. The central entity of their model is the 'trip' or a movement from one location to another. Other data such as the trip location (spatial), trip timing (temporal), the trip maker (a person within a household of other trip makers) and relevant trip attributes are linked to the trip. Trip locations are represented as paths through a network maintained using a variable-length segmentation model. While effective at maintaining data on multi-stop/multi-purpose trips, this data model does not contain the support for activity data required by time geography and activity theory. In these theories, activities and projects in space and time drive travel and telecommunication demands. Wang & Cheng (2001) formulated a STA data model that encompassed activities and projects to a greater degree (Miller 2005). The next subsection returns to the main principle of conceptual design of the activity-based spatio-temporal model presented in Wang & Cheng (2001).

3.2 *A mobility-oriented data model*

From the viewpoint of mobility status, activity patterns can be described by two types of state: Stay_At or Travel_Between. Therefore, the activities of a person can be considered as the interaction of a person with locations, either Stay_At or Travel_Between locations (as illustrated in Fig. 1). If we use mobility to represent the characteristics of travel-between, and no-mobility to describe the characteristics of stays-at, this view can be considered as a mobility-oriented view. This view treats activities as dynamic and occurring within the largely static transportation space. It deals explicitly with the moving behaviour of discrete objects (individuals). This view can also be used to describe the dynamics of travel vehicles, such as cars, buses and airplanes.

The stepwise type of spatio-temporal behaviour or dynamics, (for example, the dynamics underlying activity patterns), are revealed as the change in spatial positions, i.e. the moving of human objects in a city. Thus, the change of locations should be emphasized. Since this kind of dynamics is not always continuous, the time (period) in which the change of location happened should be recorded explicitly. Further, the time scale is not fixed in the sense that activity patterns can be defined for a day or a week. Moreover, the activity patterns are influenced by the physical and institutional environmental constraints. The physical constraints are spatial constraints, which include the information of the home and work address, the

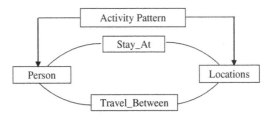

Figure 1.   A mobility-oriented view of an activity of a person (Wang and Cheng 2001).

locations of shopping centres and the network of roads. The institutional constraints require that the information about the opening hours, facilities of locations and household information of the persons should also be represented in the model. All these specialities make the spatio-temporal data model for activity-based modelling more complicated. We will follow the mobility-oriented view discussed above to design our model.

As discussed above, individuals' activity patterns can be considered as a series of Stay_At or Travel_Between in space during a particular time period. Thus, an activity pattern can be presented as a function of Stay_At and Travel_Between in the following formula:

$$Activity\ Pattern = \{Stay\_At(1), Travel\_Between(1), \dots, Stay\_At(i),$$

$$Travel\_Between(i), \dots, Travel\_Between(n-1), Stay(n)\} \tag{1}$$

$$Stay\_At(i) = f(SL_i, ST_i^s, ST_i^e, Aim) \tag{2}$$

$$Travel\_Between(i) = f(TL_i^s, TL_i^e, TT_i^s, TT_i^e, Path) \tag{3}$$

where SLi represents the location of Stay_At(i), and ST_i^s and ST_i^e represent the starting time and ending time of Stay_At(i), respectively; Aim represents the activity of this stay, such as working or playing sport; TL_i^s and TL_i^e represent the starting location and the end location of Travel_Between(i), respectively (i.e. the destination of the previous activity and the destination of the next activity). Similarly, TT_i^s and TT_i^s represent the starting time and end time of Travel_Between(i). Finally, path is the transport route of Travel_Between(i).

The spatial and temporal constraints can be represented by Equations 4 to 7.

$$SL_i = TS_i^s \quad (i = 1, \dots, n-1) \tag{4}$$

$$SL_{i+1} = TL_i^e \quad (i = 1, \dots, n-1) \tag{5}$$

$$ST_i^e = TT_i^s \quad (i = 1, \dots, n-1) \tag{6}$$

$$ST_{i+1}^s = TT_i^e \quad (i = 1, \dots, n-1) \tag{7}$$

Equation 4 means that the location of a stay is the start location of the next Travel_Between and Equation 5 means that the end location of Travel_Between(i) is the location of Stay_At(i + 1). Equation 6 states that the end time of Stay_At(i) is the start time of Travel_Between(i + 1), whose end time is the start time of the Stay_At(i + 1) (Equation 7). There are cases when two stays are at the same location and there is no Travel_Between between them. These equations still work for such cases because we may define $SL_i = SL_{i+1}$. However, there might be other spatio-temporal constraints, such as the time of playing sport should be within the opening hours of a sports studio, which is not discussed here since these constraints are more related to the methods of modelling.

The relationships between elements expressed in the equations can be represented in a hierarchical structure as illustrated in Figure 2. Since activity-based modelling normally uses household as the analytical unit, because interactions between household members are important factors determining activity engagement, we include household and person as two elements in the diagram, in addition to the major elements of activity patterns. It shows that a Household consists of several persons. Furthermore, each Person has a planned activity programme, which is then realized as an activity pattern. The fourth layer of the diagram illustrates the two major elements (Stay_At and Travel_Between) consisting of the activity pattern and their topological relationships. Under the element of Stay_At, the location and duration (which is defined by the starting time and the end time) of the stay is indicated. The element of Travel_Between is defined by from where to where the travel takes place, how long the travel lasts (which is defined by the start and end time of the travel), what transport mode is used, and which path it traverses. All Stay_At and Travel_Between are connected by topological relations between time and locations. For example, Location (i) of Stay_At(i) is the start location of Travel_Between(i) (Equation 4) and the end location of Travel_Between(i) – Location(i + 1), is the location of Stay_At(i + 1) (Equation (5)).

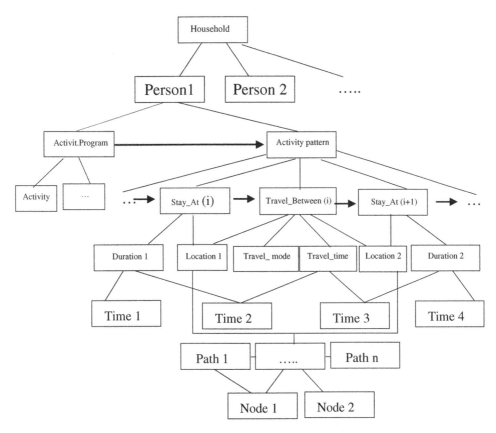

Figure 2. A conceptual framework of the hierarchy relationships between elements in mobility-oriented activity-based modelling (Wang & Cheng 2001).

Based upon the mobility-oriented view, the next section will discuss the extended model based upon the model presented here to support the transmission of epidemic diseases spreading through the common activities involved by two or more persons, either staying in the same place or travelling by the same means of transport.

4 AN EXTENDED ACTIVITY-BASED DATA MODEL FOR EPIDEMIC TRANSMISSION

The conceptual framework of epidemic disease transmission sets forth the guidelines for the design of the data model.

(1) An individual infection process—status of individuals

Since infection history is a sequence of distinct periods, it is appropriate to represent the individual infection as a series of discrete events and periods (see Fig. 3). The discrete periods indicate the infection status of an individual and are part of the individual's characteristics.

There are five statuses of an individual, corresponding to five periods, triggered by four events:

☐ Normal (or health): the individual is normal and is not infected;
☐ Latent: the individual receives the infection, but does not emit infectious material;
☐ Infectious: the individual is in the period of emitting infectious material;
☐ Recovery: the symptom appears and the individual stays in the hospital or is isolated until recovery;
☐ Carrying anti-body: the patient recovers and produces anti-body to the infection.

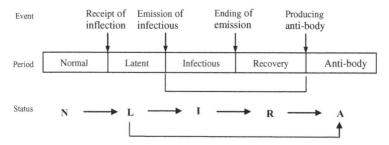

Figure 3.   An individual infection process and status.

These five states are represented as N, L, I, R and A in Figure 3. When an individual is in the infectious period, if contact is made with others, others might be infected.

(2) Mobility of individuals

The mobility and activity of individuals describe the interaction of individuals with the environment. The activities of a person can be considered as the interaction of a person with locations: either *Stay_At* or *Travel_Between* locations. If we use mobility to represent the characteristics of *Travel_Between*, and no-mobility to describe the characteristics of *Stays_At*, this view can be considered as a mobility-oriented view. This view treats activities as dynamic, occurring within the largely static transportation space. It deals explicitly with the moving behaviour of discrete objects (individuals).

Under the element of *Stay_At*, the location and duration (which is defined by the starting time and the end time) of the stay are indicated. The element of *Travel_Between* is defined by from where to where the travel takes place, how long the travel lasts (which is defined by the start and end time of the travel), what transport mode is used, and which path is traversed. All *Stay_At* and *Travel_Between* are connected by topological relations between time and locations (see Fig. 2).

(3) Infection transmission—constraints for interaction

The infection transmits through the direct contact of individuals. The interaction of individuals is realized through common activities, either "Stay_At" the same location at the same time, or "Travel_Between" locations by the same travel model at the same time. When a person Pj, who is in a normal status ("N"), interacts with a person Pi who is in infectious status ("I"), the infection will be transmitted form Pi to Pj. This process of transmission can be formalized as:

$$\text{If} \quad \text{Status(Pi)} = \text{"I" and Status(Pj)} = \text{"N"} \tag{8}$$

$$\text{and} \quad Activities(Pi) \cap Activities(Pj) \neq \Phi \tag{9}$$

$$\text{then Transmit(P}i, \text{P}j)$$

Equation 8 represents the constraint for individuals' status and Equation 9 represents the spatio-temporal constraint of the interaction.

Equation 8 can be further divided into two cases. Case one refers to the situation where two individuals are in a stationary mode, i.e. staying at a same location. Case two refers to the situation where two individual are in moving mode, i.e. travelling together between locations.

For the case of a stationary model, i.e. Stay_At, the constraint *Activities* $(Pi) \cap A(Pj) \neq \Phi$ requires that

$$L_1 = L_2 \qquad \text{and}$$
$$Ts_i \leq Te_j \leq Te_i \leq Te_j \ \text{or}$$
$$Ts_j \leq Ts_i \leq Te_j \leq Te_i \ \text{or}$$
$$Ts_i \leq Ts_j \leq Te_j \leq Te_i \ \text{or}$$
$$Ts_j \leq Ts_i \leq Te_i \leq Te_j. \tag{10}$$

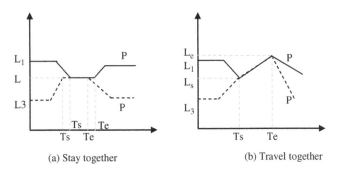

(a) Stay together   (b) Travel together

Figure 4.   Two cases of interaction of two persons.

The contact time of the two individuals for these four situations are as follows, respectively

$$\text{Contact\_time}(i,j) = \begin{cases} Te_i - Te_j \\ Te_j - Ts_i \\ Te_j - Ts_j \\ Te_i - Ts_i \end{cases}. \tag{11}$$

For the case of moving mode, i.e. Travel_Between, $Activities\ (Pi) \cap A(Pj) \neq \Phi$ requires that

$$Ls_i = Ls_j \text{ and}$$
$$Le_i = Le_j \text{ and}$$
$$Ts_i = Ts_j \text{ and}$$
$$Te_i = Te_j. \tag{12}$$

The contact time of the two individuals for this case is

$$\text{Contact\_time}(i,j) = Te - Ts \tag{13}$$

These two cases are illustrated in Figure 4, with Figure 4a representing one situation of "Stay_At" and Figure 4b representing the situation of "Travel_Between".

(4) Incorporating infectious magnitude

Taking the infectious magnitude into account, extract information should be provided for the description of the physical environment and personal status.

For the properties of Household, the population density (Pdensity) of the area where the household is located could be considered as the indicator of the infectious magnitude. The higher the density of population, the higher the infectious magnitude will be.

For the properties of Person, age can be considered as an indication of being infected. For different ages, the chance of being infected is different. Therefore, the date of birth should be provided in order to derive the information on age.

In the properties of Location, different location types (Ltype) correspond to different infectious magnitudes. So does the activity type (Atype) of Activity, the travel model (Tmode) of Travel_between and the stay type (Stype) of Stay_At.

Therefore, the following objects are essential for the database, and are defined by the properties in the brackets. The infection magnitude related properties are in italic.

HOUSEHOLD(Hid, Pnum, Income, Home Address, *Population density*);
PERSON(Pid, Name, Salary, Role, Working place, Occupation, *Date of birth*, *Infectious Status*);

141

LOCATION(Location address, *Location type*, Opening time, Closing time, Facility)
ACTIVITY(Aid, *Activity type*, Earliest starting time, Latest starting time, Duration)
TRAVEL-BETWEEN(Location1, Location2, Stating time, Ending time, *Travel mode*)
STAY_AT(Location, Starting time, Ending time, S*tay type*)

(5) Infectious probability
The infectious probability of an individual can be calculated by the following formula.

$$\text{Infectious probability} = f(\text{age, environmental factor, contact time}) \qquad (14)$$

Here, the environmental factor refers to the *Population density, Activity type, Stay type, Location type*, or *Travel mode*.

5 CASE STUDY

We used the case presented in Wang & Cheng (2001) to test our model. Table 1 presents the information on six families. The information about the husband and wife of these six families is presented in Table 2.

The husbands' and wives' daily activities are collected and listed in Tables 3 and 4.

In order to incorporate infectious factors into the transmission process, the following tables are defined. Table 5 represents the activity/location/stay type with the infectious magnitude. Table 6 presents the relationship of the travel model with the infectious magnitude, Table 7 presents the age with the infectious magnitude and Table 8 presents the population density with the infectious magnitude.

Table 1. Information on families.

HID	PERSON_NUM	INCOME	AddressID	Pop_Density
1	3	70,000	Parkview Garden	3
2	4	70,000	Xiao Shun Chun	3
3	2	60,000	Ma An Shan	3
4	4	120,000	Royal Ascot	4
5	4	80,000	University	2
6	4	100,000	Mei Lin Chun	4

Table 2. Table of PERSON for the six families.

PID	Name	HID	ROLE	OCCPUATION	Working_Address	Date_Of_Birth
1	Winter	1	Husband	Teacher	Baptist University	19660701
2	Nancy	1	Wife	Researcher	University	19690808
3	Jone	2	Husband	Teacher	University	19600305
4	Krystal	2	Wife	Sectary	University	19610506
5	Samen	3	Husband	Teacher	Baptist University	19550403
6	Ingeberg	3	Wife	Housewife	At-home	19600908
7	Mike	4	Husband	Teacher	Baptist University	19480101
8	Linda	4	Wife	Housewife	At-home	19500705
9	Lincon	5	Husband	Teacher	University	19780604
10	Karain	5	Wife	Student	University	19800912
11	Peter	6	Husband	Teacher	Baptist University	19640321
12	Amy	6	Wife	Business	Central	19651126

Here we try to simulate the transformation of the disease within these six families. To begin with, one infected individual is introduced into the population. In our case, Person 2 (Nancy) is infected. The infection for the rest of the population depends on three operational steps. The first is to identify the individuals who contact this infected individual based upon their activities (presented by the Tables Stay_At and Travel_Between). In the second step, those individuals are assigned an infected status according to infection

Table 3.   Part of table of *STAY_AT* (Wang & Cheng 2001).

SAnum	Pid	Laddress	Stype	Ts	Te
1	1	Parkview Garden	4	7:00:00 AM	8:00:00 AM
2	1	Kowloon Tong	1	8:30:00 AM	12:00:00 PM
3	1	Kowloon Tong	3	12:00:00 PM	1:00:00 PM
4	1	Kowloon Tong	1	1:00:00 PM	7:00:00 PM
5	1	Parkview Garden	4	8:00:00 PM	11:59:00 PM
6	1	Festival Walk	2	7:10:00 PM	7:50:00 PM
7	2	University	1	9:00:00 AM	12:00:00 PM
8	2	University	5	12:00:00 PM	12:30:00 PM
9	2	Parkview Garden	4	7:00:00 AM	8:30:00 AM
10	2	University	1	12:30:00 PM	5:30:00 PM

Table 4.   Part of table of *TRAVEL_BETWEEN* (Wang & Cheng 2001).

TBnum	Pid	Laddress1	Laddress2	Ts	Te	Tmode	BSAnum	ASAnum
1	1	Parkview Garden	Kowloon Tang	8:00:00 AM	8:30:00 AM	1	1	2
2	2	Parkview Garden	University	8:30:00 AM	9:00:00 AM	1	9	7
3	1	Kowloon Tang	Festival Walk	7:00:00 PM	7:10:00 PM	4	4	6
4	1	Festival Walk	Parkview Garden	7:50:00 PM	8:00:00 PM	1	6	5
5	2	University	Parkview Garden	5:30:00 PM	6:00:00 PM	1	10	11
6	3	Xiaochui Chun	University	8:40:00 AM	9:00:00 AM	3	12	19
7	4	Xiaochui Chun	Kowloon Tang	8:00:00 AM	8:30:00 AM	1	18	
8	4	Kowloon Tang	University	8:30:00 AM	9:00:00 AM	1		13
9	4	University	Xiaochui Chun	12:30:00 PM	1:00:00 PM	1	14	15
10	4	Xiaochui Chun	Sha Tin	3:00:00 PM	3:30:00 PM	1	15	16
11	4	Sha Tin	Xiaochui Chun	5:00:00 PM	5:30:00 PM	6	16	17
...	...	...	...	...	...	...	...	...

Table 5.   Table of Activity/Location types/ Stay type.

Atypeid	Type	Infectious Magnitude
1	Work Place	5
2	Shopping Centre	4
3	Sports Centre	3
4	Home	5
5	Railway Station	4
6	Restaurant	3
7	Theatre	3
8	School	5

Table 6. Table of *Travel_ Mode.*

TMode	TRAVEL	Infectious
1	Train	3
2	Bus	4
3	Car	5
4	Walking	1
5	Bike	1
6	Airplane	4

Table 7. Table of the infectious magnitude with age.

AgeClass	Age	Infectious Magnitude
1	0–2	5
2	3–12	4
3	12–30	2
4	30–40	1
5	40–60	3
6	>60	5

Table 8. Table of the infectious magnitude with population density.

DClass	Density (per km^2)	Infectious Magnitude
1	0–100	1
2	101–1000	2
3	1001–3,000	3
4	3,001–5,000	4
5	>5,000	5

probability calculated using Equation (3). The third step identifies those already infected individuals who will contact other individuals and continue the spread of infection throughout the population. To simplify the calculation, the simulation uses the infection probabilities listed in the above tables (Tables 5–8) for the susceptible and a one-day latent period and a four-day infection period for the infection.

Table 9 presents the transmission of the disease within these six families for 14 days (here only the activities of "Stay_At" are used in the case study since the programming for the activity of "Travel_Between" is still under development). It shows that in Day 1 Person 2 is infected and is infectious. Person 2 transferred the disease to Persons 1 and 3 in Day 2, so that they were in the Latent period. In Day 3, Person 1 transferred the disease to Person 5, and Person 3 transferred the disease to Person 4. So Persons 4 and 5 were in the Latent period in Day 4. In Day 5, Person 2 showed the symptoms of the disease and was transferred to hospital in the recovery period. Persons 1, 3, 4 and 5 were in the infectious period in this day, Day 6. In Day 7, Persons 1, 2 and 3 were in the recovery period and Persons 4 and 5 are infectious, as in Day 6. In Day 9, Persons 4 and 5 joined the others in the recovery period. The transmission stopped in Day 9.

Based upon Table 9, we can identify the status of individuals so that the total number of infectious individuals can be summarized. We can also trace the transmission from person to person, and trace the transmission from place to place based upon the locations associated with the activities of these persons. In the case of an isolation policy being implemented to stop the spread of the disease, we can locate the individuals to be potentially infected, due to their contact with the infectious individuals.

Table 9. Transmission of disease from person to person.

	Day 1	Day 2	Day 3	Day 4	Day 5	Day 6	Day 7	Day 8	Day 9
Infectious	2	2	2	2	1	1	4	4	
			1	1	3	3	5	5	
			3	3	4	4			
					5	5			
Latent		1		4					
		3		5					
Recovery					2	2	1	1	1
							2	2	2
							3	3	3
									4
									5

6 CONCLUSIONS

This activity-based model conceptualizes the spatial and temporal interactions of travel and activity behaviours using the concept of mobility. The activity patterns are conceptualized as a sequence of staying at or travelling between activity locations. Based upon the mobility-oriented view, the transmission of epidemic disease is modelled as spreading through the common activities of two or more persons, either staying in the same place or travelling by the same means of transport. The model can support the predication and tracing of the spread of epidemics by identifying spatial and temporal opportunities for activity participation, which will facilitate epidemiological studies and policy making to control epidemics.

Although the case used to illustrate the application of the extended data mode is simple, it has all the essential data that one may encounter in epidemic transmission modelling. As for the infectious probabilities, only age is considered as an associated probability of infection in the case study since the health history is confidential, though it can be included easily as another property of Person (see Table 2). Furthermore, only activities of "Stay_At" have been programmed for the case study. Adding the activities of "Travel_between" is definitely just trivial. However, the searching efficiency of the algorithm needs to be improved to deal with larger amounts of data in the database.

As for further research, virtual simulation of transmission presented dynamically on maps will be of great interest. This will make possible interactive movies that can be viewed at all intermediate time points and that would support queries regarding relationships among health outcomes in space and through time.

ACKNOWLEDGEMENTS

This study is funded by the Chinese Ministry of Education (985 Program, Project no. 105203200400006) and the Hong Kong Polytechnic University (Project no. G-YW92).

REFERENCES

Adams, L.A., Barth-Jones, D.C., Chick, S.E. & Koopman, J.S. 1998. Simulations to evaluate HIV vaccine trial designs. *Simulation* 71: 228–241.

Bian, L. 2004. A conceptual framework for an individual-based spatially explicit epidemiological model. *Environment and Planning B* Vol. 31: 381–395.

Baily, N.T.J. 1975. *The Mathematical Theory of Infectious Diseases and its Applications*. New York: Hafner Press.

Bian, L. 2000. Object-oriented representation for modeling mobile objects in an aquatic environment. *International Journal of Geographical Information Science* 14: 603–623.

Boulos, M.N.K. 2004. Descriptive review of geographic mapping of severe acute respiratory syndrome (SARS) on the internet. *International Journal of Health Geographics* 3:2.

Bousquet, F., Page, C.L., Bakam, I. & Takfoyan, A. 2001. Multiagent simulations of hunting wild meat in a village in eastern Cameroon. *Ecological Modeling* 138: 331–346.

Burrough, P.A. & Frank, A.U. 1995. Concepts and paradigms in spatial information: are current geographical information systems truly generic? *International Journal of Geographical Information Systems* 9: 101–116.

Cliff, A.D. & Haggett, P. 1990. Epidemic control and critical community size: spatial aspects of eliminating communicable diseases in human populations. In R.W. Thomas (ed.), *London Papers in Regional Science 21. Spatial Epidemiology*: 93–110. London: Pion.

Miller, H.J. 2004. Necessary space-time conditions for human interaction. *Environment and Planning B* 32: 381–401.

Miller, H.J. 2005. "What about people in geographic information science?' In P. Fisher and D. Unwin (eds), *Re-Presenting Geographic Information Systems*: 215–242.

Shaw, S.L. & Wang, D. 2000. Handling disaggregate spatiotemporal travel data in GIS. *GeoInformatica* 4: 161–178.

Wang, D. & Cheng, T. 2001. A spatio-temporal data model for activity-based transport demand modelling. *International Journal of Geographical Information Science* 15: 561–586.

Advances in Spatio-Temporal Analysis – Tang et al. (eds)
© 2008 Taylor & Francis Group, London, ISBN 978-0-415-40630-7

Visual exploration of time series of remote sensing data

Ulanbek D. Turdukulov & Menno J. Kraak
ITC, GIP Department, Enschede, The Netherlands

Valentyn Tolpekin
ITC, EOS Department, Enschede, The Netherlands

ABSTRACT: The approach described here facilitates the exploration of a time series of remote sensing data by offering a set of interactive visual functions that allows the study of the behaviour of dynamic phenomena and events, and the evolution of phenomena over time. Essentially, we simulate the human visual system by decomposing each spatial region into a number of 2D Gaussian functions. The algorithm compares the attributes of each function in successive frames and finds the continuous paths of the spatial object through time. These paths describe the characteristics of the object in each time step and for each object certain interesting 'events' can be described. Examples of events are: continuation, appearance/disappearance and split/merge of the phenomena. Based on the object paths, a visualization environment with 'temporal functionality' is created with a wide range of tools to support interactive exploration of events and the object's evolution. In particular, the event graph is proposed that, in combination with other visualizations, will make the process of detection and exploration of dynamic phenomena independent of the perception and experience of the observer.

1 INTRODUCTION

One of the challenging research areas within GIS is the study and representation of dynamic phenomena. A commonly used technique for monitoring dynamics in many environmental applications is remote sensing. Recent developments in remote sensing devices generate data with higher spectral, spatial and temporal resolutions. Further, remote sensing data are often pre-processed by the supplier and coupled with methods and techniques to extract various bio-physical indicators (e.g. water quality parameters, vegetation characteristics, soil properties, climatic parameters). Exploration and analysis of these fast growing spatio-temporal datasets is not a trivial task considering the amount of data being generated today, with remote sensing data alone projected to yield 50 gigabytes of data per hour (Wachowicz 2000).

The purpose of exploration is to find patterns, trends and relationships in remote sensing data for generating hypotheses and understanding the dynamics of spatial objects. One way of supporting this exploratory process is by developing visualization methods. Visualization enables the researcher to gain insights into large spatio-temporal datasets, leading to better understanding of the phenomena and their modelling.

Generally, visualization includes two important aspects: graphical representation of spatial information and support for the user's visual interaction with this information. A visual exploratory process is characterized by highly interactive and private use of representations (MacEachren & Kraak 2001). Currently, there are three main types of representation used to depict the temporal aspects of spatial data in GIS: a single static map, a series of static maps, and animations. A single map/image represents a snapshot in time and, taken together, the maps make up an event. The main problem associated with static maps is that the visualization is restricted to only a few time slices—the dynamism of the phenomena is not maintained and interaction with the spatial data is limited.

Animation is often used for large time series to integrate many maps and to add dynamics and interaction to the representation of time. It can be very useful to clarify trends and processes, as well as to provide

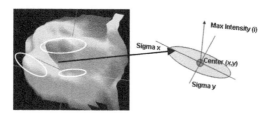

Figure 1.   3D view of hypothetical dataset with the fitted 2D Gaussian functions on the left. Properties of the Gaussian function are shown on the right.

insight into spatial relations (Kraak 2000). While animations can be effective in perceiving changes in a single spatial object, they tend to become overwhelming when representing the relationships and evolution of several objects. This is due to the fact that the human visual system has limited bandwidth and, because images fade quickly, it requires an effort to maintain them (Gahegan 1999). Therefore, even a playback animation, with the user controlling the flow, has two major drawbacks known as change blindness and inattentional blindness (Rensink 2002). Since dynamics are rarely explored only with respect to the evolution of a single object, many changes go unnoticed by the users of animation (for some amazing examples, see URL: Visual Cognition Lab).

One way to improve existing visualizations is by providing computational support to the human exploratory tasks. Exploratory tasks can be broadly divided into two stages: identifying each spatial object in each time step and comparing characteristics of each object for further understanding of the dynamics. Object tracking methods, essentially simulating a human visual tracking process, have a relatively long history in the field of scientific visualization and signal processing. While object tracking is still a field of continuing research, the purpose of this paper is to present an approach for coupling object tracking and visualization in a single exploratory environment and to describe the representations and functionality resulting from such a combination. Therefore, in the following sections, we first discuss the object tracking method, followed by a visualization example where we illustrate the principles and interactive support of different graphs designed for visual exploration of events and evolutionary stages of dynamic spatial phenomena.

2   BASIC DEFINITIONS

Each domain has its own set of interesting objects or features. These are usually defined as regions of interest in the remote sensing datasets that satisfy certain constraints: for example, an area of low NDVI values may define a stressed vegetation (Samtaney et al. 1994). In the following, we consider a scalar function (corresponding to a single band image) defined on a 2D set of image pixels. The choice of the function depends on the application domain. Our basic assumption is that this function can be decomposed into a superposition of elementary template functions having simple and well-defined form. Each template function corresponds to a single spatial object and may have different properties (spatial extent and magnitude) defined by the function parameters. The choice of a shape for the template function depends on the process at hand. The most natural function form has a maximum value in the centre of the object and decreases as the distance from the centre increases. An important decision should be made as to whether to choose a template function with finite or infinite support (corresponding to overlapping or non-overlapping spatial objects being modelled) for a given application. Overlapping functions should be used when spatially continuous physical problems are studied. In what follows, we model spatial objects with a 2D Gaussian function, which is a commonly used model in geosciences (Fig. 1). Although a Gaussian function has infinite support, it decays quickly outside the region defined by the spread of the function.

The advantages of using a template function to model a spatial phenomenon are manifold: firstly, it reduces complex datasets to a description of the overlapping functions (in terms of function parameters); secondly, the overlapping functions can describe regions with complex shapes; and thirdly, and most importantly, it leads to a robust comparison and consequently to tracking of the functions over time.

Thus, a set of spatially overlapping 2D Gaussian functions represents the physical phenomenon in the spatial region. The evolution of the spatial region can be described by following its Gaussian functions.

3   EXTRACTING AND MODELLING OBJECTS FROM RS DATA

We translated the problem of spatial object extraction into a problem of decomposition of a complex function into a superposition of template functions having simple form. If the shape of the template function is non-linear (as in the case of the Gaussian function chosen in this paper) then such decomposition is a non-trivial problem. The approach described here is based on a template function of Gaussian shape, but it might be easily adapted for other functions having continuous derivatives.

We perform the function decomposition in an iterative way. First, the approximate location of the highest magnitude template function is found (by searching for a pixel with highest magnitude). Then a region of points is sampled around this seed point in such a way that the pixel magnitude in the sampled region is decreasing with increasing distance from the seed point. This ensures that there is no contribution from different functions in the sampled region.

The sampled region is used to fit the profile with a Gaussian function. The fitting is done in two steps. First, one-dimensional fitting with a one-dimensional Gaussian function is performed separately in the vertical and horizontal cut of the sampling region:

$$A(x) = C_0 + C_1 \exp\left(-\frac{1}{2}\left(\frac{x - x_0}{\sigma_x}\right)^2\right) \tag{1}$$

$$A(y) = D_0 + D_1 \exp\left(-\frac{1}{2}\left(\frac{y - y_0}{\sigma_y}\right)^2\right) \tag{2}$$

Equation 1

Here, $A = A(x,y)$ is the image magnitude at the pixel with coordinates (x,y); $A(x)$ and $A(y)$ are correspondingly the horizontal and the vertical cross-sections of the sampled region; $x0$ and $y0$ are the coordinates of the seed point; σx and σy are the parameters describing the extent of the Gaussian profile in the horizontal and vertical directions.

The results of this fitting are used to find the approximate parameters of the two dimensional Gaussian function parameters. The second step does the fitting of the whole sampled region with a 2D Gaussian profile:

$$A(x, y) = A_0 + A_1 \exp\left(-\frac{1}{2}\left[\left(\frac{x'}{a}\right)^2 + \left(\frac{y'}{b}\right)^2\right]\right) \tag{3}$$

where x' and y' are the transformed coordinates given by

$$x' = (x - x_0)\cos\theta - (y - y_0)\sin\theta$$
$$y' = (x - x_0)\sin\theta + (y - y_0)\cos\theta$$

a and b denote the extent of the function in the principal axes and θ corresponds to the tilt of the principal axes to the original coordinates (x,y).

The function fitting is performed using the steepest gradient descent method (Press et al. 1992). A check is performed in order to detect artefact objects, such as those too small or too large or with poor quality fit (controlled by the fit error). The thresholds are determined in an empirical way for each application domain. If an artefact object is detected, it is destroyed by setting all pixel values in the sampled region to zero. Otherwise, the parameters of the fit are stored and the contribution of the object is subtracted from the image. This ensures proper treatment of overlapping objects.

Then the process of object detection and extraction is repeated for the next object. Since the algorithm removes the highest magnitude object in each step of the iterative procedure, the result is a sequence of objects sorted by magnitude in descending order. We can set the lowest object magnitude as the threshold for the termination of the process.

As a result of the object extraction, each image corresponding to a single time point is replaced by a set of records containing the parameters of 2D Gaussian objects. This significantly simplifies the information contained in the image and can be used as an input to the object tracking process.

4 TRACKING OBJECTS AND DETECTING EVENTS

The result of the extraction step is a number of features, each described by a template function at each time step. At this stage, it is not known yet if two features in successive images are actually one (moving) object in time. That is exactly the purpose of tracking—to solve a correspondence (identity) problem and to establish a path of each spatial object over time. The underlying assumption of our tracking process is that the objects representing the phenomenon and their parameters evolve consistently between the successive time steps.

The tracking process consists of two main steps: initialization and prediction-verification. The initialization phase is performed on the first three successive images to form the initial paths of the objects.

In order to initialize a pair of features in successive images for their future correspondence, the following characteristics are used: location, spread and magnitude of each feature. Given the feature in image (i), the search for the corresponding feature in image $(i + 1)$ starts from the closest until it reaches the limit of the search distance, which is proportional to the spread of the feature in image (i). Each candidate feature found within the search distance in image $(i + 1)$ is compared to the given feature in image (i) and a similarity measure is calculated. Similarity is expressed in terms of a probability and the identity is assigned to the function with the highest probability.

$$P(k,l) = P_{magn}(k,l)^* P_a(k,l)^* P_b(k,l) \qquad (4)$$

P indicates the similarity probability (the value ranges between 0 and 1) of two features k and l based on their magnitude A1 (see Eq. 2), and spreads in the directions a and b as given below:

$$P_{magn}(k,l) = \frac{2^* \sqrt{A_{1k} * A_{1l}}}{A_{1k} + A_{1l}} \qquad (5)$$

$$P_a(k,l) = \frac{2^* \sqrt{a_k * A_l}}{a_k + a_l} \qquad (6)$$

$$P_b(k,l) = \frac{2^* \sqrt{b_k * b_l}}{b_k + b_l} \qquad (7)$$

The initialization phase serves as a basis for a prediction-verification phase of the tracking process. To predict the state of each feature (described by location, spread and magnitude) for the successive time step, we use a P-th order autoregressive model for stationary time series, widely used in time series analysis and known as a forecasting process. A P-th order autoregressive model relates a forecasted value of the time series to a linear combination of past P values. Then, the forecast of the feature state is compared to the actual feature state extracted from the corresponding image. Comparison is according to Equation 4.

The output of the tracking process is the path of each feature in time. These paths are used for detecting events on the level of spatial regions—spatial objects.

Events are stages in the evolution of the phenomena. There exist many different types of events depending on the application, the type of spatial objects, and on the user's interest (Reinders 2001). Our reasoning on events is based on the identity of each feature and the total mass of the region formed by combination

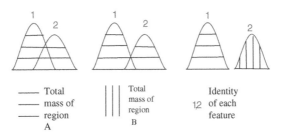

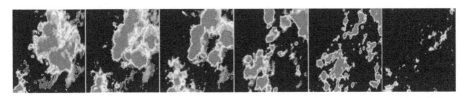

Figure 2. An example of a split event in the history of region A. Initially, region A is a composition of two features # 1 and 2 (left and middle image). As the region evolves, the feature #2 splits from the region A and forms a new region B (right image).

Figure 3. Subset of the example dataset, showing cloud movement at every hour starting at 10–00h GMT.

of given features. We detect a continuation event if the spatial region on image (i) consists of identical features as the region on image (i + 1) and if the total mass of feature identities composing the region on image (i + 1) is more than 50 per cent of the total mass of the region in the previous image. Then it is presumed that the region in the current image is a continuation of the region from the previous image and both regions form a path of the same spatial object. Similarly, a split is defined if the Gaussian identity loses connectivity with the spatial region (see Fig. 2). A merge is defined as being opposite to the split. Appearance (disappearance) is defined if the region in the current (previous) image fits none of the above criteria or the appearance (disappearance) of an isolated Gaussian function occurs.

Output of the tracking phase is the paths of spatial objects, which are essentially tree-like structures, carrying spatial, temporal and relationship information necessary for creating the exploratory visualization environment.

5 VISUALIZING THE EVOLUTION OF THE SPATIAL OBJECT

Though the approach is designed to be generic, we show an example of cloud movement in five time steps of one hour. The data used for the study were obtained from the METEOSAT Second Generation satellite with a temporal resolution of 15 minutes. The spatial resolution of the satellite is 3 km and the image has the following dimensions: 550*500 pixels and 20 time steps (a subset of the image is shown in Figure 3).

Extraction of clouds was achieved by slicing the band IR 10.8 with a threshold value of −40°C (240°K), as the top cloud temperature is lower than that of the earth's surface. Cloud data are often used for rainfall estimation, based on their duration and temperature gradient over time (the clouds with large temperature gradients are more likely to be convective, indicating a high probability of rain).

The patterns of these cloud regions were extracted and tracked over time to form the object paths, which are input to the visual exploratory system.

The design of the visual exploratory system largely depends on the task being supported and it is still a largely intuitive and ad hoc process (Ferreira de Olivera & Levkowitz 2003). In practice, however, the exploration tasks often vary depending on the perspective of the expert or the aspect of a phenomenon

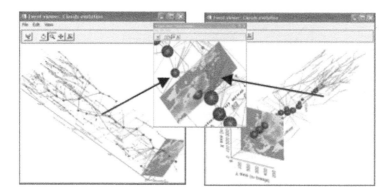

Figure 4. The proposed event graph, combined with iconic representation. On the left, the radius of icons is logarithmically plotted, while on the right the radius is directly proportional to the size of the region. In the middle, a zoomed section of the event graph.

being studied (Yuan 1997). Therefore, the best approach to the design of an exploratory environment is the combination of multiple linked views to show different aspects of object evolution.

5.1 Multiple linking representations

In the visual environment, linking allows changes in one representation to be reflected in all of them. The purpose of linking multiple views is to facilitate comparison. Currently, our exploratory environment consists of three main linked representations: an event graph, an animation and an attribute graph.

The purpose of the event graph is to show the evolution of tracked spatial objects. Essentially, our event graph is organized as a space-time cube: the bottom of the cube represents geographic space and the events and regions are drawn along the vertical time axis. Time, always present in the space-time cube, automatically introduces dynamics (Kraak 2003). The event graph includes iconic representations based on spatial attributes (position and size) of the path information and time information (Fig. 4). The event graph is realized in object graphics, meaning that the three-dimensional cube can be manipulated in space to find the best possible view (known as a focusing technique).

Each object path (consisting of several regions) has its unique colour. The radius of the icons is proportional to one of the attributes of the region (in Fig. 4 it is proportional to the size). Events are represented by connected tubes of different colour (e.g. red is a continuation event, green—merge and blue—split).

The event graph represents the generalized and abstract view of the evolution of spatial objects. In order to be effective in visual exploration, the event graph has to be linked to other spatial representations, in particular to image animation. While images are animated, the event graph has its sliding geographical space along the vertical axis synchronized to the individual images in the animation.

5.2 Functionality of linked representations

Besides the functionality already mentioned (linking views and focusing), we implemented an object brushing tool. With this tool, the users can set queries by the direct manipulation of icons on the event graph. The brushing tool works according to the filtering principle: all objects that were not selected during the user action are removed from the graphical displays. Combined with the linking technique, it allows the effect of querying in the other linked views to be observed—in animations and attribute graphs. In animations, the trajectories and spatial extent of the selected objects can be examined, while attribute graphs can display various characteristics of the objects—rate of change of area (position, etc.) in time. In the future, we intend to employ dynamic map symbols in animations for displaying trajectories and attributes of moving regions and to extend the brushing tool to spatio-temporal domains. This will allow the user to highlight specific individual time moments or areas through time or space controls, and view states of the objects at these moments or changes taking place between two time moments.

6 DISCUSSION

Conceptually, the combination of methods is generic: the identification of spatial regions can gain from research in signal processing and computer vision. The result of these identifications is often represented as a raster surface, indicating either some bio- or geo-physical indicators, or classification maps with the associated membership values. Thus, the algorithm presented can be applied to any of these surfaces if the static properties of the phenomena can be approximated by a Gaussian function.

Regarding the computation efficiency, the processing of the presented images took about 20 minutes: 19.5 minutes to fit the template functions and 10 seconds to perform the tracking and visualization on an Intel® Pentium® M processor with 1.7 GHz and 1GB of RAM. The computational efficiency can vary greatly depending on the amount of noise in the images, the pre-processing steps applied and on the fitting criteria of template functions used (i.e. minimum peak intensity threshold and the error of fit).

Extracting and tracking features preserves the structure of spatial regions and at the same time reduces the datasets: each region can be described in terms of its relations (events and feature identities) and attributes (position, size, total mass, etc.). This improves the rendering capacity and adds interactivity and dynamism to the resulting visualizations. Regarding the attributes, the combination can be easily extended to add in other measurements of location and spatial distribution depending on the purpose of the application and the events the user wants to represent. Also, functionality towards data mining and knowledge construction can be easily added (for example, to find objects with large temperature gradients on Figure 3, indicating convective types of cloud).

Most of all, the main advantage of using template functions is related to the nature of remote sensing images and spatio-temporal modelling aspects of the approach: since it is largely based on a prediction-verification scheme, it is more efficient in handling missing values that are often present in remote sensing datasets (e.g. due to cloud cover). Despite the conceptual advantages, the detailed evaluation of the algorithm, its accuracy and relative performance to other tracking algorithms needs further investigation.

7 CONCLUSION

The described approach uses advances in the fields of scientific visualization and computer vision and can be seen as part of an ongoing trend to integrate scientific visualization methods and methods for the exploration of spatial data. This trend has given name to the new discipline—geovisualization—and its current research agenda includes both representation aspects of geospatial information and integration of visual with computational methods of knowledge construction (URL: ICA Visualization 2003).

The object tracking algorithm essentially simulates a low level human visual task of identification of spatial patterns. Contrary to the human visual tracking process, the proposed method detects changes in spatial objects regardless of the number of changes present in the scene and does not depend on perception and the experience of the observer.

Thus, with proper visualizations, the user can pay more attention to higher order visual tasks—comparison and its related activities: hypothesis generation, hypothesis confirmation and further analysis. In particular, the event graph can show the essence of the object's evolution and history. Built in three-dimensional space, it utilizes the graphical variables (position, form, colour, orientation, size etc.) to portray both events and stages of the object's evolution. Synchronized with other spatial representations, the behaviour of dynamic phenomena and events, and the evolution of the phenomena can be explored interactively.

REFERENCES

Ferreira de Olivera, M.C. & Levkowitz, H. 2003. From visual exploration to visual data mining. *IEEE Transactions on Visualization and Computer Graphics* 9(3): 378–394.
Gahegan, M. 1999. Four barriers to the development of effective exploratory visualization tools for the geosciences. *International Journal of Geographic Information Science* 13(4): 289–309.

Kraak, M-J. 2000. Visualisation of the time dimension. In L. Heres (ed.), *Time in GIS: Issues in spatio-temporal modelling* 47: 27–35. Delft: Netherlands Geodetic Commission.

Kraak, M-J. 2003. The space-time cube revisited from a geovisualization perspective. *Proceedings of 21st International Cartographic Conference*: 1988–1995. Durban: International Cartographic Association.

MacEachren, A.M. & Kraak, M-J. 2001. Research challenges in geovisualization. *Cartography and Geographic Information Systems* 28(1): 3–12.

Press, W.H., Flannery, B.P., Teukolsky, S.A. & Vetterling, W.T. 1992. *Numerical recipes in Pascal: the art of scientific computing*. Cambridge, New York: Cambridge University Press.

Reinders, K.; Post, F.H. & Spoelder, H.J. 2001. Visualization of Time-Dependent Data using Feature Tracking. *Visual Computer* 17: 55–71.

Rensink, R.A. 2002. Change detection. *Annual Review of Psychology* 53: 245–277.

Samtaney, R., Silver, D., Zabusky, N. & Cao, J. 1994. Visualizing features and tracking their evolution. *Computer* 27(7): 20–27.

URL: ICA Visualization. 2003. http://kartoweb.itc.nl/icavis/agenda/index.html (accessed 31 May 2005).

URL: Visual Cognition Lab. 2003. http://viscog.beckman.uiuc.edu/djs_lab/demos.html (accessed 31 May 2005).

Wachowicz, M. 2000. The role of geographic visualization and knowledge discovery in spatio-temporal modelling. In H. Heres (ed.), *Time in GIS: Issues in spatio-temporal modelling, Publications in Geodesy* 47: 13–26.

Yuan, M. 1997. Use of Knowledge Acquisition to Build Wildfire representation in Geographic Information Systems. *International Journal of Geographic Information Science* 11(8): 723–745.

Advances in Spatio-Temporal Analysis – Tang et al. (eds)
© 2008 Taylor & Francis Group, London, ISBN 978-0-415-40630-7

Time series analysis of land cover change – The National Carbon Accounting System trial in Fujian province, China

Xiaoliang Wu, Peter Caccetta, Suzanne Furby, Jeremy Wallace & Min Zhu
CSIRO Mathematical and Information Sciences, Wembley, Australia

ABSTRACT: In Australia, remotely sensed Landsat data is routinely used for mapping and monitoring changes in the extent of woody perennial vegetation. Time series remotely sensed satellite imagery and ground information is used to form multi-temporal classifications of presence/absence of woody cover. Two broad-scale operational land cover change and monitoring projects were based on a series of algorithms and methods developed by the CSIRO. This paper gives an overview of these remote sensing techniques and demonstrates their use in China using a trial site within the Fujian province.

1 INTRODUCTION

In Australia, broad-scale operational programmes for mapping and monitoring the extent of woody perennial vegetation are based on time series of Landsat imagery. One is the National Carbon Accounting System (NCAS) Landcover Change Project, which covers the entire Australian continent (Furby 2002). Another is the Land Monitor Project, which covers 24 million hectares in Western Australia (Caccetta et al. 2000). These operational programmes are based on a series of algorithms and methods developed by the Commonwealth Scientific and Industrial Research Organisation, Australia. These methods have recently been tested in the Fujian province, China, as a trial of their applicability in a different environment. This paper provides an overview of the remote sensing techniques and of the results of the Fujian province trial.

2 METHODOLOGY

In the NCAS and Land Monitor projects, long-term sequences of ortho-rectified and calibrated Landsat satellite data are processed to provide observations relating to land cover and change. Discriminant analysis techniques are applied to quantify spectral separation of classes of interest (in this case woody versus non-woody). Samples of woody and non-woody training and validation data are obtained from interpretation of high-resolution aerial/satellite imagery and ground truth. Results are used to develop classifiers that are applied to the sequence of individual images. Finally, these classifications are combined using spatial/temporal models to reduce errors in maps of forest cover and change. Results are used in Australia's national carbon accounting and for resource management and planning.

The image processing steps are as follows:

(1) Selection of Landsat scenes and registration of time series Landsat data;
(2) Calibration of Landsat data to a common reference;
(3) Processing of the calibrated data to remove 'corrupted' data, which include dropouts, data affected by fire, smoke and cloud;
(4) Terrain illumination correction of the calibrated data to adjust for differential viewing geometry;
(5) Stratification of the data into 'zones', where land cover types within a zone have similar spectral properties;
(6) Discriminant analysis of ground data and spectral data to determine a single-date classifier to produce a series of classification maps; and

(7) Specification of a joint model for multi-temporal classification. Greater classification accuracies are obtained after applying a joint model.

Steps 3 and 5 are self explanatory. The other processing steps, as implemented in the CSIRO methods, are briefly described in the following subsections.

2.1 *Image rectification*

The two steps in establishing a rectified sequence of Landsat imagery are: (1) establish a common ortho-rectified base mosaic of Landsat data; and (2) ortho-rectify temporal sequences of images to the common base.

Accurate ortho-rectification was achieved using a rigorous earth-orbital model. PCI OrthoEngine software was used for this purpose. Once the ortho-rectified base was established, ground control points (GCP) were automatically matched using a cross-correlation technique. This approach improves efficiency and the accuracy of the results. For quality assurance, visual inspection and numerical summaries based on cross-correlation feature matching are used to assess the accuracy of ortho-rectification of the time series images.

2.2 *Image calibration*

Ideally, all images would be calibrated to standard reflectance units. However, when comparing images to detect change, it is sufficient to convert raw digital counts to be consistent with a chosen reference image.

Three calibration steps are applied in the radiometric correction procedure for Landsat imagery:

(1) Top-of-Atmosphere (TOA) reflectance calibration (also called sun angle and distance correction);
(2) Bi-directional Reflectance Distribution Function (BRDF) calibration (Danaher et al. 2001); and
(3) Terrain illumination correction.

Each step is briefly described in the following subsections.

2.2.1 *Top-of-Atmosphere (TOA) reflectance calibration*
The TOA calibration is to correct the reflectance differences caused by the solar distance and angle. The sun zenith and azimuth angles for each pixel and the distance from the scene centre to the sun are calculated and the reflectance correction is then calculated for each band as described in Vermote et al. (1994).

2.2.2 *BRDF Calibration*
Angular effects across the Landsat image result in BRDF effects that are relatively small, but significant in the context of broad-scale monitoring. The BRDF correction is a simple linear function of scan angle that is applied to each band. A two-kernel empirical BRDF model was used to correct the remaining scene-to-scene differences. Simple variations of Walthall's model, as described by Danaher et al. (2001), were used in the BRDF calibration approach. The model is a three-parameter model (see Equation 3 in Danaher et al. (2001)), where the three parameters were calculated by solving equations based on the image overlap areas, and the same parameters were applied to all scenes (Wu et al. 2001).

2.2.3 *Terrain Illumination Correction*
This third step is required where there are significant terrain illumination effects, resulting in bright and dark sides of hills and mountains. This is particularly important for time series imagery where terrain effects vary with different dates. The details of the terrain illumination correction used can be found in Wu et al. (2004), and is based on the C-correction (Teillet et al. 1982) and incorporates a ray-tracing algorithm for identifying true shadow. A high-resolution digital elevation model (DEM) is required to achieve adequate removal of terrain effects.

2.3 *Woody classification*

Canonical Variate Analysis (CVA) (Campbell & Atchley 1981, McKay & Campbell 1982a, b) and related procedures are used to define optimal discriminant indices for each zone. The index-threshold classification approach (Furby 2002) is used to produce the forest or woody probability maps for each date within the sequence.

2.4 *Spatial temporal models*

The time series of classification maps is derived from images of varying quality and spectral discrimination. To improve classification accuracy, joint models are used that incorporate error rates of the initial classifications as well as temporal and spatial rules.

The sequence of cover class probabilities from all dates is analysed using a joint model approach (Caccetta 1997, Kiiveri & Caccetta 1998). This approach uses the probabilities from neighbouring dates to modify the probabilities of each pixel. The effect of the method is that it 'smoothes out' sudden changes (e.g. from cultivation), and reduces uncertainty and errors in the individual dates. The result is a series of modified probability maps for each date (the forest probability maps before applying the joint model are called 'the prior forest probability map' while the probability map produced after applying the joint model is called 'the modified forest probability map').

The joint model is described as follows. To begin, we need to define some notation. For pixel $k = 1 \cdots m$ and time slice $i = 1 \cdots n$ we write $y_{ik}, y'_{ik}, l_{ik}, l'_{ik}$ and z_{ik} to represent Landsat coverage, index image, initial interpretation, 'true' class label and stratification zone respectively. We use the notation $r(k)$ to denote the 8 pixels adjacent to k. The joint model for the observed and unobserved images can be written as

$$p(y'_{1k} \cdots y'_{nk}, l'_{1k} \cdots l'_{nk}, l'_{1k} \cdots l_{nk}, l'_{1r}(k) \cdots l'_{nr(k)}, z_{1k} \cdots z_{nk})$$

$$= \Pi_{i=1 \cdots n} p(y'_{ik}|z_{ik}, l_{ik}) p(l_{ik}|z_{ik}, l'_{ik}) p(l'_{ik}|l'_{i-1,k}, l'_{ir(k)}) p(z_{ik}) \tag{1}$$

where $p(l'_{1k}|l'_{0k}, l'_{ir(k)})$ is defined as $p(l'_{1k}|l'_{ir(k)})$.

The 'true' class labels are obtained from the joint distribution of $l'_{1k} \cdots l'_{nk}$ given the observable images by a cyclic ascent algorithm as in (Caccetta 1997, pp187–192) and (Kiiveri & Caccetta 1998).

Useful properties of the approach include:

(1) Propagation of uncertainties in inputs and calculation of uncertainties in outputs;
(2) Production of hard and soft maps;
(3) Handling of missing data by using all available information to make predictions; and
(4) Existence of well-developed statistical tools for parameter estimation.

2.5 *Land cover change maps*

From the modified probability maps produced from the joint model, 'yes/no' woody masks for each date were formed, and then compared to form change maps. The land cover change maps were provided for further statistical analysis of changes and for carbon modelling.

3 FUJIAN TRIAL

As a part of an Australia-China project organized by the Australian Greenhouse Office (AGO) and the Chinese Academy of Forestry, a trial using these methods for monitoring forest change in China was performed for a test area in the Fujian province, China. The aim of the trial was to assess the application of the methods in China. This section outlines the data processing and results for land cover change in the Fujian trial. Some issues are discussed and some recommendations are made.

3.1 *The area and satellite image data*

The trial region in the Fujian province, covered by a single Landsat scene, is shown as the rectangular region in Figure 1. A time-series of eight Landsat TM images spanning 16 years were carefully chosen for the Fujian trial region – 1988, 1989, 1992, 1995, 1998, 2001, 2003 and 2004. The scene seasons are around summer and autumn (from September to November). These scenes were purchased from the Remote Sensing Ground Station (RSGS), Beijing, China.

Figure 1. Coverage of Fujian trial region, China (Map source: http://encarta.msn.com).

3.2 *Digital Elevation Model (DEM)*

Since a high-resolution DEM (25 metres, the same as the ortho-rectified images' pixel size) was not available for the project, the SRTM DEM (Shuttle Radar Topography Mission, U.S. Geological Survey) of the trial region was used instead. A sun-shaded sub-region of the SRTM DEM is shown in Figure 2(c).

Some pre-processing steps were applied to the SRTM DEM in order to meet our purposes:

(1) the projection was converted to the UTM Projection: NUTM 50, Datum: WGS84, the same as the Landsat TM images,
(2) DEM grid cells were resampled from 90 m to 25 m using the bilinear interpolation method, and
(3) there are some empty cells in the SRTM DEM caused by terrain effects; these were assigned values based on interpolating the elevation values from neighbouring cells.

3.3 *Ground Control Points (GCP)*

An ortho-rectified Landsat 7 ETM+ scene (21 October 2001) obtained from the Global Land Cover Facility–GLCF (http://glcf.umiacs.umd.edu/index.shtml) was chosen as the rectification reference for the Fujian trial. Ninety-nine GCPs were then collected from the reference image. The GLCF ortho-rectified scene itself was not used in subsequent analysis as it was generated using the nearest-neighbour interpolation method (alternative resampling methods such as the 16-point Kaiser-Damped Sinc convolution techniques are recommended).

3.4 *Woody sites – ground information*

Usually ground data, high-resolution aerial/satellite images and local knowledge are required to identify tree types on images inside the trial region. Dr Hong Yan from the Chinese Academy of Forestry assisted in identifying the tree types on the 2001 image. Some Chinese Fir sample sites were also provided by Dr Trevor Booth and Dr Keryn Paul from CSIRO Forestry and Forest Products, and their colleagues in the Chinese Academy of Forestry.

3.5 *Methodology*

The steps used to produce the series of vegetation maps were described in the previous section. A few comments related to Fujian Landsat data processing are presented below.

(1) The SRTM DEM was used for the Fujian Landsat ortho-rectification and terrain illumination correction. Ortho-rectified sub-region images are shown in Figure 2(a) (2001) and Figure 2(b) (2004).
(2) The CVA analysis from the woody/non-woody samples results in using two indices: $-3*$band3 + band4 (index 1) and $-3*$band3 $-$ band4 + band5 (index 2). Thresholds are used to divide the index scores into values that are certainly woody (100% probability), certainly non-woody (0% probability), and uncertain cover in between (0–100% probability) scaled linearly with distance relative to the thresholds. The threshold values for these indices are known to vary spatially. A threshold matching technique was

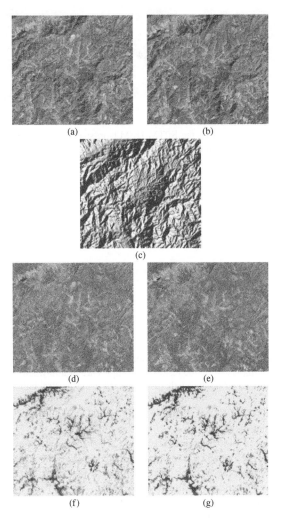

Figure 2. A sample area that demonstrates reasonably good results in the trial region. (a) and (b) show the images before applying terrain-correction ((a) is year 2001 image and (b) is year 2004 image). (c) shows the SRTM DEM (sun-shaded) corresponding to the same area. (d) and (e) show images after applying terrain-correction. (f) shows the prior forest probability maps (2001 in red and 2004 in green) and (g) shows the modified forest probability maps (2001 in red and 2004 in green). Most of the 'false change' in (f) has been removed by the time-series processing.

applied to automatically adjust these thresholds for other time slice maps based on the base map (year 2001 in this case). This automatic matching improves consistency and saves labour-intensive work.

3.6 *Results from the Fujian trial*

Ortho-rectification plays a critical role in the whole processing. If time series images are mis-registered, the same pixel on another year's image will be shifted and it will cause severe problems for tracking land cover change through time. The average GCP residuals from earth-orbital model fitting for all eight time series images are less than a third of the pixel size (<8 m).

Terrain illumination correction has removed the large terrain effects, as can be seen from the images shown in Figure 2(d) and Figure 2(e). However, many finer terrain effects, especially in local valleys on hillsides, are still visible and remain on the image after applying terrain-correction (Fig. 3). This is due to

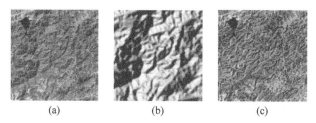

(a)   (b)   (c)

Figure 3. A sample area that illustrates that the SRTM DEM (sun-shaded) with 90 m resolution is inadequate for undulating or fragmented terrain such as this area. Some shadows/terrain effects are still visible and remain on the image after applying terrain-correction: (a) the un-corrected image; (b) the corresponding SRTM DEM; and (c) the corrected image.

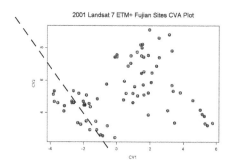

Figure 4. Ordination plot from CVA analysis for 2001 image using woody/non-woody samples. These dots in the bottom-left part are the woody samples while the rest of the samples are of non-woody cover types. This plot clearly shows that woody and non-woody cover can be easily discriminated using spectral indices.

the coarse resolution of the SRTM DEM. A high-resolution DEM would improve the correction and the results in regions with undulating terrain.

The CVA analysis ordination plot for the 2001 image (Fig. 4) clearly shows that woody/non-woody cover is well discriminated in the Fujian trial region using the Landsat TM data. The prior probability map obtained for each year shows the indices derived from CVA analysis incorporated with the index-threshold approach work very well for the time series images (Fig. 2(f)).

The joint model computation produces reasonable modified forest probability maps from 1988 to 2004. The joint model smoothed individual-date classification error and even removed some of the remaining terrain-induced error effects on the prior probability maps (Fig. 2(g)). One consequence of the joint model approach is that long term land cover change (such as clearing of forest for agriculture) is more accurately mapped, but 'transient' land cover change, which only appears in one image date, can be 'over-smoothed'. For example, the information on clearing of forest, which then re-grows before the next image, may be lost in the joint model processing.

3.7 *Accuracy and limitations*

The typical results (e.g. Fig. 2) obtained from the Fujian trial clearly demonstrate that our remote sensing techniques work well in the trial area, using almost identical procedures to those which are applied successfully in Australia. However, similar to the situation in Australia, a few issues remain when applying our remote sensing techniques.

The land cover change maps are derived from reflectance signals detected by Landsat TM, and depend on a contrast between woody and other cover types (soil, crops, bare rocks, etc.). The thresholds for classification of woody cover have been derived from interpretation and comparison/matching with the base map (year 2001).

This classification as woody relies on the spectral contrasts of cover types resulting from physical differences on the ground, and effectively requires a certain density of vegetation. Hence thin, scattered

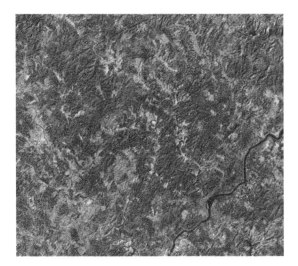

Figure 5. This image illustrates typical land cover changes in the Fujian region from three dates: 1998, 2001 and 2004 in blue, red and green, respectively. Landsat TM band 5 was used for these years.

vegetation with a high proportion of soil background may be omitted. In particular, bare or very thin areas within woody vegetation will not be classified as 'woody cover'. Common examples are tracks, rocks, fire-scars and salt-affected vegetation. Hence, the areas mapped as woody at particular dates will not necessarily correspond to administrative definitions of reserves, etc. Certain dense but highly reflective vegetation types may also be omitted, but no cases of such omissions are known.

There is a time lag in detection of re-vegetated areas, which varies with region and vegetation type. Re-vegetated areas will not be mapped until the vegetation achieves a sufficient density. Hence, some recent, slow-growing or sparse re-vegetated areas will not be detected until they are present for several image dates.

Errors of commission may occur when other land covers give a similar spectral response to woody cover. The temporal smoothing of the joint model removes most of the transient cultivation effects that might cause these errors. However, there are cases where errors of commission may remain after the joint model processing. Examples include cleared areas with persistently dark soil, and also lake fringes and normally dry lake surfaces where changes in water level have dramatic effects on the cover and reflectance. Errors that are incurred in these areas may result in incorrect mapping of change in lake edge vegetation. However, these errors have not been removed by manual digitizing as some may be real land cover change.

Various regions show that reasonable results were produced from the joint model even when some terrain effects still remain on the prior forest probability maps, while the results also show the vegetation re-growth issue mentioned in the above paragraphs.

Figure 5 presents one type of visual display of land cover change in the Fujian region formed from three dates: 1998, 2001 and 2004 in blue, red and green, respectively. Landsat TM band 5 was used for these years. Areas displayed in shades of grey have undergone little or no change, whereas brightly coloured areas have changed. The colours give an indication of the timing of the change.

3.8 *Conclusions and recommendations*

The results obtained from the Fujian trial clearly demonstrate that our remote sensing techniques work reasonably well in the Fujian trial region. A few comments when our processing procedure was applied to the Fujian trial follow.

(1) Default BRDF kernel coefficients were used to calibrate all images because of the lack of enough images to estimate the BRDF kernel coefficients. It is desirable to use locally estimated BRDF kernel coefficients if there are enough overlapping images.

(2) The SRTM DEM is sufficient for performing ortho-rectification but the resolution is not high enough to achieve ideal terrain illumination corrections.

(3) The joint model performed well in removing single-date classification error and some of the remaining terrain effects from the prior probability maps.

(4) The joint model applied in this study 'over-smoothed' in areas of transient land cover change, such as forest areas that were cleared then re-grew or were replanted before the next image date.

Some recommendations:

(1) In areas affected by terrain, a high-resolution DEM is required to ensure images are properly ortho-rectified and terrain corrected.

(2) Advanced models, such as a second-order joint model, may be required to better track the changes through time.

(3) More frequent time slices may be required to improve the results for detection of 'transient' land cover change. One-year intervals between images may be a solution.

ACKNOWLEDGEMENTS

We are grateful to the Australian Greenhouse Office (AGO) for their financial support of the Australia-China project. We greatly appreciate the help of Dr Hong Yan (the Chinese Academy of Forestry – CAF) in identifying the tree types on the images, and of Dr Trevor Booth and Dr Keryn Paul (CSIRO Forestry and Forest Products) for additional ground sites.

We would like to thank USGS and NASA for making the SRTM DEM and the GLCF Landsat data available for public use. Our thanks also extend to Mr Hong Su (the Remote Sensing Ground Station, Beijing, China) for helping us to obtain the appropriate product level of Landsat TM scenes.

REFERENCES

Caccetta, P.A. 1997. *Remote Sensing, Geographic Information Systems (GIS) and knowledge-based methods for monitoring land condition.* PhD. Thesis, School of Computing, Curtin University of Technology.

Caccetta, P.A., Campbell, N.A., Evans, F.H., Furby, S.L., Kiiveri, H.T. & Wallace, J.F. 2000. Mapping and monitoring land use and condition change in the South-West of Western Australia using remote sensing and other data. *Proceedings of the Europa 2000 Conference, Barcelona.*

Campbell, N.A. & Atchley, W.R. 1981. The geometry of canonical variate analysis. *Systematic Zoology* 30: 268–280.

Danaher, T., Wu, X. & Campbell, N.A. 2001. Bi–directional Reflectance Distribution Function Approaches to Radiometric Calibration of Landsat TM imagery. *IGARSS 2001 Conference Proceedings.*

Furby, S.L. 2002. Land Cover Change: Specification for Remote Sensing Analysis. *National Carbon Accounting System – Technical Report No. 9.* Australian Greenhouse Office.

Kiiveri, H.T. & Caccetta, P.A. 1998. Image fusion with conditional probability networks for monitoring salinisation of farmland. *Digital Signal Processing* 8,4: 225–230.

McKay, R.J. & Campbell, N.A. 1982a. Variable selection techniques in discriminant analysis. I. Description. *British Journal of Mathematical and Statistical Psychology* 35: 1–29.

McKay, R. J. & Campbell, N. A. 1982b. Variable selection techniques in discriminant analysis. II. Allocation. *British Journal of Mathematical and Statistical Psychology* 35: 30–41.

Shuttle Radar Topography Mission. http://srtm.usgs.gov (accessed 8 may 2005).

Teillet, P.M., Guindon, B. & Goodenough, D.G. 1982. On the slope–aspect correction of multispectral scanner data. *Canadian Journal of Remote Sensing* 8: 84–106.

The Global Land Cover Facility. http://glcf.umiacs.umd.edu/index.shtml (accessed 8 may 2005).

Vermote, E., Tanré, D., Deuzé, J.L., Herman, M. & Morcrette, J.J. 1994. S*econd simulation of the satellite signal in the solar spectrum (6S), 6S User Guide Version 0.*

Wu, X., Danaher, T., Wallace, J.F. & Campbell, N.A. 2001. A BRDF–Corrected Landsat 7 Mosaic of the Australian Continent. *IGARSS 2001 Conference Proceedings.*

Wu, X., Furby, S.L. & Wallace, J.F. 2004. An Approach for Terrain Illumination Correction. *The 12th Australasian Remote Sensing and Photogrammetry Conference Proceedings, Fremantle, Western Australia.*

Advances in Spatio-Temporal Analysis – Tang et al. (eds)
© 2008 Taylor & Francis Group, London, ISBN 978-0-415-40630-7

3D modelling of groundwater based on volume visualization technology

Zhiyong Xu
Key Laboratory of GIS, Ministry of Education, School of Resources and Environment Science, Wuhan University, Wuhan, China

Xiaofang Wu
College of Information, South China Agriculture University, Guangzhou, China

Guorui Zhu & Huiwu Yan
Key Laboratory of GIS, Ministry of Education, School of Resources and Environment Science, Wuhan University, Wuhan, China

ABSTRACT: At present, groundwater is being heavily extracted, which causes a series of grave environmental problems such as, land subsidence, seawater inbreak and deterioration of water quality. It is very necessary to study the groundwater resource and make good, rational use of it. Therefore, volume visualization technology is introduced to the field of groundwater resource in this paper. When using volume visualization technology, some scattered points are first sampled in finite space and each point will have one or more properties, which may be its physical property, chemical attributes and so on. Then, using the physical and chemical property data, the inverse distance to a power is adopted and improved to build the 3D model of the existing groundwater conditions. Numerical simulation technology is used to calculate the depression of the groundwater cone, and the velocity and direction of groundwater flow. The finite difference method is adopted to build the model of the groundwater flow field. Lastly, the time-space distribution and dynamic change characteristic of the hydrogeologic layer and its inner physical and chemical attributes are represented by a hybrid volume-rendering technique. The research provides the scientific foundation for decision support for the extraction of groundwater.

1 INTRODUCTION

Groundwater lies under the ground, so it cannot be observed directly. Only by means of hydrogeologic reconnaissance and groundwater dynamic monitoring can we discover the present condition and movement of groundwater. However, owing to the limitations of the reconnaissance budget, the degree of hydrogeologic reconnaissance is very limited; therefore we cannot directly gain the information on hydrogeologic volume between the reconnaissance wells but can only gather the information from the experience of hydrogeologic workers. However, there are errors in recognizing the hydrogeologic condition and there are blind-areas, all adversely affecting the veracity of the hydrogeologic research result. Therefore, how the present scientific techniques are used to show the existent condition of groundwater, its moving regulation and dynamic characteristics and provide the scientific foundation for hydrogeologic research becomes one of the subjects in the present hydrogeologic field requiring urgent attention.

Because of these problems, many experts have researched this topic. For example, MODFLOW (McDonald & Harbaugh 1988) was developed for simulating unconfined aquifers but it is limited to aquifer systems in which there is a minimal vertical component of flow. IVM (Interactive Volume Modelling), which was developed by the Dynamic Graphic Company in America, supports the modelling, visualization and interaction of objects in 3D space. Its advantage is that it represents the continuously changing phenomena in 3D space, but it has difficulty in representing the non-continuously changing phenomena. SGM (Stratigraphic Geo-cellular Modelling), which was developed by the Stratamodel Company in America, disposes the bore data and describes the geologic conformation and scatter attributes according to the boundary of stratum, but it is limited to representing the continuously changing non-layered phenomena.

3D geo-modelling is one of the main components of the all the software mentioned above. The traditional methods of 3D geo-modelling include digitized modelling and geometric modelling. Digitized modelling can describe the attributes of inner physics and chemistry in the geologic body. Geometric modelling describes the spatial shape. Therefore, the regular method is to synthesize the digitized and geometric modelling in 3D geo-modelling, where the attributes of the geologic body are represented by digitized modelling and the spatial geometric shape is described by geometric modelling. Therefore, the present research problem is how to construct a model that can not only describe the spatial geometric shape, but also represent the inner physical and chemical attributes and avoid the problem of separation between the spatial geometric shape and the inner attributes of the geologic body.

Therefore, the volume visualization technique is introduced in this paper. The volume visualization technique includes the representation, analysis, operation and visualization of volume data concerning the inner information of objects, and its main function is to disclose the complicated inner structure of objects. The volume data is gained by sampling in finite space, and each value from a sampling point will be of one or more of its properties, representing its physical properties. It is a true 3D data model, includes the inner information of the objects, and is very suitable for representing the true 3D geo-phenomena. Compared with traditional 3D modelling, the volume visualization technique is made up of the consecutive geometry model. In addition, the object is described not by the polygon and boundary, but by the voxel, which includes the inner and external information on the object. So, the volume visualization technique is used to construct and show the hydrogeologic layer and its inner physical and chemical attributes such as intrinsic permeability, porosity, water quality, groundwater temperature field, and applies the hybrid volume-rendering technique to represent the time-space distribution and dynamic change characteristics.

In this paper, the research region examined is Chang Zhou, Wu Jin district, where groundwater over-extraction, groundwater cone of depression and land subsidence are all serious problems. In the research region, the groundwater resource is very abundant, so its macroscopical characteristic and trend analysis are very remarkable. To study the 3D visualization of the groundwater in detail, the authors divide the groundwater 3D data into two classes: the scalar field and vector field. Section 2 focuses primarily on the 3D modelling of scalar field (Existing Condition). 3D modelling of the vector field will be discussed in Section 3. Section 4 analyses the research result and proposes the method of spatial data mining. Finally, the conclusion is given in Section 5.

2 3D MODELLING OF EXISTENT CONDITION

In the hydrogeologic field, the geometry model is built using the data from probing and boring. The volume data is gained by voxelling the geometry model, which is implemented by mathematical interpolation. The present hydrogeologic condition can be divided into three hydrogeologic phenomena: one is the phenomenon of homogeneity change characteristic, the second is the phenomenon of horizontal change characteristic, and the third is the phenomenon of 3D change characteristic. Because the data of these phenomena are scattered, the method of improved inverse distance to a power is applied to build by interpolation the 3D volume model of the hydrogeologic scalar field.

In the next section, the inverse distance to a power is introduced; its shortcomings are analysed and revised. Finally, it is applied to the 3D modelling of the existing hydrogeologic condition.

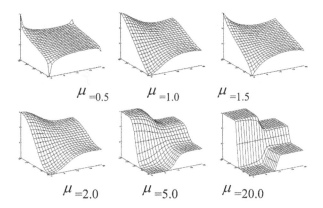

μ =0.5 μ =1.0 μ =1.5

μ =2.0 μ =5.0 μ =20.0

Figure 1. Different interpolated surfaces based on inverse distance to a power.

2.1 The improved inverse distance to a power

2.1.1 Basic principle of inverse distance to a power

The basic idea of inverse distance to a power is: supposing that there are spatial point sets $\{(x_i, y_i, z_i) | i = 1, \cdots\cdots, N\}$, the projection of point (x_i, y_i, z_i) in the x-y plane is (x_i, y_i), named D_i. The z value of any point $P(x, y)$ in the plane is influenced by the z_i in D_i. What is more, the influence of z_i is opposite to the distance between P and D_i:$d[P, D_i]$, which can be abbreviated to d_i. The z value of point P is represented by Formula 1:

$$f(P) = \begin{cases} \dfrac{\sum_{i=1}^{N}(d_i)^{-\mu}z_i}{\sum_{i=1}^{N}(d_i)^{-\mu}} & d_i \neq 0 \text{ for each } D_i \\ z_i & \text{When } d_i = 0 \end{cases} \tag{1}$$

μ is the exponent and should be more than zero, and is very important to the result. Figure 1 is the resulting graph of $\mu = 0.5, 1.0, 1.5, 2.0, 5, 20$. Certainly, when $\mu = 2.0$, the result is the best, and operation speed is fastest because it need not extract the root.

From Formula 1, it can be seen that inverse distance to a power is very simple, but it has many shortcomings:

(1) When the number of points is very large, the calculation of $z = f(P)$ is huge and difficult.
(2) It considers only the influence of distance, not the direction. Sometimes, the direction is very important in interpolation.
(3) The surface is not limited by derivatives in each known point.
(4) When $d_i \approx 0$, the calculation is very difficult because the error is very sensitive, especially when the denominator is close to zero.

Each of the shortcomings can be addressed. Neighbouring points can be selected to solve the first and second problem; the third problem can be solved by estimating derivatives; and the fourth problem can be solved by reducing the calculation error.

2.1.2 Selecting neighbouring points

From Formula 1, we find that only the points nearby P are important when calculating the z value of P. The points that are far away from P have very little influence on P. So computing time can be saved if only points nearby P are used in Formula 1.

Voronoi is a good method for representing the vicinity relationship. So Voronoi is used to get the neighbour points of P. At the same time, the neighbour points of P are classified because each class has a different influence on P. When building the Voronoi map, the neighbour of P is named as the Class-First-Neighbour-Point (CFNP), the neighbour of CFNP is named as the Class-Second-Neighbour-Point (CSNP), and then the Class-N-Neighbour-Point can be obtained.

Obviously, the influence of CFNP on P is larger than that of CSNP. The experiment shows that both the linear model and the exponential model from CSNP to CFNP can represent the degree of influence. Generally, of all neighbour degrees, only CFNP and CSNP are used in the formula.

Let the neighbour of P be C'_p (or C'). If the neighbour degree is considered, the new weight function is $W_i = (d_i)^{-2} w_i$, and the z value of P is as follows :

$$f(P) = \begin{cases} \dfrac{\sum\limits_{D_j \in C'} W_i z_i}{\sum\limits_{D_j \in C'} W_i} & d_i \neq 0 \text{ for each } D_i \\[2ex] z_i & \text{When } d_i = 0 \end{cases} \tag{2}$$

2.1.3 Estimate derivative

It is a character of inverse distance to a power that the grads value of each point is zero. In order to change the character, a small increment is added to the parameter z_i in the Formula 2 at the vicinity of each point, so that the surface has a reasonable derivative at each point of D_i. Firstly, the derivative is displayed using differences in both the X and Y directions.

$$A_i = \frac{\sum\limits_{D_j \in C''_i} W_j \frac{(z_j - z_i)(x_j - x_i)}{(d[D_j, D_i])^2}}{\sum\limits_{D_j \in C''_i} W_j} \quad B_i = \frac{\sum\limits_{D_j \in C''_i} W_j \frac{(z_j - z_i)(y_j - y_i)}{(d[D_j, D_i])^2}}{\sum\limits_{D_j \in C''_i} W_j} \tag{3}$$

$$\Delta z_i = [A_i(x - x_i) + B_i(y - y_i)] \frac{v}{v + d_i} \tag{4}$$

The character of the expression $v/(v + d_i)$. is the same as d_i^{-1}. The result of the expression decreases from from 1 to 0 simultaneously with d_i increasing from 0 to the larger value, which can control the influence of distant points. But it is possible that the value of Δz_i is very large. So v is formulated as follows:

$$v = \frac{0.1[\max\{z_i\} - \min\{z_i\}]}{[\max\limits_i\{(A_i^2 + B_i^2)\}]^{\frac{1}{2}}} \tag{5}$$

We can be sure that: $\qquad |\Delta z_i| \leq 0.1[\max\{z_i\} - \min\{z_i\}] \tag{6}$

Obviously, the derivative of D_i:

$$\left. \begin{array}{l} \dfrac{\partial}{\partial x}(\Delta z_i)\Big|_{\substack{x = x_i \\ y = y_i}} = A_i \\[3ex] \dfrac{\partial}{\partial y}(\Delta z_i)\Big|_{\substack{x = x_i \\ y = y_i}} = B_i \end{array} \right\} \tag{7}$$

This is the result that we need.

$$f(P) = \begin{cases} \dfrac{\sum\limits_{D_j \in C'} W_i(z_i + \Delta z_i)}{\sum\limits_{D_j \in C'} W_i} & d_i \neq 0 \text{ for each } D_i \\[2ex] z_i & \text{when } d_i = 0 \end{cases} \tag{8}$$

166

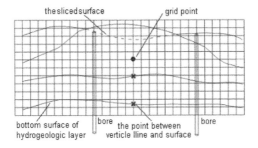

Figure 2. 3D model of phenomenon of homogeneity change characteristic.

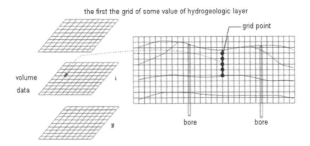

Figure 3. 3D model of horizontal change characteristic.

2.1.4 Reduction of calculation error

When p is near to some values of D_i, rounding or truncation will result in large errors. So, the ε vicinity of D_i is built.

$$\lim_{P \to D_i} f(P) = z_i \tag{9}$$

If p is in a certain ε vicinity, let $f(P) = z_i$. If p is in ε vicinity of many points, $f(P)$ equals the average z value of these points. Suppose $N(P)$ is the set of these points, the following is the improved formula.

$$f(P) = \begin{cases} \dfrac{\sum\limits_{D_i \in C'} W_i(z_i + \Delta z_i)}{\sum\limits_{D_i \in C'} W_i} & d_i \neq 0 \text{ for each } D_i \\[2em] \dfrac{\sum\limits_{D_i \in N(P)} z_i}{\sum\limits_{D_i \in N(P)} 1} & \text{when } d_i = 0 \end{cases} \tag{10}$$

2.2 3D modelling of scalar field

2.2.1 Phenomenon of homogeneity change characteristic

The spatial distribution of the hydrogeologic layer has generally a layered character. In the vertical direction, the hydrogeologic layer is divided into several different layers (such as aquifers). There is a typical interface between one layer and its neighbour and the inner part of every layer is homogeneous. Therefore, the spatial distribution structure of a hydrogeologic layer is set up by building the model of the interface between the layers.

There are four steps in setting up the architectural model. Firstly: boundary interpolation. A series of control points is added on the boundary or outside the research region so as to discover the spatial distribution of the hydrogeologic layer in the research region. The boundary points are always generated by means of improved inverse distance to a power. Secondly, the improved inverse distance to a power is used to generate the bottom surface of the hydrogeologic layer. Thirdly, the bottom surface is overlaid,

167

then the surfaces are pruned in terms of the incision and stagger relations between the surfaces, and the 3D geologic model of the hydrogeologic layer is formed. Lastly, this model is united by stitching the top surface and the bottom surface, and building the cube model by interpolation (Nalder & Wein 1998).

2.2.2 Phenomenon of horizontal change characteristic

Strictly speaking, the porosity, water quality, pondage rate, intrinsic permeability and feedwater degree are the continual changing geo-phenomena in 3D space. In order to gain the geologic information, firstly the rock core is gained by boring, and then the rock sample is analysed by physical and chemical experiments. From every bore and rock data, the attribute value of one level position in the rock layer is described.

The spatial distributing character of the horizontal geo-phenomenon is not changing in the vertical direction, but continually distributes in the horizontal direction. Therefore, the value of any point in the hydrogeologic layer is generated by an interpolation algorithm; then the layered model is built. The volume model of the whole region is formed last.

In this paper, the extent of sampling data in the research region is taken as the bound box of the volume model. The bound box is divided into NX × NY × NZ. NX, NY, NZ are the grid numbers of the X, Y, Z directions. The interpolation of grid data is implemented by the interpolation algorithm of inverse distance to a power; then the 3D volume model is formed (Decencière Etienne & Meyer 1998).

2.2.3 Phenomenon of 3D change characteristic

The groundwater temperature and its physical and chemical attributes are the continually changing hydrogeologic phenomenon in 3D space. In the long term, the true 3D display and water temperature surveying of any position have great shortcomings because of the limits of surveying conditions and visualization technology, which can be resolved by the volume visualization technology.

By means of the improved inverse distance to a power, the attribute value of each voxel is gained by interpolation. Then the 3D spatial model of change characteristic is composed of all the voxels (Deutsch 1996).

The volume visualization of the hydrogeologic layer and its physical and chemical attributes are shown in Figure 4 below.

3 MODELLING OF GROUNDWATER FLOW FIELD

In order to quantify the groundwater resource of the research region and control the geologic disaster efficiently, the groundwater flow field must be represented. Two aspects are important (Hin & Post 1993): one is to represent the development of the groundwater cone of depression in the aquifer; the other is to represent the changing regulation of the velocity of flow and the direction of flow with the change of time (Max et al. 1993).

At present, numerical simulation technology is used to calculate the groundwater cone of depression, velocity and direction of groundwater flow. Changzhou city in China was selected as the research region, which lies in the downriver south side of Changjiang. See also Figure 5.

The width of the region is about 55 km and its length is about 81 km. The hydrogeologic layer is divided into eight layers: diving aquifer, the waterproof layer, the confined aquifer, the waterproof layer, the confined aquifer, the waterproof layer, the confined aquifer, the waterproof layer. In this paper, the typical confined aquifer is taken as the research example, and the 3D mathematic model of groundwater flow of

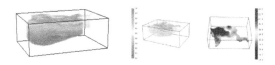

Figure 4. Volume visualization results of three phenomena.

the confined aquifer is set up according to the equilibrium principle and Darcy law.

$$\begin{cases} \frac{\partial}{\partial x}(K_{xx}\frac{\partial H}{\partial x}) + \frac{\partial}{\partial y}(K_{yy}\frac{\partial H}{\partial y}) + \frac{\partial}{\partial z}(K_{zz}\frac{\partial H}{\partial z}) + W = \mu_s\frac{\partial H}{\partial t} \ (x,y,z) \in D & \\ H(x,y,z,t)|_{t=0} = H_0(x,y,z) & \text{initial condition} \\ H(x,y,z,t) = \varphi_1(x,y,z) & AB \\ H(x,y,z,t) = \varphi_2(x,y,z,t) & CD \\ \partial H/\partial n = 0 & BC, DA \end{cases} \qquad (11)$$

K_{xx}, K_{yy} and K_{zz} are the penetrative parameters, their unit is LT^{-1}.

H:water level (L)

W:flux per volume (T^{-1})

μ_s:feedwater rate of the multi-bore medium, which represents the feedwater degree of inter space medium per volume when the water level depresses one per cent.

t:time(T)

D:seepage region

This is a complex equation, which involves time and spatial factors. To solve this equation, the derivative is replaced by difference, and the finite difference method is used normally. The famous groundwater simulative software: MODFLOW and the 3D finite difference method are used to obtain the numerical solution for the above model. First, the time and space dispersing section and difference functions are built, and the hydrogeologic conditions and parameters are manipulated. The SOR is applied to gain the data of water level, water depth and water flux of the confined aquifer from 1999 to 2000. Then the groundwater flow model is set up; at the same time, the value of H in special time is gained, so the cone of depression model is generated by means of a boundary-confined DEM. The space is dispersed with the regular grid and the time is dispersed by equal isometry, when the equation is solved. Then the volume data of the flow field can be gained. Figure 6 is the result for the groundwater cone and Figure 7 is the visualization of the groundwater effluent seepage field.

4 SPATIAL DATA PROBE

In groundwater resources research, geological engineers need to understand the hydrogeological section with respect to certain directions, and comprehensively research it from various viewpoints in order to discover the hydrogeological regulation.

The basic demand of a space data probe is to obtain quickly some attribute information at any place. Obviously, this involves the problem of space interpolation. The improved inverse distance to a power can be adopted to interpolate and get attribute information at any point.

The method of slice on volume is that many regular grid points are distributed on the selected plane, and the attributes of these points are gained by interpolation based on the improved inverse distance to a power. Figure 8 is the flow field of one particular slice.

Figure 5. Map of the confined aquifer of the research region.

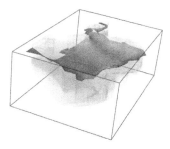

Figure 6. Groundwater cone of depression.

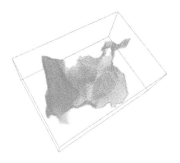

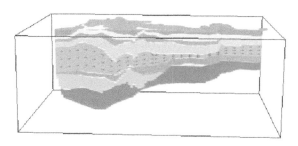

Figure 7. Map of the confined
aquifer of the research region.

Figure 8.   Groundwater cone of depression.

Considering the geographic map of the research region, the authors found that the centres of the cones of depression were mostly distributed in the region of developed industry and dense population along the Shanghai-Ningbo railroad. Therefore, we can come to the conclusion that the groundwater cone of depression is caused by groundwater over-extraction. In addition, the distributed field of ground depression accords with the groundwater cone of depression in the confined aquifer. Therefore, groundwater extraction from the confined aquifer should be controlled in order to avoid the problem becoming worse.

5   CONCLUSION

In this paper, the improved inverse distance to a power is used to build the 3D model of scalar fields. Then, the 3D flow model was built up based on the finite difference method. The space data probe is discussed. Lastly, the research result will help us to understand and recognize the condition of groundwater and promote research into the development of groundwater resources.

REFERENCES

Decencière Etienne, F.C. de & Meyer, F. 1998. Applications of kriging to image sequence coding. *Signal Processing: Image Communication* 13(3): 227–249.

Deutsch, C.V. 1996. Correcting for negative weights in ordinary kriging. *Computers & Geosciences* 22(6): 765–773.

Griffith, D.A. 1983. The boundary value problem in spatial statistical analysis. *Journal of Regional Science* 23: 377–387.

Hin, A.J.S. & Post, F.H. 1993. Visualization of turbulent flow with particles. *IEEE Proceedings of Visualization'93*: 46–51.

Jimenez–Espinosa, R. & Chica-Olmo, M. 1999. Application of geostatistics to identify gold-rich areas in the Finisterre–Fervenza region, NW Spain. *Applied Geochemistry* 14: 133–145.

Klassen, R. & Harrington, S. 1991. A Technique for visualizing 2D slice of 3D vector field. *Proceedings Visualization'91*. IEEE Computer Society Press.

Max, N., Becker, B. & Crawfis, R. 1993. Flow volume for interactive vector field. *IEEE Proceedings of Visualization' 93*: 19–24.

McDonald, M.G. & Harbaugh, A.W. 1988. A Modular Three–Dimensional Finite Difference Groundwater Flow Model. *USGS–TWRI Book 6, Chapter A1.*

Nalder, I.A. & Wein, R.W. 1998. Test of a new method in the Canadian boreal forest. *Agricultural and Forest Meteorology* 92(4): 211–225.

Rosenblum, L.J. 1994. Research issues in scientific visualization. *IEEE Computer Graphics and Applications* 3: 61–85.

Spatial reasoning and data mining

Advances in Spatio-Temporal Analysis – Tang et al. (eds)
© 2008 Taylor & Francis Group, London, ISBN 978-0-415-40630-7

Spatial data mining and knowledge discovery

Deren Li
National Laboratory for Information Engineering in Surveying, Mapping and Remote Sensing,
Wuhan University, Wuhan, China

Shuliang Wang
National Laboratory for Information Engineering in Surveying, Mapping and Remote Sensing,
Wuhan University, Wuhan, China
International School of Software, Wuhan University, Wuhan, China

ABSTRACT: Growing attention is being paid to spatial data mining and knowledge discovery (SDMKD). This paper presents the principles of SDMKD, proposes three new techniques and gives their applicability and examples. First, the motivation of SDMKD is briefly discussed. Secondly, the intention and extension of the SDMKD concept are presented. Thirdly, three new techniques are proposed in this section: SDMKD-based image classification that integrates spatial inductive learning from a GIS database and Bayesian classification; a cloud model that integrates randomness and fuzziness; and data field that radiates the energy of observed data from the universe of the observed sample to the universe of discourse. Fourthly, applicability and examples are studied for three cases. The first is remote sensing classification, the second is landslide-monitoring data mining and the third is uncertain reasoning. Finally, conclusions are drawn and discussed.

1 MOTIVATIONS

The technical progress in computerized data acquisition and storage results in the growth of vast databases. With this continuous increase and accumulation, the huge amount of computerized data has far exceeded the ability of humans to completely interpret and use it (Li et al. 2006). This phenomenon may be more serious in geo-spatial science. In order to understand and make full use of these data repositories, a few techniques have been tried, e.g. expert systems, database management systems, spatial data analysis, machine learning and artificial intelligence. In 1989, knowledge discovery in databases was proposed. In 1995, data mining also appeared. As both data mining and knowledge discovery in databases virtually point to the same techniques, people started to refer to them together, i.e. data mining and knowledge discovery (DMKD). As 80 per cent of data are geo-referenced, it was necessary to consider the spatial characteristics in DMKD and to develop a further branch in geo-spatial science, i.e. SDMKD (Li & Cheng 1994, Ester et al. 2000).

Spatial data are more complex, more changeable and larger in volume than datasets of common affairs. The spatial dimension means that each item of data has a spatial reference (Haining 2003), defining where each entity occurs on the continuous surface, or where the spatial-referenced relationship exists between two neighbouring entities. Spatial data include not only positional data and attribute data, but also spatial relationships among spatial entities (Wang 2005). Moreover, spatial data structure is more complex than the tables in an ordinary relational database. Besides tabular data, there are vector and raster graphic data in spatial databases and the features of graphic data are not explicitly stored in the database. However, contemporary GIS have only basic analysis functions, the results of which are explicit. Further, it is under the assumption of dependency and on the basis of the sampled data that geo-statistics make estimates at unsampled locations or make maps of the attributes. Because the discovered spatial knowledge can

Table 1. Spatial data mining and knowledge discovery from various points of view.

Viewpoints	Spatial data mining and knowledge discovery
Discipline	A interdisciplinary subject, and its theories and techniques are linked with database, computer, statistics, cognitive science, artificial intelligence, mathematics, machine learning, network, data mining, knowledge discovery database, data analysis, pattern recognition, etc.
Analysis	Discover unknown and useful rules from huge amounts of data via a set of interactive, repetitive, associative and data-oriented manipulations
Logic	An advanced technique of deductive spatial reasoning. It is discovery, not proof. The knowledge is in the context of the mined data.
Cognitive science	An inductive process that goes from concrete data to abstract patterns, from special phenomena to general rules.
Objective data	Data forms: vector, raster and vector-raster Data structures: hierarchy, relation, net and object-oriented Spatial and non-spatial data contents: positions, attributes, texts, images, graphics, databases, file system, log files, voices, Web and multimedia.
Systematic information	Original data in database, cleaned data in data warehouse, detailed commands from users, background knowledge from applicable fields.
Methodology	Match the multidisciplinary philosophy of human thinking that suitably deals with the complexity, uncertainty and variety when briefing data and representing rules.
Application	All spatial data-referenced fields and decision-making processes, e.g. GIS, remote sensing, GPS (global positioning system), transportation, police, medicine, transportation, navigation, robotics, etc.

support and improve spatial data-referenced decision-making, growing attention is being paid to the study, development and application of SDMKD (Han & Kamber 2001, Miller & Han 2001, Li et al. 2001, 2002).

This paper proposes the concepts, techniques and applications of SDMKD. Section 2 describes the concept of SDMKD, paying attention to the knowledge to be discovered, the discovery mechanism and the mining granularity. Section 3 presents the techniques to be used in SDMKD. After the existing techniques are reviewed, three new techniques are proposed. Section 4 gives the applicability and examples of SDMKD. Finally, the conclusions are presented in Section 5.

2 CONCEPTS

Spatial data mining and knowledge discovery (SDMKD) is the efficient extraction of hidden, implicit, interesting, previously unknown, potentially useful, ultimately understandable, spatial or non-spatial knowledge (rules, regularities, patterns, constraints) from incomplete, noisy, fuzzy, random and practical data in large spatial databases. It is a confluence of database technologies, artificial intelligence, machine learning, probabilistic statistics, visualization, information science, pattern recognition and other disciplines. Understood from different viewpoints (Table 1), SDMKD shows many new interdisciplinary characteristics.

2.1 Mechanism

SDMKD is a process of discovering a form of rules, plus exceptions, at hierarchal view-angles with various thresholds, e.g. drilling, dicing and pivoting on multi-dimensional databases, spatial data warehousing, generalizing, characterizing and classifying entities, summarizing and contrasting data characteristics, descriptive rules, predicting future trends and so on (Han & Kamber 2001). It is also a supportable

Table 2. Main spatial knowledge to be discovered.

Knowledge	Description	Examples
Association rule	A logical association among different sets of spatial entities that associate one or more spatial objects with other spatial objects. Study the frequency of items occurring together in transactional databases.	Rain (x, pour) $\Rightarrow$ Landslide (x, happen), interestingness is 98 per cent, support is 76 per cent, and confidence is 51 per cent.
Characteristics rule	A common character of a kind of spatial entity, or several kinds of spatial entities. A kind of tested knowledge for summarizing similar features of objects in a target class.	Characterize similar ground objects in a large set of remote sensing images.
Discriminate rule	A special rule that distinguishes one spatial entity from other spatial entities. Different spatial characteristics rules. Comparison of general features of objects between a target class and a contrasting class.	Compare land price in suburban areas and land price in urban centre.
Clustering rule	A segmentation rule that groups a set of spatial entities together by the virtue of their similarity or proximity to each other, when what groups and how many groups will be clustered are unknown in advance. Organize data in unsupervised clusters based on attribute values.	Group crime locations to find distribution patterns.
Classification rule	A rule that defines whether a spatial entity belongs to a particular class or set when the number and details of the classes are already known. Organize data in given/ supervised classes based on attribute values.	Classify remotely sensed images based on spectrum and GIS data.
Serial rules	A spatio-temporal constrained rule that relates spatial entities in time continuously, or the function dependency among the parameters. Analyse the trends, deviations, regression, sequential pattern, and similar sequences.	In summer, landslide disasters often happen. Land price is a function of influencing factors and time.
Predictive rule	An inner trend that forecasts future values of some spatial variables when the temporal or spatial centre is moved. Predict some unknown or missing attribute values based on other seasonal or periodical information.	Forecast the movement trend of landslides based on available monitoring data.
Exceptions	Outliers that are isolated from common rules or deviate very much from other data observations	A monitoring point with much bigger movement.

process of spatial decision making. There are two mining granularities: spatial object granularity and pixel granularity (Li et al. 2005).

SDMKD may be mainly partitioned into three major steps: data preparation (positioning mining objective, collecting background knowledge, cleaning spatial data); data mining (decreasing data dimensions, selecting mining techniques, discovering knowledge); and knowledge application (interpretation, evaluation and application of the discovered knowledge).

In order to discover confidential knowledge, it is common to use more than one technique to mine the data sets at the same time. And it is also good to select the mining techniques on the basis of the given mining task and the knowledge to be discovered.

2.2 *Knowledge to be discovered*

Knowledge is more generalized, condensed and understandable than data. Common knowledge is summarized and generalized from huge amounts of spatial data. While the amount of spatial data is huge, the number of spatial rules is very small. The more generalized the knowledge, the bigger the contrast. There are many kinds of knowledge that can be mined from large spatial data sets (Miller & Han 2001, Wang 2002) (See Table 2).

In Table 2, these kinds of rules are not isolated, and they often benefit from each other and various forms, such as linguistic concepts, characteristic tables, predication logic, semantic networks, object orientation and visualization, can represent the discovered knowledge. Very complex non-linear knowledge may be depicted by a group of rules.

Knowledge is rule plus exception. A spatial rule is a pattern showing the intersection of two or more spatial objects or space-depending attributes according to a particular spacing or set of arrangements (Ester 2000). In addition to the rules, during the discovering process of description or prediction, there may be some exceptions (also named outliers) that deviate very much from other data observations (Shekhar et al. 2003). These identify and explain exceptions (surprises). For example, spatial trend predictive modelling first discovers the centres that are local maximal of some non-spatial attribute, then determine the (theoretical) trend of some non-spatial attribute when moving away from the centres. Finally, a few deviations are found in that some data are far from the theoretical trend. These deviations may arouse the suspicion that they are noise, or generated by a different mechanism. How are these outliers explained? Traditionally, the detection of outliers has been studied via statistics, and a number of discordancy tests have been developed. Most of them treat outliers as "noise" and they try to eliminate the effects of outliers by removing them or by developing some outlier-resistant method (Hawkins 1980). In fact, these outliers prove the rules. In the context of data mining, they are meaningful input signals rather than noise. In some cases, outliers represent unique characteristics of the objects that are important to an organization. Therefore, a piece of generic knowledge is virtually in the form of rule plus exception.

3 TECHNIQUES FOR SDMKD

Because SDMKD is an interdisciplinary subject, there are various techniques associated with the different types of knowledge mentioned above (Li et al. 2002). They may include probability theory, evidence theory, spatial statistics, fuzzy sets, cloud model, rough sets, neural network, genetic algorithms, decision tree, exploratory learning, inductive learning, visualization, spatial online analytical mining (SOLAM), outlier detection, etc. The main techniques are briefly described in Table 3.

Some of the techniques in Table 3 have been further developed and applied; for example, the algorithms in spatial inductive learning include AQ11 and AQ15 by Michalski, AE1 and AE9 by Hong, CLS by Hunt, ID3, C4.5 and C5.0 by Quinlan, and CN2 by Clark, etc. (Di 2001). The implementation of data mining in spatial databases still needs to be further studied. The following are our proposed techniques – SDMKD-based image classification, cloud model and data field.

3.1 SDMKD-based image classification

Based on the integration of remote sensing and GIS (Li & Guan 2002), this subsection presents an approach that combines spatial inductive learning with Bayesian image classification in a loose manner, taking the learning tuple as the mining granularity for subdividing learning knowledge classes into subclasses, i.e. pixel granularity and polygon granularity, and selects class probability values of the Bayesian classification (shape features, locations and elevations) as the learning attributes. GIS data are used in training area selection for the Bayesian classification, generating learning data of two granularities and testing area selection for classification accuracy evaluation. The ground control points for image rectification are also chosen from GIS data. It implements inductive learning in spatial data mining via the C5.0 algorithm on the basis of learning granularities. Figure 1 shows the principle of the method.

In Figure 1, the remote sensing images are classified, initially by the Bayesian method before using the knowledge, and the probabilities of each pixel to every class are retained. Secondly, inductive learning is conducted by the learning attributes. Learning with probability simultaneously makes use of the spectral information of a pixel and the statistical information of a class, since the probability values are derived from both of them. Thirdly, knowledge on the attributes of general geometric features, spatial distribution patterns and spatial relationships are further discovered from the GIS database, e.g. the polygons of different

Table 3. Techniques to be used in SDMKD.

Techniques	Description
Probability theory	Mine spatial data with randomness on the basis of stochastic probabilities. The knowledge is represented as a conditional probability in the contexts of given conditions and a certain hypothesis of truth (Arthurs 1965). Sometimes, probability and mathematical statistics are also called.
Spatial statistics	Discover sequential geometric rules from disorder in data via covariance structure and variation functions in the contexts of adequate samples and background knowledge (Cressie 1991). Clustering analysis is such a technique.
Evidence theory	Mine spatial data via a belief function and a possibility function. It is an extension of probability theory and suitable for stochastic uncertainty based SDMKD (Shafer 1976).
Fuzzy sets	Mine spatial data with fuzziness on the basis of a fuzzy membership function that depicts an inaccurate probability, by using fuzzy comprehensive evaluation, fuzzy clustering analysis, fuzzy control, fuzzy pattern recognition, etc. (Zadeh 1965).
Rough sets	Mine spatial data with incomplete uncertainties via a pair of lower and upper approximations (Pawlak 1991). Rough sets-based SDMKD is also a process of intelligent decision-making under the umbrella of spatial data.
Neural network	Mine spatial data via a non-linear, self-learning, self-suitable, parallel and dynamic system composed of many linked neurons in a network. The set of neurons collectively finds out rules by continuously learning and training samples in the network (Gallant 1993).
Genetic algorithms	Search the optimized rules from spatial data via three algorithms simulating the replication, crossover and aberrance of biological evolution (Buckless & Petry 1994).
Decision tree	Reasoning the rules via rolling down and drilling up a tree-structured map, of which a root node is the mining task, item and branch nodes are mining processes, and leaf nodes are exact data sets. After pruning, the hierarchical patterns are uncovered (Quinlan 1986).
Exploratory learning	Focusing on data characteristics by analysing topological relationships, overlaying map-layers, matching images, buffering features (points, lines, polygons) and optimizing road (Dasu 2003).
Spatial inductive learning	Comes from machine learning. Summarize and generalize spatial data in the context of a given background that is from users or a task of SDMKD. The algorithms require that the training data be composed of several tuples with various attributes. And one of the attributes of each tuple is the class label (Muggleton 1990).
Visualization	Visually mine spatial data by computerized visualization techniques that make abstract data and complicated algorithms change into concrete graphics, images, animations, etc., which the user may sense directly by eye (Soukup & Davidson 2002).
SOLAM	Mine data via online analytical processing and a spatial data warehouse. Based on multi-dimensional view and Web, it is a tested mining technique that highlights executive efficiency and timely responsibility to commands (Han 1998).
Outlier detection	Extract the interesting exceptions from spatial data via statistics, clustering, classification and regression, as well as the common rules (Shekhar et al. 2003).

classes. For example, the water areas in the classification image are converted from pixels to polygons by raster to vector conversion, and then the location and shape features of these polygons are calculated. Finally, the polygons are subdivided into subclasses by deductive reasoning based on knowledge, e.g. class water is subdivided into subclasses such as river, lake, reservoir and pond. In Figure 2, the final classification results are obtained by post-processing of the initial classification results by deductive reasoning. Except for the class label attribute, the attributes for deductive reasoning are the same as that in inductive learning. The knowledge discovered by the C5.0 algorithm is a group of classification rules and a default class, and each rule has a confidence value between 0 and 1. According to how the rule is activated when the attribute

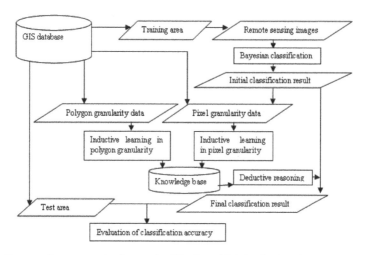

Figure 1.   Flow diagram of remote sensing image classification with inductive learning.

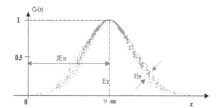

Figure 2.   Three numerical characteristics.

values match the conditions of this rule, the deductive reasoning adopts four strategies: (1) If only one rule is activated, then the final class is the same as this rule; (2) If several rules are activated, then the final class is the same as the rule with the maximum confidence; (3) If several rules are activated and the confidence values are the same, then the final class is the same as the rule with the maximum coverage of learning samples; and (4) If no rule is activated, then the final class is the default class.

3.2   Cloud model

The cloud model is a model of the uncertainty transition between qualitative and quantitative analysis, i.e. a mathematical model of the uncertainty transition between a linguistic term of a qualitative concept and its numerical representation. A piece of cloud is made up of lots of cloud drops, the shape visible as a whole, but fuzzy in detail, which is similar to a natural cloud in the sky. So the terminology cloud is used to name the uncertainty transition model proposed here. Any one of the cloud drops is a stochastic mapping in the universe of discourse from a qualitative concept, i.e. a specified realization with uncertain factors. With the cloud model, the mapping from the universe of discourse to the interval [0, 1] is a one-point to multi-point transition, i.e. pieces of cloud drops instead of a membership curve. Also, the degree to which any cloud drop represents the qualitative concept can be specified. Cloud models may mine spatial data with both fuzzy and stochastic uncertainties, and the discovered knowledge is close to human thinking. Now, in geo-spatial science, the cloud model has been further explored to spatial intelligent queries, image interpretation, land price discovery, factors selection, mechanism of spatial data mining and landslide monitoring (Li & Du 2005, Wang 2002, etc.).

The cloud model integrates well the fuzziness and randomness in a unified way via three numerical characteristics, Expected value (Ex), Entropy (En) and Hyper-Entropy (He). In the universe of discourse,

Ex is the position corresponding to the centre of the cloud gravity, whose elements are fully compatible with the spatial linguistic concept; En is a measure of the concept coverage, i.e. a measure of the spatial fuzziness, which indicates how many elements could be accepted by the spatial linguistic concept; and He is a measure of the dispersion of the cloud drops, which can also be considered as the entropy of En. The extreme case, {Ex, 0, 0}, denotes the concept of a deterministic datum where both the entropy and hyper entropy equal zero. The greater the number of cloud drops, the more deterministic the concept. Figure 2 shows the three numerical characteristics of the linguistic term "displacement is 9 mm around". Given three numerical characteristics Ex, En and He, the cloud generator can produce as many drops of the cloud as you would like.

The above three visualization methods are all implemented with the forward cloud generator in the context of the given {Ex, En, He}. Despite of the uncertainty in the algorithm, the positions of cloud drops produced each time are deterministic. Each cloud drop produced by the cloud generator is plotted deterministically according to its position. On the other hand, it is an elementary issue in spatial data mining that the spatial concept is always constructed from the given spatial data, and spatial data mining aims to discover spatial knowledge represented by a cloud from the database. That is, the backward cloud generator is also necessary. It can be used to perform the transition from data to linguistic terms, and may mine the integrity {Ex, En, He} of cloud drops specified by many precise data points. Under the umbrella of mathematics, the normal cloud model is common, and the functional cloud model is more interesting. Because it is common and useful to represent spatial linguistic atoms (Li et al. 2001), the normal compatibility cloud will be taken as an example to study the forward and backward cloud generators.

The input to the normal forward cloud generator is three numerical characteristics of a linguistic term, (Ex, En, He), and the number of cloud-drops to be generated, N, while the output is the quantitative positions of N cloud drops in the data space and the degree to which each cloud-drop can represent the linguistic term. The algorithm in detail is:

[1] Produce a normally distributed random number En' with mean En and standard deviation He;
[2] Produce a normally distributed random number x with mean Ex and standard deviation En';
[3] Calculate $y = e^{-\dfrac{(x - Ex)^2}{2(En')^2}}$;
[4] Drop (x_i, y_i) is a cloud-drop in the universe of discourse;
[5] Repeat step [1]–[4] until N cloud-drops are generated.

Simultaneously, the input of the backward normal cloud generator is the quantitative positions of N cloud-drops, $x_i(i = 1, \ldots, N)$, and the certainty degree that each cloud-drop can represent a linguistic term, $y_i(i = 1, \ldots, N)$, while the output is the three numerical characteristics, Ex, En, He, of the linguistic term represented by the N cloud-drops. The algorithm in detail is:

[1] Calculate the mean value of $x_i(i = 1, \ldots, N)$, $Ex = \dfrac{1}{N} \sum\limits_{i=1}^{N} x_i$;

[2] For each pair of (x_i, y_i), calculate $En = \sqrt{\dfrac{(x_i - Ex)}{2ln y_i}}$;

[3] Calculate the mean value of $En_i(i = 1, \ldots, N)$, $En = \dfrac{1}{N} \sum\limits_{i=1}^{N} En_i$;

[4] Calculate the standard deviation of En_i, $He = \sqrt{\dfrac{1}{N} \sum\limits_{i=1}^{N} (En_i - En)^3}$

With the given algorithms of the forward and backward cloud generators, it is easy to build the mapping relationship inseparably and interdependently between the qualitative concept and quantitative data. The cloud model lessens the weakness of rigid specification and too much certainty, which comes into conflict with the human recognition process, and appearing in commonly used transition models. Moreover, it performs the interchangeable transition between qualitative concept and quantitative data through the use of strict mathematical functions, and the preservation of the uncertainty in transition makes the cloud model well suited to the needs of real life situations. Obviously, the cloud model is not a simple combination of probability methods and fuzzy methods.

3.3 *Data field*

The obtained spatial data are comparatively incomplete. Each datum in the concept space has its own contribution to forming the conception and the concept hierarchy. So it is necessary for the observed data to radiate their data energies from the sample space to their parent space. In order to describe data radiation, data field is proposed.

Spatial data radiate energies into the data field. The power of the data field may be measured by its potential with a field function. This is similar to the electric charges that contribute to the formation of an electric field, in that every electric charge has an effect on the electric potential everywhere in the electric field. So the function of the data field can be derived from physical fields. The potential of a point in the number universe is the sum of all data potentials.

$$p = k \cdot e^{-\sum\limits_{i=1}^{N} \frac{r_i^2}{\rho_i}}$$

where, k is a constant of radiation gene, r_i is the distance from the point to the position of the ith observed data, ρ_i is the certainty of the ith data, and N is the amount of the data. With a higher certainty, the data may have a greater contribution to the potential in concept space. In addition, space between the neighbour isopotential, computerized grid density of Descartes coordinate, etc. may also make their contributions to the data field.

4 APPLICABILITY AND EXAMPLES

SDMKD may be applied in many spatial data referenced fields. Our three case studies are: remote sensing classification, landslide monitoring data mining, and spatial uncertainty reasoning.

4.1 *Remote sensing image classification*

A land use classification experiment was performed in the Beijing area using a SPOT multi-spectral image and a 1:100,000 scale land use database. Discovering knowledge from the GIS database and remote sensing image data can improve land use classification. Two kinds of knowledge on land use and elevation will be discovered from the GIS database, which are then applied to subdivide water and green patches, respectively. Pixel granularity is used to subdivide green patches and polygon granularity is used to subdivide water.

The original image is 2412 by 2399 pixels and has three bands, and was obtained in 1996. The land use database was built before 1996 and has land use, contours, roads and annotation layers (Fig. 3). The original image is stretched and rectified to fit the GIS data. The image is 2834 by 2824 pixels after rectification and is used as the source image for classification. ArcView 3.0a, ENVI 3.0 and See5.1.10, which is a development based on the C5.0 algorithm by the Rulequest Cooperation, were used in the study. We also developed several programs for data processing and format conversion using Microsoft Visual C++6.0.

For the sake of comparison, only the Bayesian method was applied to classify the image at first. The rectified image is overlaid with land use data layers, and the training and test areas are interactively selected. Then the image is classified into 8 classes (water, paddy, irrigated field, dry land, vegetable field, garden, forest and residential areas). As shown in the confusion matrix (Table 5), the overall accuracy is 77.6199 per cent. Water, paddy, irrigated field, residential area and vegetable field are classified with high accuracy. The vegetable field is easily distinguished from other green patches because it is lighter than the others. Dry land, garden and forest are seriously confused and the accuracy is 65.58 per cent, 48.913 per cent and 59.754 per cent, respectively. Some forest shadows are misclassified as water.

Spatial inductive learning is used in two respects to improve the Bayesian method of land use classification; one is to discover rules to subdivide water by polygon granularity, and the other to discover rules to reclassify dry land, garden and forest by pixel granularity. The land use layer (polygon) and contour layer

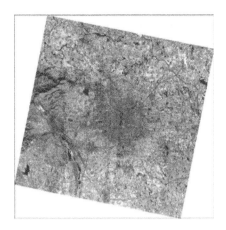

Figure 3.   Original SPOT image (resampled).

Table 4.   Rules discovered by inductive learning.

```
Rule 1:  (cover 19)                Rule 6:   (cover 213)
         compactness > 7.190882              Ycoord > 4428958
         height = lt50                       compactness ⇐ 7.190882
         -> class 71 [0.952]                 height = lt50
                                             -> class 74 [0.986]
Rule 2:  (cover 5)
         Xcoord > 453423.5         Rule 7:   (cover 281)
         Xcoord ⇐ 455898.7                   Xcoord > 451894.7
         Ycoord > 4414676                    compactness ⇐ 7.190882
         Ycoord ⇐ 4428958                    -> class 74 [0.975]
         compactness > 2.409397
         compactness ⇐ 7.190882    Rule 8:   (cover 38)
         -> class 72 [0.857]                 area ⇐ 500000
                                             height = 50-100
                                             -> class 74 [0.950]
Rule 3:  (cover 33)
         Xcoord ⇐ 455898.7
         Ycoord > 4414676          Rule 9:   (cover 85)
         Ycoord ⇐ 4428958                    height = gt200
         compactness ⇐ 7.190882              -> class 99 [0.989]
         height = lt50
         -> class 72 [0.771]       Rule 10:  (cover 7)
                                             height = 100-200
                                             -> class 99 [0.778]
Rule 4:  (cover 4)
         area > 500000
         height = 50-100           Default class: 74
         -> class 73 [0.667]
Rule 5:  (cover 144)               ----------------
         Ycoord ⇒ 4414676          Evaluation (604 cases):
         compactness ⇒ 7.190882
         height = lt50             Errors 7(1.2%)
         ->   class 74 [0.993]
```

Table 5. Confusion matrix of Bayesian classification.

| Classified | Real class | | | | | | | | |
	Water	Paddy	Irrigated field	Dry land	Vegetable field	Garden	Forest	Residential area	Sum
Water	3.900	0.003	0.020	0.013	0.002	0.021	2.303	0.535	6.797
Paddy	0.004	8.496	0.087	0.151	0.141	0.140	0.103	0.712	9.835
Irrigated field	0.003	0.016	10.423	0.026	0.012	0.076	0.013	0.623	11.192
Dry land	0.063	0.48	0.172	1.709	0.361	2.226	2.292	1.080	8.384
Vegetable field	0.001	0.087	0.002	0.114	3.974	0.634	0.435	0.219	5.465
Garden	0.010	0.009	0.002	0.325	0.263	4.422	4.571	0.065	9.666
Forest	0.214	0.006	0.000	0.271	0.045	1.354	15.671	0.642	18.202
Residential area	0.132	0.039	0.127	0.080	0.049	0.168	0.839	29.024	30.459
Sum	4.328	9.135	10.834	2.689	4.846	9.041	26.227	32.901	100
Accuracy (%)	90.113	93.010	96.204	63.580	81.994	48.913	59.754	88.217	
Overall accuracy = 77.6199%				Kappa coefficient = 0.7474					

(line) are selected for these purposes. Because there are few contours and elevation points, it is difficult to interpolate a DEM accurately. Instead, the contours are converted to height zones, such as <50 m, 50–100 m, 100–200 m and >200 m, which are represented by polygons.

In the learning for the subdivision of water, several attributes of the polygons in the land use layer are calculated as condition attributes, such as area, location of the centre, compactness (perimeter^2/(4π•area)), height zone, etc. The classes are river (code 71), lake (72), reservoir (73), pond (74) and forest shadow (99). 604 water polygons were learned, and 10 rules were discovered (Table 5).

In Table 5, only 1.2 per cent of samples are misclassified in the learning stage; thus the learning accuracy is 98.8 per cent. These rules reveal the spatial distribution patterns and general shape of features, etc. For example, rule 1 states "If compactness of a water polygon is greater than 7.190882, and locates in the height zone <50 m, then it is a river". Here, the compactness measure plays a key role in identifying the river from other water types. Rule 2 identifies lakes by location and compactness; rule 9 and rule 10 distinguish forest shadows from water by height, and so on.

In the learning for reclassifying dry land, garden and forest, the condition attributes are image coordinates, heights and the probability values of the three classes that were produced by the Bayesian classification. One per cent (2909) of samples were selected randomly from the vast amount of pixels. 63 rules were discovered and the learning accuracy is 97.9 per cent. The test accuracy is 94.4 per cent, which was evaluated by another one per cent of randomly selected samples. These rules are omitted here because of paper size limitation.

After inductive learning, the Bayesian classified image was reclassified by deductive reasoning based on the discovered rules. Because the Bayesian method cannot subdivide water, only the rules to identify forest shadows from water are used in order to compare the result with the Bayesian classification. The final class is determined by the maximum confidence principle (See Fig. 4).

Accuracy evaluation is accomplished using the same test areas as used in the Bayesian classification. The confusion matrix is shown in Table 6. The overall accuracy of the final result is 88.8751 per cent. The accuracy of dry land, garden and forest is 69.811 per cent, 78.561 per cent and 91.81 per cent, respectively. Comparing the final result with the result produced only by Bayesian classification, the overall accuracy had increased by 11.2552 per cent and the accuracy of dry land, garden and forest had increased by 6.231 per cent, 29.648 per cent and 32.056 per cent, respectively.

The land use classification result shows that the overall accuracy had increased by more than 11 per cent, and the accuracy of some classes, such as garden and forest, are further increased to approximately 30 per

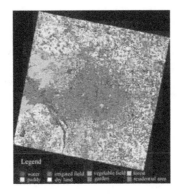

Figure 4. Final image classification (resampled).

Table 6. Confusion matrix of Bayesian classification combined with inductive learning.

Classified	Real class Water	Paddy	Irrigated field	Dry land	Vegetable field	Garden	Forest	Residential area	Sum
Water	3.900	0.003	0.020	0.012	0.002	0.019	0.139	0.535	4.631
Paddy	0.004	8.496	0.087	0.151	0.141	0.14	0.103	0.712	9.835
Irrigated field	0.003	0.016	10.423	0.026	0.012	0.076	0.013	0.623	11.192
Dry land	0.063	0.480	0.172	1.877	0.361	0.205	0.149	1.080	4.386
Vegetable field	0.001	0.087	0.002	0.114	3.974	0.634	0.435	0.219	5.465
Garden	0.009	0.009	0.002	0.210	0.263	7.102	0.470	0.065	8.131
Forest	0.215	0.006	0.000	0.218	0.045	0.696	24.079	0.642	25.899
Residential area	0.132	0.039	0.127	0.080	0.049	0.168	0.839	29.024	30.46
Sum	4.328	9.135	10.834	2.689	4.846	9.041	26.227	32.901	100
Accuracy (%)	90.113	93.01	96.204	69.811	81.994	78.561	91.81	88.217	
Overall accuracy = 88.8751%				Kappa coefficient = 0.8719					

cent. First, this indicates that the proposed SDMKD-based image classification approach of integrating the Bayesian classification with inductive learning not only improves classification accuracy greatly, but also extends the classification by subdividing some classes with the discovered knowledge. The approach is implemented feasibly and effectively. Secondly, spatial inductive learning in GIS databases can resolve the problems of spectral confusion to a great extent. SDMKD is very helpful in improving the Bayesian classification, and using probability values generates more accurate learning results than using the pixel values directly. Thirdly, utilizing the knowledge discovered from GIS databases is likely to be more beneficial in improving remote sensing image classification than the conventional way in which GIS data are utilized directly in pre- or post-processing of image classification.

4.2 Landslide monitoring

The Baota landslide is located in Yunyang, Chongqing, China. Monitoring started in June 1997. Up to now, this database on the displacements has amounted to 1 Gigabytes, and all the attributes are numerical displacements, i.e. dx, dy, and dh are the measurements of displacements in the X, Y and H directions at the landslide-monitoring points, and |dx|, |dy| and |dh| are their absolute values. In the following, it should be noted that all spatial knowledge is discovered from the databases with the properties of dx, dy, and dh, while |dx|, |dy| and |dh| are only used to visualize the results of spatial data mining. The properties of dx are the major examples. The linguistic terms of different displacements of dx, dy and dh may be depicted by the pan-concept hierarchy tree (Fig. 6) in the conceptual space, which are formed by cloud models (Fig. 11). It can be seen from Figure 6 and Figure 10 that the nodes "very small" and "small" both have the son node "9 mm around", which indicates that the pan-concept hierarchy tree is a pan-tree structure.

From the observed landslide-monitoring values, the backward cloud generator can mine Ex, En and He of the linguistic term indicating the level of that landslide displacement, i.e. gain the concept with the forward cloud generator. Then, with the three gained characteristics, the forward cloud generator can reproduce as many deterministic cloud-drops as one would like, i.e. produce synthetic values with the backward cloud generator.

According to the landslide monitoring characteristics, let the linguistic concepts of "smallest (0~9 mm), small (9~18 mm), big (18~27 mm), bigger (27~36 mm), very big (36~50 mm), extremely big (> 50 mm)" with Ex, "lowest (0~9), low (9~18), high (18~27), higher (27~36), very high (36~50), extremely big (>50)" with En, "most stable (0~9), stable (9~18), unstable (18~27), more unstable (27~36), very unstable (36~50), extremely unstable (>50)" with He, respectively, depicting the movements, scattering levels and stabilities of the displacements. Further, let the |dx|-axis, |dy|-axis respectively depict the absolute displacement values of the landslide monitoring points. The certainty of the cloud drop $(dx_i, C_T(dx_i))$, $C_T(dx_i)$ is also defined as,

$$C_T(dx_i) = \frac{dx_i - \min(dx)}{\max(dx) - \min(dx)}$$

where, $\max(dx)$ and $\min(dx)$ are the maximum and minimum of $dx = \{dx_1, dx_2, \ldots, dx_i, \ldots, dx_n\}$. Then the rules for monitoring the Baota landslide in the X direction can be discovered from the databases in the conceptual space.

BT11: the displacements are big south, high scattered and unstable;
BT12: the displacements are big south, high scattered and very unstable;
BT13: the displacements are small south, lower scattered and more stable;
BT14: the displacements are smaller south, lower scattered and more stable;
BT21: the displacements are extremely big south, extremely high scattered and extremely unstable;
BT22: the displacements are bigger south, high scattered and unstable;
BT23: the displacements are big south, high scattered and extremely unstable;
BT24: the displacements are big south, high scattered and more unstable;
BT31: the displacements are very big south, higher scattered and very unstable;
BT32: the displacements are big south, low scattered and more unstable;
BT33: the displacements are big south, high scattered and very unstable; and
BT34: the displacements are big south, high scattered and more unstable.

Figure 5 visualizes the above rules, where different rules are represented by ellipses with different colours, and the numbers in the ellipses denote the number of the rules. The generalized result at a higher hierarchy than that of Figure 4 in the feature space is the displacement rule for the whole landslide, i.e. "the whole displacement of the Baota landslide is bigger south (to Yangtze River), higher scattered and extremely unstable". Because large amounts of consecutive data are replaced by discrete linguistic terms, the efficiency of spatial data mining can be improved. Meanwhile, the final result mined will be stable due to the randomness and fuzziness of the concept indicated by the cloud model.

All the above landslide monitoring points form the potential field and the isopotential lines in the feature space. Intuitively, these points can be grouped naturally into clusters. These clusters represent different kinds of spatial objects recorded in the database, and naturally form the cluster graph. Figure 5(a) shows the potential of all points.

In Figure 6(a), all points' potentials form the potential field and the isopotential lines. Seen from this figure, when the hierarchy jumps up from Level 1 to Level 5, i.e. from the fine granularity world to

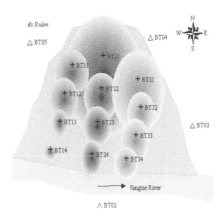

Figure 5.   Spatial rules at the Baota landslide monitoring points.

the coarse granularity world, these landslide monitoring points can be intuitively grouped into different clusters at different hierarchies of various levels. That is,

[1] No clusters at the hierarchy of Level 1;
[2] Four clusters at the hierarchy of Level 2. These are cluster BT14, cluster A (BT13, BT23, BT24, BT32, BT34), cluster B (BT11, BT12, BT22, BT31, BT33) and cluster BT21;
[3] Three clusters at the hierarchy of Level 3. These are cluster BT14, cluster (A, B) and cluster BT21;
[4] Two clusters at the hierarchy of Level 4, namely cluster (BT14, (A, B)) and cluster BT21; and
[5] One cluster at the hierarchy of Level 5. That is cluster ((BT14, (A, B)), BT21).

These denote [1] the displacements of landslide monitoring points are separate at the lowest hierarchy; [2] at the lower hierarchy, the displacements of landslide monitoring points (BT13, BT23, BT24, BT32, BT34) have the same trend of "the displacements are small", and (BT11, BT12, BT22, BT31, BT33) are all "the displacements are big", while BT14, BT21 show different trends with both of them and each other, i.e. the exceptions, "the displacement of BT14 is smaller", "the displacement of BT21 is extremely big"; [3] when the hierarchy becomes high, the displacements of landslide monitoring points (BT13, BT23, BT24, BT32, BT34) and (BT11, BT12, BT22, BT31, BT33) have the same trend of "the displacements are small"; however, BT14, BT21 still cannot be grouped into this trend; [4] when the hierarchy gets higher, the displacements of landslide monitoring point BT14 can be grouped into the same trend as (BT13, BT23, BT24, BT32, BT34) and (BT11, BT12, BT22, BT31, BT33), that is "the displacements are small"; however, BT21 is still an outlier; [5] the displacements of landslide monitoring points are unified at the highest hierarchy, that is, the landslide is moving.

These show different "rules plus exceptions" in the fine granularity world to the coarse granularity world. That is to say, the clustering or associations between attributes at different cognitive levels make many combinations, "rules plus exceptions", showing the discovered knowledge with different information granularities. When the exceptions BT14, BT21 are granted and eliminated, the rules and the clustering process will be more obvious (Fig. 6(b)). Simultaneously, these clusters represent different kinds of landslide monitoring points recorded in the database. They can naturally form the cluster spectrum figures as shown in Figure 6(c) and Figure 6(d). As seen from these two figures, the displacements of points (BT13, BT23, BT24, BT32, BT34) and (BT13, BT23, BT24, BT32, BT34) first compose two new classes, cluster A and cluster B, then the two new classes compose a larger class with cluster BT14, and they finally compose the largest class with cluster BT21, during the process of which the mechanism of spatial data mining is still "rules plus exceptions". That is, different users may discover knowledge from the displacement database of Baota landslide with different demands.

185

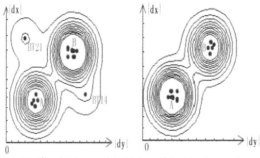

(a) All points' potential   (b) Potential without exceptions

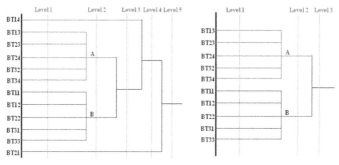

(c) All points' cluster graph   (d) Cluster graph without exceptions

Figure 6.   The clusters and cluster graphs of Landslide monitoring points.

When the Committee of Yangtze River investigated the region of the Yunyang Baota landslide, they found that the landslide had moved towards the Yangtze River. By landslide monitoring point BT21, a small sized landslide had taken place. Now there are still two pieces of large rift, especially the wall rift of farmer G. Q. Zhang's house being nearly 15 mm. These actual results match the discovered spatial knowledge very well, which indicates that the techniques of cloud model-based spatial data mining are practical and creditable.

4.3   Uncertainty reasoning

Uncertainty reasoning may include one-rule reasoning and multi-rule reasoning.

4.3.1   One-rule reasoning

If there is only one factor in the rule antecedent, we call the rule a one-factor rule. Figure 6 is a one-factor, one-rule generator for the rule "If A, then B". CGA is the X-conditional cloud generator for linguistic term A, and CGB is the Y conditional cloud generator for linguistic term B. Given a certain input x, CGA generates random values μ_i. These values are considered as the activation degree of the rule and input to CGB. The final outputs are cloud drops, which form a new cloud.

Combining the algorithm of the X and Y conditional cloud generators (Li 2005), we present the following algorithm for one-factor, one-rule reasoning.

[1] $En'_A = G(En_A, He_A)$ Produce random values that satisfy the normal distribution probability of mean En_A and standard deviation He_A.

[2] Calculate $\mu = \exp\left[-\dfrac{(x - Ex_A)^2}{2En'^2_A}\right]$

186

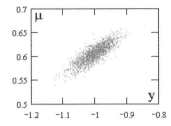

Figure 7.   Output cloud of one-rule reasoning.

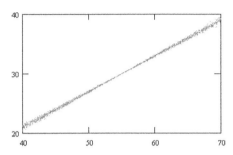

Figure 8.   Input-output response of one-rule reasoning.

[3] $En'B = G(EnB, HeB)$ Produce random values that satisfy the normal distribution probability of mean EnB and standard deviation HeB.

[4] Calculate $y = Ex_B \pm \sqrt{-2\ln(\mu)}En'_B$, let (y, μ) be cloud drops. If $x < ExA$, "-"is adopted, while id $x > ExA$, "+" is adopted.

[5] Repeat step [1] to [4], generating as many cloud drops as required.

Figure 7 is the output cloud of a one-factor, one-rule generator with one input. We can see that the cloud model based reasoning generates an uncertain result. The uncertainty of the linguistic terms in the rule is propagated during the reasoning process. Since the rule output is a cloud, we can give the final result in several forms: (1) One random value; (2) Several random values as sample results; (3) Expected value, which is the mean of many sample results; (4) Linguistic term, which is represented by a cloud model, and the parameters of the model are obtained by the inverse cloud generator method.

If we input a number of values to the one-factor rule and present the inputs and outputs in a scatter plot, we can obtain the input-output response graph of the one-factor, one-rule reasoning (Fig. 8). The graph looks like a cloud band, not a line. Closer to the expected values, the band is more focused, while further from the expected value, the band is more dispersed. This is consistent with human intuition. The above two figures and the discussion show that the cloud model based uncertain reasoning is more flexible and powerful than the conventional fuzzy reasoning method.

If the rule antecedent has two or more factors, such as "If $A_1, A_2, ..., A_n$, then B", we call the rule a multi-factor rule. In this case, a multi-dimensional cloud model represents the rule antecedent. Figure 9 is a two-factor, one-rule generator, which combines a 2-dimensional X-conditional cloud generator and a 1-dimensional Y-conditional cloud generator. It is easy to give the reasoning algorithm on the basis of the cloud generator algorithms stated in Section 3.2 and consequently, multi-factor, one-rule reasoning is conducted in a similar way.

4.3.2   Multi-rule reasoning
Usually, there are many rules in a real knowledge base. Multi-rule reasoning is frequently used in an intelligent GIS or a spatial decision support system. Figure 10 is a one-factor, multi-rule generator and the algorithm is as follows.

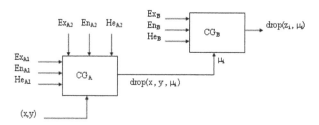

Figure 9. A two-factor, one-rule generator.

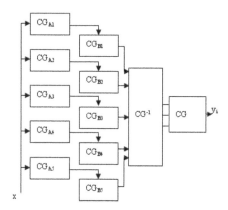

Figure 10. One-factor, five-rule generator.

The algorithms for the one-factor, multi-rule generator (Fig. 10) are as follows.

[1] Given input x, determine how many rules are activated. If $Ex_{Ai} - 3En_{Ai} < x < Ex_{Ai} + 3En_{Ai}$, then rule i is activated by x.
[2] If only one rule is activated, output a random y by the one-factor, one-rule reasoning algorithm. Go to step [4]
[3] If two or more rules are activated, firstly each rule outputs a random value by the one-factor, one-rule reasoning algorithm, and a virtual cloud is constructed by the geometric cloud generation method of Section 3.2. A cloud generator algorithm is conducted to output a final result y with the three parameters of the geometric cloud. Because the expected value of the geometric cloud is also a random value, we can take the expected value as the final result for simplicity. Go to step [4]
[4] Repeat step [1] to [3], generating as many outputs as required.

The main idea of the multi-rule reasoning algorithm is that when several rules are activated simultaneously, a virtual cloud is constructed by the geometric cloud method. Because of the property of least squares fitting, the final output is close to the rule of activated summarization. This is consistent with the human being's intuition. Figure 11 is a situation of two rules being activated and Figure 12 is the situation when three rules are activated. Only the mathematical expected curves are drawn for clarity and the dashed curves are the virtual clouds. The one-factor, multi-rule reasoning method can be easily extended to multi-factor, multi-rule reasoning on the basis of multi-dimensional cloud models. We omit the algorithm here.

4.3.3 An illustrative example
The following is an illustrative example of multi-factor, multi-rule reasoning. Suppose we have the following five rules to describe qualitively the terrain features in a GIS.

Rule 1: If location is southeast, then elevation is low.
Rule 2: If location is northeast, then elevation is low to medium.

188

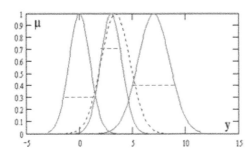

Figure 11.　Two activated rules.

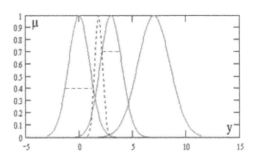

Figure 12.　Three activated rules.

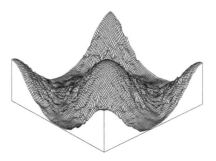

Figure 13.　2-dimensional clouds to represent the linguistic term of location.

Rule 3: If location is central, then elevation is medium.
Rule 4: If location is southwest, then elevation is medium to high.
Rule 5: If location is northwest, then elevation is high.

Rule input is the location. Because the location is determined by x and y coordinates, the linguistic terms for location are represented by 2-dimensional clouds (Fig. 13). Rule output is the elevation, which is represented by 1-dimensional clouds (Fig. 14).

Figure 15 is the input-output response surface of the rules. The surface is an uncertain surface and the surface is uneven (rough). Closer to the centre of the rules (the expected values of the clouds), the surface is more smooth showing the smaller uncertainty; while at the overlapping areas of the rules, the surface is more rough showing larger uncertainty. This proves that the multi-factor, multi-rule reasoning also represents and propagates the uncertainty of the rules as did the one-factor, one-rule reasoning.

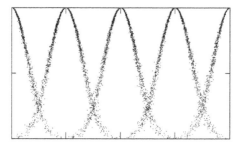

Figure 14.   1-dimensional clouds to represent the linguistic term of elevation.

Figure 15.   The input-output response surface of the rules.

5   CONCLUSIONS AND DISCUSSION

After the concept and principle of spatial data mining and knowledge discovery (SDMKD) were outlined, this paper proposed three mining techniques – SDMKD-based image classification, cloud model and data field – together with applicable examples, i.e. remote sensing image classification, Baota landslide monitoring data mining and spatial uncertain reasoning.

The technical progress in computerized spatial data acquisition and storage has resulted in the growth of vast databases, which caused a branch of data mining, SDMKD, to be developed in geo-spatial science. This paper takes the mechanism of SDMKD as a process of discovering a form of rules plus exceptions at hierarchal view-angles with various thresholds.

SDMKD-based image classification integrates spatial inductive learning from a GIS database with Bayesian classification. In the experimental results of remote sensing image classification, the overall accuracy was increased more than 11 per cent, and the accuracy of some classes, such as garden and forest, was further increased to approximately 30 per cent. The intelligent integration of GIS and remote sensing encourages the discovery of knowledge from spatial databases and applies this knowledge in image interpretation for spatial data updating. The applications of inductive learning in other image data sources, e.g. TM, SAR, and the applications of other SDMKD techniques in remote sensing image classification, may be valuable future directions of further study.

Cloud models integrate the fuzziness and randomness in a unified way via the algorithms of forward and backward cloud generators in the contexts of three numerical characteristics, {Ex, En, He}. It takes advantage of human natural language, and might search for the qualitative concept described by natural language to generalize a given set of quantitative datum with the same feature category. Moreover, the cloud model can act as an uncertainty transition between a qualitative concept and its quantitative data. With this method, it is easy to build the mapping relationship inseparably and interdependently between qualitative concepts and quantitative data, and the final discovered knowledge with its hierarchy can match different demands from different levels of user. A series of uncertainty reasoning algorithms were presented based on the algorithms of cloud generators and conditional cloud generators.

Data field radiates the energy of observed data from the universe of sample to the universe of discourse, considering the function of each data item in SDMKD. In the context of data field, the conceptual space represents different concepts in the same characteristic category, while the feature space depicts complicated spatial objects with multi-properties. A concept hierarchy may be automatically shaped with the isopotentials of all data, along with the clusters and clustering hierarchy of all spatial entities. The clustering or associations between attributes at different cognitive levels might make many combinations.

The experimental results on the monitoring of the Baota landslide show that spatial data mining with data field and cloud model can reduce the task complexity, improve the implementation efficiency, and enhance the comprehension of the discovered spatial knowledge.

ACKNOWLEDGEMENTS

This study is supported by 973 Program (2006CB701305), National Excellent Doctoral Thesis Fund, the State Key Laboratory of Software Engineering Fund (WSKLSE05-15), and the Ministry Key GIS Lab Fund.

REFERENCES

Arthurs, A.M. 1965. *Probability theory*. London: Dover Publications.
Buckless, B.P. & Petry, F.E. 1994. *Genetic Algorithms*. Los Alamitos, California: IEEE Computer Press.
Cressie, N. 1991. *Statistics for Spatial Data*. New York: John Wiley and Sons.
Dasu, T. 2003. *Exploratory Data Mining And Data Cleaning*. New York: John Wiley & Sons, Inc.
Di, K.C. 2001. *Spatial Data Mining and Knowledge Discovery*. Wuhan: Wuhan University Press.
Ester, M., Frommelt, A., Kriegel, H.P. & Sander, J., 2000. Spatial data mining: databases primitives, algorithms and efficient DBMS support. *Data Mining and Knowledge Discovery* 4: 193–216.
Gallant, S.I. 1993. *Neural Network Learning and Expert Systems*. Cambridge: MIT Press.
Han, J. & Kamber, M. 2001. *Data Mining: Concepts and Techniques*. San Francisco: Academic Press.
Han, J. 1998. *Towards on-line analytical mining in large databases*. ACM SIGMOD–Record.
Haining, R. 2003. *Spatial Data Analysis: Theory and Practice*. Cambridge: Cambridge University Press.
Hawkins, D. 1980. *Identifications of outliers*. London: Chapman and Hall.
Hong, J.R. 1997. *Inductive Learning – algorithm, theory and application*. Beijing: Science Press.
Li, D.R. & Guan, Z.Q. 2002. *Integration and realization of spatial information system*. Wuhan: Wuhan University Press.
Li, D.R. & Cheng, T. 1994. KDG: Knowledge Discovery from GIS – Propositions on the Use of KDD in an Intelligent GIS. In *Proc. ACTES, The Canadian Conf. on GIS*. 1001–1012.
Li D.R., Wang, S.L., Shi, W.Z. & Wang, X.Z., 2001. On spatial data mining and knowledge discovery (SDMKD). *Geomatics and Information Science of Wuhan University* 26(6): 491–499.
Li, D.R., Wang, S.L., Li, D.Y. & Wang, X.Z., 2002. Theories and technologies of spatial data mining and knowledge discovery. *Geomatics and Information Science of Wuhan University* 27(3): 221–233.
Li, D.R., Wang, S.L. & Li, D.Y. 2005. *Theories and Applications of Spatial Data Mining*. Beijing: Science Press.
Li, D.Y. & Du, Y. 2005. *Artificial Intelligence with Uncertainty*. Beijing: National Defense Industry Press.
Li, X., Zhang, S.C. & Wang, S.L. 2006. Advances in Data Mining Applications. *Special Issue of IJDWM (International Journal of Data Warehouse and Mining)* 2(3): i–iii.
Miller, H.J. & Han, J. 2001. *Geographic Data Mining and Knowledge Discovery*. London and New York: Taylor and Francis.
Muggleton, S. 1990. *Inductive Acquisition of Expert Knowledge*. Wokingham, England: Turing Institute Press in association with Addison-Wesley.
Pawlak, Z. 1991. *Rough Sets: Theoretical Aspects Of Reasoning About Data*. London: Kluwer Academic Publishers.
Quinlan, J. 1986. Introduction of decision trees. *Machine Learning* (5): 239–266.
Shafer, G. 1976. *A Mathematical Theory of Evidence*. Princeton: Princeton University Press.

Shekhar, S., Lu C.T. & Zhang, P. 2003. A Unified Approach to Detecting Spatial Outliers. *GeoInformatica* 7(2): 139–166.

Soukup, T. & Davidson, I. 2002. *Visual Data Mining: Techniques and Tools for Data Visualization and Mining*. New York: Wiley Publishing, Inc.

Wang, S.L. 2002. *Data Field and Cloud Model –Based Spatial Data Mining and Knowledge Discovery*. Ph.D. Thesis. Wuhan: Wuhan University.

Wang, S.L., Shi, W.Z., Yuan, H.N. & Chen G.Q., 2005. Attribute uncertainty in GIS data. *Lecture Notes in Computer Science* 3614: 614–623.

Zadeh, L.A. 1975. The concept of linguistic variable ant its application to approximate reasoning. *Information Science* 8: 199–249.

Advances in Spatio-Temporal Analysis – Tang et al. (eds)
© 2008 Taylor & Francis Group, London, ISBN 978-0-415-40630-7

Genetic neural network based data mining and its application in police case analysis

Hanli Liu, Lin Li & Haihong Zhu
School of Resource and Environment Science, Wuhan University, Wuhan, P.R. China

ABSTRACT: This paper puts forward a method that combines the learning algorithm of BP neural networks with a genetic algorithm to train BP networks and optimize the weight values of the network in a global scale. This method offers global optimization, high accuracy and fast convergence. The data-mining model based on the genetic neural network has been widely applied to the procedure of data mining in case information in the command centre of a police office. It achieves an excellent effect in assisting the solution of cases and making good decisions. In this paper, the principles and methods of this data-mining model are described in detail. An actual case of its application is also presented that allows us to draw the conclusion that this data-mining model is scientific, efficient and practicable.

1 INTRODUCTION

1.1 *The definition of data mining*

Data mining is a procedure of information extraction and knowledge acquirement from a large quantity of data. The information and knowledge are hidden behind the data, unknown by the users but potentially useful. The data have the characteristics of incompleteness, contain noise, are fuzzy and uncertain (Hard 2003). As a kind of interdisciplinary field that synthesizes multiple disciplines including database technology, artificial intelligence, neural networks, statistics, knowledge acquisition and information extraction, data mining has nowadays become one of the foremost research directions in the international realms of information-based decision making. By analysing and synthesizing data obtained from different aspects, data mining methods dig out useful and hidden information for prediction from a large amount of data stored in databases and data warehouses. The methods that have been used include association rules, classification knowledge, clustering analysis, tendency and deviation analysis, as well as similarity analysis (Qing 2003). By searching for valuable information from the analysis results, users can employ the information to guide their business actions, administrative actions and their scientific research. All of these provide new opportunities and challenges to the development of all kinds of fields related to data processing.

1.2 *The application of data mining*

Applied to the procedure of data analysis, data process, data utilization and decision making in many social departments, data warehouse and data mining technologies assist these departments to make scientific and reasonable decisions (Berson & Smith 1999). This is meaningful for the development of our society and economy. Data mining can be applied to various realms. For instance, many sales departments use data mining technology to determine the distribution or geographical position of their sales network and to discover potential customer groups, thus allowing adjustment to their sale strategies. In insurance companies, stock companies, banks and credit card companies, data mining technology is applied to detect deceptive actions by customers in order to reduce commercial fraud. Data mining has also been widely applied to medical treatments, genetic engineering and many other fields. In recent years, with the increase in information held in police departments in China, data mining technology has been applied to police departments, especially in the command centres of police offices. This paper mainly discusses the principle and the practical application of a genetic neural network based data mining model in case analysis in a police office.

2 THE MEANING AND METHOD OF DATA MINING IN THE COMMAND CENTRE OF A POLICE OFFICE

2.1 *The meaning of data mining in the command centre of a police office*

Every day in the command centre of police offices, much case information is received by various means. The information is input to databases to form a large amount of case data. These data are archived annually and periodically to form a resource of historical cases. By processing and analysing these historical cases, police officers can gain some experience and learn lessons to help them solve cases and make decisions in the future. Therefore, in order to assist police departments to solve cases rapidly and make decisions efficiently, we should synthesize and organize these historical data, use proper data mining models to discover potential and useful knowledge from the data, and then make predictions and analyse the important factors in the data, including the rate of crime, the constitution of crime population, the crime age structure, the area distribution of crime, the developing tendency of crime, the means and approaches of crime, where criminals hide, and so on. At present, all of these have become urgent tasks that police offices need to accomplish by their data processing procedures.

2.2 *Two steps of data mining*

The data mining procedure in the command centre of a police office includes two main steps:

(1) First, filtering, selecting, cleaning and synthesizing the archived historical case information, and then performing transformation, if necessary, and finally loading data into a data warehouse.
(2) Choosing appropriate models and algorithms of data mining to dig out the potential knowledge from the data. By much analysis and comparison among various data mining models, we selected the error back propagation (BP) neural network as the general purpose calculation model in our data mining. We train the neural network with a supervised learning method and combine the BP algorithm with a genetic algorithm to optimize the weight values of the BP network (Li 2004). Further, we apply the trained model to the prediction, classification and rule extraction of the case information.

3 DATA MINING MODEL OF NEURAL NETWORK

3.1 *Common methods of data mining*

At present, data mining methods include statistical methods, association discovery, clustering analysis, classification and regression, OLAP (On Line Analytical Processing), query tools, EIS (Executive Information System), neural networks, genetic algorithms, etc. Because of its high durability against noisy data, its good generalization ability, high accuracy and low error rate, the neural network model possesses great advantages compared to other data mining methods (Xu 2004). It has now become a popular method in the field of data mining.

3.2 *Data mining model of BP neural network*

BP neural network is a kind of feedforward network and is now widely used. Generally, it has a multi-layer structure that consists of at least three layers. The network includes one input layer, one output layer and one or more hidden layers. There are full connections between neurons in adjacent layers and no connections between neurons in the same layer (Zhong 1993). Based on a set of training samples and a set of test data, a BP neural network trains its neurons and completes the procedure of learning. The BP algorithm is suitable for a data mining environment in which it is impossible to solve problems using ordinary methods. We need to use complex functions with several variables to complete non-linear calculations and accomplish

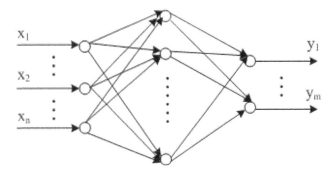

Figure 1. The structure of a BP neural network.

the semi-structural and non-structural decision-making supporting procedure. So for the command centre of a police office, we choose the BP neural network as the data mining model.

The basic structure of a BP neural network is shown in Figure 1.

The learning procedure of the neural network can be divided into three phases:

(1) The first is a forward propagation phase in which a specified input pattern has been past through the network from the input layer, through the hidden layers to the output layer and becomes an output pattern.

(2) The second is an error back propagation phase. In this phase, the BP algorithm compares the real output and the expected output to calculate the error values. After that, it propagates the error values from the output layer through the hidden layers to the input layer. The connection weights will be altered during this phase.

(3) These two phases proceed repeatedly and alternately to complete the memory training of the network until the network convergences and the global error tends to a minimum (Xu 2003).

3.3 Learning algorithm of the neural network

In the practical application of data mining, we use the three-layer BP neural network model that includes one input layer, one output layer and a single hidden layer. We choose the differentiable Sigmoid function as its activation function. The function is defined in formula (1):

$$f(x) = \frac{1}{1 + e^{-x}} \tag{1}$$

The learning algorithm of the BP neural network is described as follows:

(1) Setting the initial weight values $W(0)$: Generally we generate small random and non-zero floating numbers in $[0, 1]$ as the initial weight values.

(2) Choosing certain pairs of input and output samples and calculating the outputs of the network. The input samples are $x_s = (x_{1s}, x_{2s}, \ldots, x_{ns})$. The output samples are $o_s = (o_{1s}, o_{2s}, \ldots, o_{ms})$, $s = 1, 2, \ldots, L$. L is the number of input samples. When the input sample is the sth sample, the output of the ith neutron is y_{is}:

$$y_{is}(t) = f\left(\sum_j w_{ij}(t)x_{js}\right) \tag{2}$$

where x_{js} is the jth input of the ith neutron when the input sample is the sth sample and t is the training times.

195

(3) Calculating the global error of the network. When the input sample is the sth sample, E_s is the error of the network. The formula for calculating E_s is:

$$E_s(t) = \frac{1}{2} \sum_k (o_{ks} - y_{ks}(t))^2$$

$$= \frac{1}{2} \sum_k e_{ks}^2(t) \tag{3}$$

where k represents the kth neutron of the output layer. $y_{ks}(t)$ is the output of the network when the input sample is the sth sample and the weight values have been adjusted t times. After training the network t times and based on all of the L groups of samples, the global error of all of these samples is:

$$G(t) = \sum_s E_s(t) \tag{4}$$

(4) Determining if the algorithm ends.

$$G(t) \leq \varepsilon \tag{5}$$

when the condition of formula (5) is satisfied, the algorithm ends. ε is the limit value of error that is specified beforehand. $\varepsilon > 0$.

(5) Calculating the error of back propagation and adjusting the weights. The gradient descent algorithm has been used to calculate the adjustment values of weights. The formula is as follows:

$$w_{ij}(t+1) = w_{ij}(t) - \eta \frac{\partial G(t)}{\partial w_{ij}(t)}$$

$$= w_{ij}(t) - \eta \sum_s \frac{\partial E_s(t)}{\partial w_{ij}(t)} \tag{6}$$

where η is the learning rate of the network and also the step of the weight adjustment.

3.4 The problems of the BP Network and their solutions

Because we use the gradient descent algorithm to calculate the values of weights, the BP neural network still encounters problems such as local minimum, slow convergence speed and convergence instability in its training procedure (Wang 2005). We combine two methods to solve these problems. One solution is to improve the BP network algorithm. By adding a steep factor or acceleration factor in its activation function, the speed of convergence can be increased. In addition, by compressing the weight values when they are too large, network paralysis can be avoided. The improved activation function is defined by formula (7):

$$f_{a,b,\lambda}(x) = \frac{1}{1 + e^{(x-b)/\lambda}} + a \tag{7}$$

where a is a deviation parameter, b is a position parameter and λ is the steep factor.

Another solution is possible because the genetic algorithm is a concurrence global search algorithm. Because of its excellent performance in global optimization, we can combine the genetic algorithm with the BP network to optimize the connection weights of the BP network. Finally, we can use the BP algorithm for accurate prediction or classification.

4 UTILIZING THE GENETIC ALGORITHM TO OPTIMIZE THE BP NEURAL NETWORK

4.1 The principle of the genetic algorithm

A genetic algorithm is a kind of search and optimization model built by simulating the lengthy evolution period of heredity selection and natural elimination of a biological colony (Goldberg 1992). It is an algorithm of global probability search. It does not depend on gradient data and does not need the differentiability of the function that will be solved and only needs that the function can be solved under the condition of constraint. A genetic algorithm has the powerful ability of macro scope search and is suitable for global optimization. So by using a genetic algorithm to optimize the weights of the BP neural network, we can eliminate the problems of the BP network and enhance the generalization performance of the network (Zhou 2005).

The individuals in genetic space are chromosomes. The basic constitution factors are genes. The position of a gene in an individual is called the locus. A set of individuals constructs a population. The fitness represents the evaluation of adaptability of individuals to the environment (Goldberg 1989).

The elementary operation of a genetic algorithm consists of three operators: selection, crossover and mutation (Srinivas et al. 1994). Select is also called copy or reproduction. By calculating the fitness f_i of individuals, we select high quality individuals with high fitness, copy them to a new population and eliminate the individuals with low fitness from the new population. Generally used strategies of selection include roulette wheel, expectation value, paired competition and retaining high quality individuals. After selection, a crossover operator puts individuals into matched pools and randomly makes individuals into pairs to form a parent generation. Then, with a crossover probability and the specified method of crossover, it exchanges part of the genes of individual pairs to generate new individual pairs in the child generation. The common methods of crossover are one point crossover, multi point crossover and average crossover. Finally, with a specified mutation rate, a mutation operator substitutes genes with their opposite genes in some loci to generate new individuals.

4.2 The calculating steps of the genetic algorithm

The methods and steps of utilizing a genetic algorithm to optimize the weights of the BP network are described as follows:

(1) First, k groups of weights are given at random and respectively assigned to k sets of BP networks. By training the networks, k groups of new weights are calculated and adjusted. They constitute the original solution space.

(2) Using real number coding methods, the weights are encoded into decimals and used as chromosomes. k groups of chromosomes comprise a population. So the original solution space has been mapped to the search space of the genetic algorithm. The length of gene string after encoding is $L = n \times h + h \times m$, where n is the number of neutrons in the input layer, h is the number of neutrons in the hidden layer, m is the number of neutrons in the output layer.

(3) Using the maximum optimization method, the fitness function can be determined. The formula for the fitness function is:

$$f = \frac{1}{2G} = \frac{1}{2\sum\limits_{j=1}^{s}\sum\limits_{i=1}^{m}(o_{ij} - y_{ij})^2} \tag{8}$$

where s is the total number of samples. m is the number of neutrons in the output layer. G is the global error of all of s numbers of samples. o_{ij} is the expected output of the network and y_{ij} is the real output of the network.

(4) The weights are optimized using the genetic algorithm. We calculate the fitness and perform the selection using the roulette wheel method. After that, we copy the individuals with high fitness into the next generation. After selection, we crossover the individuals with probability P_c. Because we encode weights into decimals, the operator of crossover should be altered. If the crossover is performed between the ith individual and the $(i+1)$th individual, the operator is as follows:

$$\begin{cases} X_i^{t+1} = c_i \cdot X_i^t + (1 - c_i) \cdot X_{i+1}^t \\ X_{i+1}^{t+1} = (1 - c_i) \cdot X_i^t + c_i \cdot X_{i+1}^t \end{cases} \tag{9}$$

where X_i^t, X_{i+1}^t is a pair of individuals before crossover, X_i^{t+1}, X_{i+1}^{t+1} is a pair of individuals after crossover. c_i is a random datum of uniform distribution in $[0, 1]$. With probability P_m, we mutate the individuals after crossover. When we mutate the ith individual, the operator is:

$$X_i^{t+1} = X_i^t + c_i \tag{10}$$

where X_i^t is an individual before mutation, X_i^{t+1} is an individual after mutation, c_i is a random datum of uniform distribution in $[u_{min} - \delta_1 - X_i^t, u_{max} + \delta_2 + X_i^t]$, where u_{min} and u_{max} are respectively the minimum value and maximum value of weights in the BP networks. δ_1 and δ_2 are used as adjustment parameters. After these operations are performed once, a new population is generated. By repeating the procedures of selection, crossover and mutation, the weight combination is adjusted close enough to the most optimized weight combination (Heckerling & Gerber 2004).

(5) Finally, through the BP networks, the weights can be adjusted delicately. Now the whole procedure of optimization ends.

With respect to every kind of prediction and analysis problem in the course of data mining, we extract proper sets of training samples and test data, train mature neural network models with the above-mentioned methods and apply the models to future case analysis and prediction.

5 A REAL INSTANCE OF APPLICATION

In this section, we give a real application of data mining in the command centre of a police office as an example. In this example, we analyse people's gender, age, education degree, crime history, experiences, personal features, habits, social relations, economical incomes, family environment and marital status. To some extent, we find that these factors affect people's social actions that may lead people to commit a crime. Using these factors as input variables, we utilize a genetic neural network to predict the current crime possibility of these people.

5.1 Clean data in database

In the first step, we fill up the missing data, smooth the noisy data in the database and solve the problems of the same names but with different meanings and different names with the same meaning. And then, we load related data into the data warehouse.

5.2 Select training samples for BP networks

Because in the case of an archive database, the information is arranged in order of time, representative data can be obtained by random sampling. To obtain the training sample set, we selected 5,000 records from the data warehouse. In addition, we extracted another 2,000 records as the test sample set.

5.3 Normalize samples

The most important input variables of the BP network include gender, age, education degree, crime history, salary level and bad habits. The output of samples is the status (Yes or No) of whether these people commit a crime at present. The output of the BP network is the probability of people's present status (percentage of crime probability) of crime. Table 1 gives a list of the first 10 samples out of the total of 5,000 training samples.

By normalizing the above input and output variables, the range of values of these variables has been mapped to the range of $[0, 1]$. The mapping relationship is given as follows:

Gender

Male: 1.0;   Female: 0.0

Table 1. Values of input variables.

No.	Sex	Age	Education degree	Crime history	Salary level	Bad habits	Present status of crime
1	M	25	Secondary school	Yes	300–800	No	Yes
2	M	32	Secondary school	No	1–300	Yes	Yes
3	M	40	Primary School	Yes	300–800	Yes	Yes
4	F	30	Primary School	No	5,000–8,000	No	No
5	F	27	Secondary School	Yes	300–800	Yes	Yes
6	M	28	University	No	1,500–3,000	No	No
7	M	50	Junior University	No	800–1,500	No	No
8	M	38	Post-graduate	No	5,000–8,000	No	No
9	M	70	Primary School	No	1–300	Yes	No
10	F	35	High School	No	300–800	No	No

Age

0: 0.00; 1: 0.01; 2: 0.02; …; 100 and above: 1.0

Education degree

Illiterate: 0.0
Graduate of Primary School: 0.125
Graduate of Secondary School: 0.25
Graduate of High School: 0.375
Graduate of Junior University: 0.5
Graduate of University: 0.625
Postgraduate: 0.75
Doctor: 0.875
Post doctor: 1.0

Crime history

Yes: 1.0; No: 0.0

Salary level

None: 0.0; Below 300 Yuan: 0.125; 300–800 Yuan: 0.25;
800–1,500 Yuan: 0.375; 1,500–3,000 Yuan: 0.5;
3,000–5,000 Yuan: 0.625; 5,000–8,000 Yuan: 0.75;
8,000–15,000 Yuan: 0.875; 15,000 Yuan and above: 1.0

Bad habits

Yes: 1.0; No: 0.0

Present status of crime

Yes: 1.0; No: 0.0
In Table 2, we give a value list of the first 10 normalized samples from the total 5,000 training samples.

5.4 *Build and train the BP neural networks*

5.4.1 *Build and train BP neural networks*
Including the above six important variables, the total number of input variables is 10, so we determined that the number of neutrons in the input layer is 10 and the number of neutrons in the output layer is one.

Table 2. Normalized value of input variables.

No.	Sex	Age	Education degree	Crime history	Salary level	Bad habits	Present status of crime
1	1	0.25	0.25	0	0.25	0	1
2	1	0.32	0.25	1	0.125	1	1
3	1	0.40	0.125	1	0.25	1	1
4	0	0.45	0.125	0	0.75	0	0
5	0	0.27	0.25	1	0.25	1	1
6	1	0.28	0.625	0	0.5	0	0
7	1	0.50	0.5	0	0.375	0	0
8	1	0.38	0.75	0	0.75	0	0
9	1	0.70	0.125	0	0.125	1	0
10	0	0.35	0.375	0	0.25	0	0

According to our experience and conforming to the principle of simplifying the network structure, we set the number of neutrons in the hidden layer to 16. With the above parameters, we built 10 BP networks that have the same structure. Then we generated 10 sets of small random numbers as initial weights of these networks and used the extracted 5,000 samples as input and output samples of the networks. After that, we utilized the BP algorithm to train the networks and got 10 sets of trained weights. The training times are 5,000. After training, we tested the networks with our test sample set.

5.4.2   Calculate generalization ability

After training, we calculated the global training error and the global testing error of every BP network and gained the average global training error and global testing error of these 10 BP networks. Further, we trained the BP networks another 3,000 times and calculated the global errors again. We found that the changes in error values were very small. So, we calculated the average global errors again and ended the training. As a result, the average generalization ability of these networks is shown in Figure 2, where X is the times of training, Y is the value of error, G1 is the average global error of the training sample set and G2 is the average global error of the test sample set.

5.4.3   Training result

From Figure 2, we can see that the training result shows that the BP networks can converge rapidly after limited times of training. And indeed, every one of these networks has converged. But the global error ends up with a relatively large value. It is probable that the global errors of BP networks converge to a local minimum value, but not the global minimum value. To achieve the global minimum error and increase the prediction accuracy of the BP network, we use the genetic algorithm to optimize the weight values.

5.5   Utilize genetic algorithm to optimize the weight values

We encoded the 10 sets of trained weights as decimals and used the weights after encoding as chromosomes. 10 groups of chromosomes make up a population. Then we optimize these weights using the genetic algorithm until the weights after decoding are adjusted close enough to the most optimized weight combination.

5.6   Use the optimized weights to train the BP network again

5.6.1   Train a BP network again

Finally, we choose the chromosome in which the fitness function is maximized and decode it to form the optimized weights. Then we use one of the BP networks to adjust the optimized weights delicately. The training times and test times for this adjustment are 8,000. As a result, we gain the generalization ability of the BP network. It is shown in Figure 3.

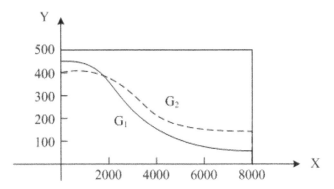

Figure 2.   The generalization ability of BP networks.

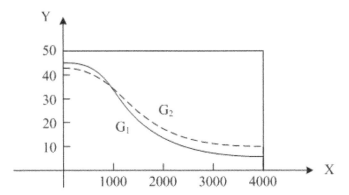

Figure 3.   The generalization ability of the BP network.

5.6.2   *Analyse training result*

Figure 3 shows that the BP network converges to a global training error and a global testing error that are much smaller than the values before optimization. The global error values are small enough and the global testing error is close enough to the global training error. From that, we can determine that the BP network converges to a global minimum. So we end the training and optimizing procedure.

5.7   *Testing and real application*

For further testing, we chose another 2,000 case samples to evaluate the real generalization ability of this genetic network model. We used the finally adjusted weights as the running weights of the BP network to predict the probability of crime that people may commit at present. The probability is the output of the BP network and is a floating point number in [0, 1] that represents the occurrence probability of events. The test result shows that the global error of the network is very small and the prediction is highly accurate. In addition, in the real work of the command centre of a police office, we used this data-mining model to find the most probable suspects in current cases. The result shows that in most cases, the prediction can help to find the potential suspects or provide some clues for case solving. So, we can draw the conclusion that the trained BP network has a high generalization ability and the data-mining model can be widely used in solving real cases. We can use this model not only to guide the monitoring and tracing of former criminals, but also to assist the finding and confirming suspects. So this data-mining model is very useful for case solving and decision-making in a police office.

5.8 *Retrospection and summary of experimental method*

In this application, we combined the advantages of the BP neural network data-mining model and the global optimization ability of a genetic algorithm. This genetic neural network model has greatly enhanced the generalization ability of the BP network and improved the accuracy of crime probability prediction. It has proved to be successful and efficient in practice.

The experimental system was realized with Visual C++ 6.0. The developing and running environment is Windows XP/NT/2000. It is a general-purpose tool and can build three-layer BP neural networks in any kind of structure and has an optimized genetic algorithm. In the future, we will improve the experimental system and put it to more practical uses.

6 CONCLUSION

The BP neural network that has been applied to data mining possesses the characteristics of high memory ability, high adaptability, accurate knowledge discovery, no restriction on the quantity of data and fast speed of calculation. Using a genetic algorithm to optimize the BP network can effectively avoid the problem of local minimum. Therefore, the genetic neural network based data-mining model has many advantages over other data mining models. In the real practice of data mining in the command centre of a police office, the advantages have been fully demonstrated. With the development of research in the future, this model will be used more widely in police offices.

REFERENCES

Berson, A. & Smith, S.J. 1999. *Data Warehousing, Data Mining, & OLAP.* New York: McGraw-Hill Book Co.
Goldberg, D.E. 1989. *Genetic Algorithms in Search, Optimization and Machine Leaning.* New York: Addison-Wesley Publishing Company.
Goldberg, D.E. 1992. Genetic Algorithms: A Bibliography. *IlliGAL Technical Report, 920008.*
Guo, Z. 2003. An Extensible System for Data Cleaning. *Computer Engineer* 29(3): 95–96.
Hard, D. 2003. *Principles of Data Mining.* Beijing: Machine Industry Publisher.
Heckerling, P.S. & Gerber, B.S. 2004. Use of Genetic Algorithms for Neural Networks to Predict Community-Acquired Pneumonia. *Artificial Intelligence in Medicine* 30(1): 71–75.
Li, M. 2002. *The Principle and Application of Genetic Algorithm.* Beijing: Science Publisher.
Li, Y. 2004. A Data Mining Architecture Based on ANN and Genetic Algorithm. *Computer Engineer* 30(6): 155–156.
Qing, G. 2003. Acquirement of Knowledge on Data Mining. *Computer Engineer* 29(21): 20–22.
Srinivas, M. & Patnaik, L.M. 1994. Genetic Algorithms: A Survey. *Computer* 27(6): 17–26.
Wang, Y. 2005. Predictive Model Based on Improved BP Neural Networks and its Application. *Computer Measurement & Control* 13(1): 39–42.
Xu, L. 2003. *Neural Network Control.* Beijing: Electronic Industry Publisher.
Xu, Z. 2004. A Data Mining Algorithm Based on the Rough Sets Theory and BP Neural Network. *Computer Engineer and Application* 31: 169–175.
Zhang, L. 1993. *The Model and Application of Artificial Neural Network.* Shanghai: Fudan University Publisher.
Zhou, A. 2005. A Genetic–Algorithm–Based Neural Network Approach for Short-Term Traffic Flow Forecasting. *Advances in Neural Networks* 3498: 965–969.

Advances in Spatio-Temporal Analysis – Tang et al. (eds)
© *2008 Taylor & Francis Group, London, ISBN 978-0-415-40630-7*

Qualitative spatial reasoning about Internal Cardinal Direction relations

Yu Liu, Yi Zhang, Xiaoming Wang & Xin Jin
Institute of Remote Sensing and Geographic Information Systems, Peking University, Beijing, P.R. China

ABSTRACT: One class of spatial relations, the Internal Cardinal Direction (ICD) relation, is introduced and investigated in this paper. Based on the ICD-9 model that includes nine atomic base relations, the characteristics and simplification rules for the ICD model are discussed first. Then the ICD relations involved in qualitative spatial reasoning, including four composition cases, are proposed formally. They are: composing two nesting ICD relations; composing two coordinate ICD relations to deduce conventional (or external) cardinal direction relations; qualitative distance relations; and topological relations. In terms of ICD relations, the container object determines the scale of spatial analysis and forms a positioning framework together with ICD relation predications. Consequently, research on ICD related reasoning should contribute to the representation and processing of survey knowledge.

1 INTRODUCTION

1.1 *Internal cardinal direction relations*

In Geographical Information Systems (GIS), Artificial Intelligence (AI) and databases, Qualitative Spatial Reasoning (QSR) has attracted much attention. Spatial relations, including topological relations, cardinal direction relations and metric relations, play essential roles in QSR. To the authors' knowledge, leaving aside temporal factors, research into QSR has until recently mainly focused on four issues. These are:

1. Formalizing one type of spatial relation and investigating its characteristics, such as conceptual neighbourhood graphs, computability, etc. (Egenhofer & Al-Taha 1992, Randell et al. 1992, Duckham & Worboys 2001, Skiadopoulos et al. 2004).
2. Composing two or more spatial relations to obtain a previously unknown relation. This aspect includes compositions of topological relations (Ligozat 1999, Renz 2002), compositions of cardinal direction relations (Frank 1996, Skiadopoulos & Koubarakis 2001, Isli et al. 2000), and compositions of topological relation with metric relations (Giritli 2003), etc.
3. Determining the position of a place according to a set of given spatial relations (Frank 1992, Clementini et al. 1997, Isli & Moratz 1999, Moratz et al. 2003), where cardinal direction relations and metric relations are often employed.
4. Establishing a calculus based on a set of well-defined relations and studying its relation-algebraic features (Ligozat 1998, Düntsch et al. 2001, Ligozat & Renz 2004). Furthermore, other SQR issues, such as the path-consistency problem and minimal labels problem, can be solved using these calculi.

In this paper, the QSR about the Internal Cardinal Direction (ICD) relation is proposed. Being different from the other types of spatial relations, ICD is used to represent the direction relations between an object and another areal object containing it. The ICD relations between a containee and a container depend on the containee's relative position to the latter.

It is well known that spatial knowledge development includes three stages, i.e. landmark knowledge, route knowledge and survey knowledge (Montello 2001). To express and transfer survey knowledge, some base landmarks are often selected first and described using ICD relations. Then the other places are determined with reference to the base landmarks using topological relations, cardinal direction relations or qualitative distance relations. A typical statement expressing survey knowledge might be "*A* locates

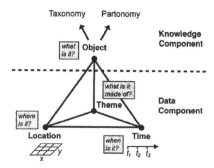

Figure 1.   Pyramid framework for spatial knowledge (Mennis et al. 2000).

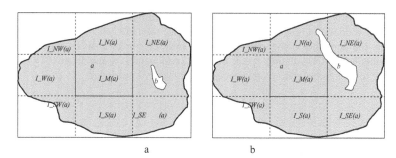

Figure 2.   ICD-9 model. (a) atomic ICD relation. (b) complex ICD relation.

in the south-east of *B*, while *C* lies to the north of *A*". In Mennis et al. (2000), a pyramid framework (Fig. 1) for modelling geographic data and geographic knowledge is developed based on geographic cognition. Following this framework, geographic knowledge includes two parts: taxonomy (superordinate-subordinate relationships) and partonomy (part-whole relations). Obviously, an ICD relation implies part-whole relations as well as partonomy knowledge. Hence, research on ICD-involved QSR should contribute to the representation of spatial knowledge.

This paper is structured as follows. First, the ICD-9 model and its characteristics are described briefly. Then some fundamental concepts are defined. Based on these concepts, we establish the simplification rules for ICD-9 in Section 2. In the third part of this paper, the ICD relation involved in qualitative spatial reasoning is discussed. A group of composition tables is presented to demonstrate the QSR about ICD relations with ICD relations, ECD (External Cardinal Direction) relations, topological relations and qualitative distance relations. We conclude the paper and present an agenda for future work in the last part.

1.2   *ICD-9 model*

Liu et al. (2005) developed three ICD models with different levels of detail. They are ICD-5, ICD-9 and ICD-13. In the ICD-9 model, regions are defined as non-empty sets of points in R^2. The reference object (or container) *a* is assumed to be a connected region. "Connected" means for every pair of points in *a* there exists a line (not necessarily a straight one) joining these two points such that all the points on the line are also in it.

As shown in Figure 2, to partition the container *a* into ICD parts, the Minimum Bounding Rectangle (MBR) of *a* is first divided into nine average tiles. These tiles intersect *a* and form corresponding parts denoted by $I\_N(a)$, $I\_NE(a)$, $I\_E(a)$, $I\_SE(a)$, $I\_S(a)$, $I\_SW(a)$, $I\_W(a)$, $I\_NW(a)$ and $I\_M(a)$. If another object *b* is located inside $I\_E(a)$, then $b\ I\_E\ a$, i.e. *b* is in the east of *a* (Fig. 2-a). The other ICD relations including I_NE (northeast), I_N (north), I_SE (southeast), I_S (south), I_SW (southwest), I_W (west),

I_NW (northwest) and *I_M* (middle) could be defined similarly. These ICD relations are atomic. However, if *b* covers more than one part of *a*, it is a complex ICD relation. Using the method proposed by Skiadopoulos and Koubarakis (2001), it can be represented by $bR_1: \ldots :R_k a$, where $2 \leq k \leq 9$, and

$$R_i \in \{I\_N, I\_NE, I\_E, I\_SE, I\_S, I\_SW, I\_W, I\_NW, I\_M\}, 1 \leq i \leq k \tag{1}$$

Hence, the ICD relation depicted in Figure 2-b is denoted by $b\ I\_N:I\_NE:I\_E:I\_M\ a$. In order to define complex ICD relations, a function λ is introduced to combine a set of basic ICD relations into a complex one. For instance, $\lambda(I\_N, I\_NE, I\_NW:I\_N) = I\_NW:I\_N:I\_NE$. If *b* is restricted to be connected, some disconnected cases, such as $\lambda\ (I\_N, I\_NE, I\_S)$, are impossible.

Referring to the research suggested in Skiadopoulos and Koubarakis (2001), we use the function δ as a shortcut to express a set of ICD relations. For arbitrary atomic cardinal direction relations $R_1, \ldots, R_k$, the notation $\delta(R_1, \ldots, R_k)$ is a shortcut for the disjunction of all valid basic cardinal direction relations that can be constructed by combining atomic relations $R_1, ldots, R_k$. For instance, $\delta(I\_SW, I\_W, I\_NW)$ stands for the disjunctive relations $\{I\_SW, I\_W, I\_NW, I\_SW:I\_W, I\_W:I\_NW, I\_SW:I\_W:I\_NW\}$. Obviously, $\delta(R_1, \ldots, R_k)$ include $2^k - 1$ basic relations.

Definition 1. Using the δ function, the set including all internal cardinal relations is denoted by ***ICD*-9**. ***ICD*-9** $= \delta(I\_N, I\_NE, I\_E, I\_SE, I\_S, I\_SW, I\_W, I\_NW, I\_M)$. It includes 511 elements.

Definition 2. $ICD(b,a)$ is a function used to test the ICD relation between *b* and *a*, i.e. $ICD(a, b) = R \Leftrightarrow aRb$.

Definition 3. If *b* and *a* have an ICD relation *R*, i.e. *bRa*, then we have: $b \subseteq R(a)$. If *R* is an atomic relation, $R(a)$ stands for only one partition cell of *a*. However, if *R* is complex, say $R_1: \ldots :R_k$, then

$$R(a) = \bigcup_{i=1}^{k} R_i(a) \tag{2}$$

In the context of ICD relations to *a*, $R(a)$ can be viewed as an upper approximation of *b*. For simplicity, if *b* is inside *a*, we use the notation $\bar{b}$ to denote the approximation of *b* based on the ICD relation between *b* and *a*.

According to the above definitions, for the given condition: $b \subseteq R(a)$, where $R = R_1: \ldots :R_k$ and $1 \leq k \leq 9$, the set of all possible ICD relations between *b* and *a* is $\delta(R_1, \ldots, R_k)$, i.e. $ICD(b,a) \in \delta(R_1, \ldots, R_k)$. Briefly, $\delta(R_1, \ldots, R_k)$ can be written as $\delta(R)$. Note some disconnected combination cases should be excluded from $\delta(R)$ if *b* is restricted to be connected. We use the function $\delta'(R)$ to represent the subset instead of $\delta(R)$. For instance, $\delta'(I\_N, I\_M, I\_S) = \{I\_N, I\_M, I\_S, I\_N:I\_M, I\_M:I\_S, I : N:I\_M:I\_S\}$, where the relation $I\_N:I\_S$ is excluded.

2 CHARACTERISTICS AND SIMPLIFICATION RULES OF ICD-9

Let R_1 and R_2 be two atomic ICD relations, according to the relations between $R_1(a)$ and $R_2(a)$, the conceptual relation between R_1 and R_2 could be determined using the following definitions.

Definition 4. R_1 and R_2 are *equal* if $R_1(a) = R_2(a)$. R_1 and R_2 are *neighbouring* if $R_1(a)$ externally meet $R_2(a)$. R_1 and R_2 are *disjoint* if $R_1(a) \cap R_2(a) = \emptyset$. In particular, R_1 and R_2 are *opposite* if R_1 and R_2 are disjoint and centrally symmetrical.

The symbols *Q*, *N*, *D*, and *O* are used to denote these four relations. For example, we have $I\_N\ Q$ I_N, $I\_S\ N\ I\_M$, $I\_W\ D\ I\_SE$, and $I\_NE\ O\ I\_SW$. A quantitative representation is introduced to define the above conceptual relations. Assuming there is a Cartesian coordinate system, the origin of which is the middle part of the container, then each tile can be represented by an ordered pair $<Q_x, Q_y>$, where

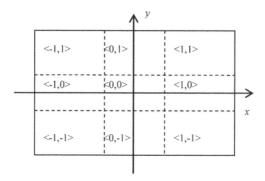

Figure 3.   Quantification of atomic ICD relations.

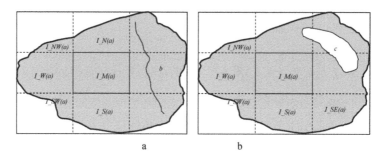

Figure 4.   Complex ICD relation that can be simplified.

$Q_x, Q_y \in \{-1, 0, 1\}$(Fig. 3). Actually, -1, 0, and 1 stand for south, middle, and north or west, middle, and east along two axes, respectively.

Theorem 1. If R_1 and R_2 are atomic and can be encoded as $<Q_{1x},Q_{1y}>$ and $<Q_{2x},Q_{2y}>$, respectively, then:
$R_1 Q R_2$ iff $(Q_{1x} = Q_{2x}) \wedge (Q_{1y} = Q_{2y})$,
$R_1 N R_2$ iff $abs(Q_{1x} - Q_{2x}) \leq 1 \wedge abs(Q_{1y} - Q_{2y}) \leq 1 \wedge \neg R_1 Q R_2$,
$R_1 D R_2$ iff $abs(Q_{1x} - Q_{2x}) = 2 \vee abs(Q_{1y} - Q_{2y}) = 2$ and
$R_1 O R_2$ iff $Q_{1x} + Q_{2x} = 0 \wedge Q_{1y} + Q_{2y} = 0 \wedge Q_{1x} \neq 0 \wedge Q_{2x} \neq 0$.

This representation method can be extended to complex ICD relations. If $R(a)$ is a rectangle, then R can be described using the range of $R(a)$. For example, $I\_N{:}I\_NE{:}I\_M{:}I\_E$ could be denoted by $<[0,1],[0,1]>$.

Using the ordered pairs of atomic ICD relations, a complex ICD relation can be simplified. As shown in Figure 4, a linear object b and an areal object c are inside a. Their corresponding ICD relations are b $I\_NE{:}I\_E{:}I\_SE$ a and c $I\_N{:}I\_NE{:}I\_E$ a. However, the more natural and geographical cognition accordant statements for describing these relations might be "b goes through the east of a" and "c is located in the northeast of a" in practice. Hence, a simplification method is necessary to simplify the complex ICD relations.

Let $R = R_1{:}R_2 \ldots {:}R_n$ be a complex ICD relation, and each involved atomic relation can be represented as $<Q_{1x},Q_{1y}>$, $<Q_{2x},Q_{2y}>$, ..., and $<Q_{nx},Q_{ny}>$. We define

$$< Q_{sx}, Q_{sy} > = < \frac{\sum_{i=1}^{n} Q_{ix}}{n}, \frac{\sum_{i=1}^{n} Q_{iy}}{n} > \qquad (3)$$

206

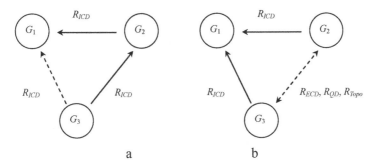

Figure 5.   ICD-involved composition of spatial relations.

as the result for simplification. For example, the pair's values for ICD relations in Figure 4 are <1,0> and <0.67,0.67>, respectively. Then, R could be simplified to an atomic ICD relation according to the minimum Euclidean distance from $<Q_{sx},Q_{sy}>$ to the pairs of all atomic relations. Following this point, $ICD(b,a)$ and $ICD(c,a)$ are simplified to I_E and I_NE. This is accordant to commonsense geographical cognition. Especially, if $ICD(b,a)$ is complex and its simplification result is I_M, the size of b is usually in the same scale as a.

3   ICD RELATED QUALITATIVE SPATIAL REASONING

It is argued by Goodchild (2001) that many geographic phenomena are scale-specific. An important characteristic of ICD relations is that the container determines the background scale for describing and analysing the entities inside it. To position an object in the container, distance relations, cardinal direction relations and topological relations are all necessary as well as ICD relations. Hence, it is necessary to study the qualitative spatial reasoning by integrating ICD relations and the other types of spatial relations.

Figure 5 demonstrates two categories of ICD relation involved compositions, where G_1, G_2, and G_3 are three spatial objects, while $RICD$, $RECD$, RQD and $RTopo$ denote ICD relations, ECD relations, qualitative distance relations and topological relations, respectively. The solid lines stand for known ICD relations, while the dashed lines represent unknown spatial relations to be inferred. Figure 5-a depicts the composition of nested ICD relations. This makes G_1, G_2, and G_3 belong to different scale levels. If G_2 and G_3 are inside the same container G_1, then the external cardinal direction relations, qualitative distance relations and topological relations can be deduced based on their ICD relations to G_1. Figure 5-b stands for such compositions.

The composition given in Figure 5-b is somewhat different from the common relation compositions, which can be defined as:$R_1 \circ R_2 = \{< x,y > |\exists t(xR_1t \wedge tR_2y)\}$, just like the case presented in Figure 5-a. However, we should still consider these categories of compositions. This is because ICD relations are not closed under inverse (Liu et al. 2005). For example, if a has a specific ICD relation with b, then the relation between b and a will not be an ICD relation any more. The characteristic of ICD is significantly different from the other types of spatial relations (e.g. topological relations); in other words, container and containee play asymmetric roles. Since the container forms an analysis context of the containees, it is more valuable to infer the relation between two objects inside the same container based on their ICD relations. For the sake of making a standard composition, we could inverse such a composition to predict the ICD relations by composing an ICD relation with an ECD (or topological, qualitative distance) relation. Besides, the unclosed property means that ICD relations themselves could not form a complete algebra. It is thus necessary to include the other types of spatial relations in the composition.

3.1   *QSR about nesting ICD relations*

The qualitative spatial reasoning of two nesting ICD relations given by the following theorem is somewhat straightforward.

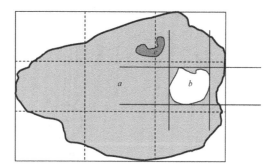

Figure 6. Composing ICD and ECD relations.

Theorem 2. Let a, b, and c be three objects that have IDC relations of R_1 and R_2 respectively, i.e. bR_1a and cR_2b. Then $R_2 \circ R_1 = \delta(R_1)$.

Proof. At first, according to definition 3,

$$bR_1a, R_1 \in ICD\_9 \Rightarrow b \subseteq R_1(a),$$

$$cR_2b, R_2 \in ICD\_9 \Rightarrow c \subseteq R_2(b) \Rightarrow c \subset R_1(a)$$

Therefore, we have

$$ICD(c,a) \in \delta(R_1)$$

Finally,

$$R_2 \circ R_1 \in \delta(R_1)$$

Theorem 2 indicates that $R_2 \circ R_1$ is independent of R_2. This is a natural result. However, we seldom use this reasoning process in practice due to different spatial scale levels. For example, if a city b is located in the north of a state a, meanwhile a building c is in the west of city b, we seldom say that "building c locates in the north of state a" for the reason of scale, although it is a correct statement.

3.2 QSR about ICD and ECD relations

Compared with ICD relations, the conventional direction relations when two geometries are disjoint can be named as external cardinal direction relations. In order to represent ECD-relations, some models were developed. They include the cone-based model, project-based model (Frank 1991), double-cross model (Freksa 1992) and MBR-based model (Goyal & Egenhofer 2000, Goyal & Egenhofer 2001, Skiadopoulos & Koubarakis 2001, Skiadopoulos et al. 2004), etc. To keep consistency with the ICD-9 model, the MBR-based ECD model is selected to represent external cardinal relations (Fig. 6). There are two main approaches for representing the MBR-based model, i.e. the matrix-based method (Goyal & Egenhofer 2000) and the formula-based method (Skiadopoulos & Koubarakis 2001). In this paper, we use the latter.

When the MBR-based ECD model is adopted, the middle part of the container in ICD relations can be viewed as another reference object for ECD relations. Meanwhile, the other eight parts have corresponding external cardinal relations with the middle part. ICD relations thus have similar characteristics to ECD relations. The differences between them include the following two aspects:

1. ICD partitions only the MBR of the reference object, while ECD partitions the whole space R^2.
2. The areas of all ICD parts are usually in similar scales (depending on the actual shape of the container). However, ECD tiles do not have such a feature. With a given reference object, the area of the middle part is fixed, and areas of the other eight parts are infinite. Moreover, areas of E_NE, E_SE, E_SW, and E_NW are higher order infinities than E_N, E_E, E_S, and E_W. This makes the MBR-based ECD model unsuitable for a relatively "small" object. As an extreme case, if the reference object degenerates to a point, then a cone-based model should be adopted instead of the MBR-based model.

Table 1. Composition table of ICD and ICD to get ECD relations.

	I_N	I_NE	I_E	I_SE	I_S	I_SW	I_W	I_NW	I_M
I_N	*	δ'(E_E, E_NE, E_SE)	E_SE	E_SE	δ'(E_S, E_SW, E_SE)	E_SW	E_SW	δ'(E_W, E_NW, E_SW)	δ'(E_S, E_SW, E_SE)
I_NE	δ'(E_W, E_NW, E_SW)	*	δ'(E_S, E_SW, E_SE)	δ'(E_S, E_SW, E_SE)	E_SW	E_SW	E_SW	δ'(E_W, E_NW, E_SW)	E_SW
I_E	E_NW	δ'(E_N, E_NE, E_NW)	*	δ'(E_S, E_SW, E_SE)	E_SW	E_SW	δ'(E_W, E_NW, E_SW)	E_NW	δ'(E_W, E_NW, E_SW)
I_SE	E_NW	δ'(E_N, E_NE, E_NW)	δ'(E_N, E_NE, E_NW)	*	δ'(E_W, E_NW, E_SW)	δ'(E_W, E_NW, E_SW)	E_NW	E_NW	E_NW
I_S	δ'(E_N, E_NE, E_NW)	E_NE	E_NE	δ'(E_S, E_SW, E_SE)	*	δ'(E_W, E_NW, E_SW)	E_NW	E_NW	δ'(E_N, E_NE, E_NW)
I_SW	E_NE	E_NE	E_NE	δ'(E_E, E_NE, E_SE)	δ'(E_E, E_NE, E_SE)	*	δ'(E_N, E_NE, E_NW)	δ'(E_N, E_NE, E_NW)	E_NE
I_W	E_NE	E_NE	δ'(E_E, E_NE, E_SE)	E_SE	E_SE	δ'(E_S, E_SW, E_SE)	*	δ'(E_N, E_NE, E_NW)	δ'(E_E, E_NE, E_SE)
I_NW	δ'(E_E, E_NE, E_SE)	δ'(E_E, E_NE, E_SE)	E_SE	E_SE	E_SE	δ'(E_S, E_SW, E_SE)	δ'(E_S, E_SW, E_SE)	*	E_SE
I_M	δ'(E_N, E_NE, E_MW)	E_NE	δ'(E_E, E_NE, E_SE)	E_SE	δ'(E_S, E_SW, E_SE)	E_SW	δ'(E_W, E_NW, E_SW)	E_NW	*

In the QSR about ICD and ECD relations, the containee objects are restricted to be connected. Hence, function δ' is used to represent the composition result. It can also be introduced to ECD relations with similar meanings. The reason for the restriction is that the spatial relations between two disconnected geometries are often complicated and less meaningful, although there is some literature addressing this issue, for example Behr and Schneider (2001).

Let b and c be two objects inside region a. They occupy corresponding parts of a, and the corresponding ICD relations are bR_1a and cR_2a. If R_1 and R_2 are atomic, the composition can be deduced via Table 1.

In the above table, the symbol "*" stands for universal relation. Obviously, if two relations are opposite, their composition result is *; on the contrary, if two relations are equal, their composition remains unchanged. The table can be easily computed. Suppose the MBR of a, b, and c be (x_{al}, y_{ab})-(x_{ar}, y_{at}), (x_{bl}, y_{bb})-(x_{br}, y_{bt}), and (x_{cl}, y_{cb})-(x_{cr}, y_{ct}), where the MBR is represented by the coordinates of its two corner points (left-bottom and right-top). Let us take the case that $b \, I\_N \, a$ and $c \, I\_NE \, a$ as an example, we have:

$$x_{al} + \frac{1}{3}(x_{ar} - x_{al}) < x_{bl} < x_{br} < x_{al} + \frac{2}{3}(x_{ar} - x_{al})$$

$$y_{ab} + \frac{2}{3}(y_{at} - y_{ab}) < y_{bb} < y_{bt} < y_{at}$$

$$x_{al} + \frac{2}{3}(x_{ar} - x_{al}) < x_{cl} < x_{cr} < x_{ar} \tag{4}$$

$$y_{ab} + \frac{2}{3}(y_{at} - y_{ab}) < y_{cb} < y_{ct} < y_{at}$$

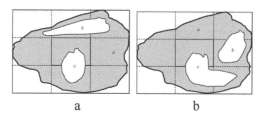

a   b

Figure 7.   Composing complex ICD relations to get ECD relations.

The relation between b's MBR and c' MBR is $x_{bl} < x_{br} < x_{cl} < x_{cr}$ along the x axis; meanwhile, the relations between y_{bb}, y_{bt} and y_{cb}, y_{ct} is undetermined. Thus, based on the MBR-based ECD model, the ECD relation between c and b might be $c\ E\_E\ b$, $c\ E\_NE\ b$, $c\ E\_SE\ b$, $c\ E\_NE:E\_E\ b$, $c\ E\_E:E\_SE\ b$, or $c\ E\_NE:E\_E:E\_SE\ b$. They can be written as $\delta'(E\_E, E\_NE, E\_SE)$.

When R_1 and R_2 are not atomic relations, i.e. b or c covers more than one ICD part, the relation composition is a little more complex. As shown in Figure 7-a, given $b\ I\_NW:I\_N:I\_NE\ a$ and $c\ I\_M:I\_S\ a$, the ECD relation between c and b could be inferred as $c\ E\_S\ b$. To obtain the composition of two complex ICD relations, the quantitative representation for the ICD relation should be employed in the following definition.

Definition 5. Let R be a complex ICD relation. $R = R_1: R_2 \ldots R_n$. Two intervals of R along x and y axes are defined as:

$$[\min(Q_{1x}, Q_{2x}, \ldots, Q_{nx}), \max(Q_{1x}, Q_{2x}, \ldots, Q_{nx})] \text{ and}$$
$$[\min(Q_{1y}, Q_{2y}, \ldots, Q_{ny}), \max(Q_{1y}, Q_{2y}, \ldots, Q_{ny})]$$

where $<Q_{ix},\ Q_{iy}>$ is the quantification of R_i $(i = 1, 2, \ldots, n)$. For example, the intervals of $I\_NW:I\_N:I\_NE:I\_E$ are $[-1, 1]$ and $[0, 1]$, respectively.

The ECD relations can be inferred based on this definition. Assuming b R1 a and c R2 a, and at least one of them is complex, each atomic ICD relation in R2 and the intervals of R1 should be considered to obtain the ECD relation between c and b. Suppose the intervals of R1 are [Fx, Tx] and [Fy, Ty], and an atomic relation R2i in R2 is denoted by <Qix, Qiy>. Qix may have six Jointly Exhaustive and Pairwise Disjoint (JEPD) relations to interval [Fx, Tx]: Qix = Fx = Tx, Qix<Fx, Qix = Fx<Tx, Fx<Qix<Tx, Fx<Qix = Tx, and Qix>Tx. They are similar to Allen's interval algebra (Allen 1983). Each relation determines the possible ECD relations along the horizontal axis. The ECD relations along the vertical axis can be inferred similarly. According to their combinations that include 6*6 = 36 cases, the ECD relations between R2i(a) that is one part of c, and b can be obtained (Table 2).

Based on Table 2, we can obtain the relation between each tile of R2(a) and b. Supposing they are S1, ..., Sn, then we can compute their Cartesian product S = S1 × S2 ... × Sn. Finally, the λ function is used to combine the elements in S and to obtain the ECD relations in $R_1 \circ R_2$. Let us take Figure 7-b as an example. The intervals of ICD(b,a) are [1,1] and [0,1], and ICD(c,a) includes three atomic relations that are quantified as <0,0>, <0,−1> and <1,−1>. Hence, we get three sets according to Table 2. They are:

$\delta'(E\_W, E\_SW) = \{E\_W, E\_SW, E\_W:E\_SW\}$,
$\{E\_SW\}$, and
$\delta'(E\_S, E\_SE) = \{E\_S, E\_SE, E\_S:E\_SE\}$.

Their Cartesian product is $\{(E\_W, E\_SW, E\_S),\ (E\_W, E\_SW, E\_SE),\ (E\_W, E\_SW, E\_S:E\_SE),$ $(E\_SW, E\_SW,\ E\_S),\ (E\_SW, E\_SW,\ E\_SE),\ (E\_SW,\ E\_SW, E\_S:E\_SE),\ (E\_W:E\_SW,\ E\_SW,\ E\_S),$ $(E\_W:E\_SW,\ E\_SW,\ E\_SE),\ (E\_W:E\_SW, E\_SW,\ E\_S:E\_SE)\}$. After adopting the λ function for each element in the above set and removing the duplicated cases, the result of composing R_1 and R_2 is $\{E\_W:E\_SW:E\_S,\ E\_W:E\_SW:E\_S:E\_SE,\ E\_SW:E\_S,\ E\_SW:E\_S:E\_SE\}$. Since c is assumed to be connected, some disconnected cases are excluded.

3.3 *QSR about ICD and qualitative distance relations*

In QSR, scale is an important concept, and refers to the size of the unit at which some problem is analysed, such as at the county or state level (Montello 2001). It is widely accepted that qualitative distance relations

210

Table 2. Composition of ECD relations along two axes.

	$Q_{ix}=F_x=T_x$	$Q_{ix}<F_x$	$Q_{ix}=F_x<T_{1x}$	$F_x<Q_{ix}<T_{1x}$	$F_x<Q_{ix}=T_x$	$Q_{ix}>T_x$
$Q_{iy}=F_y=T_y$	*	δ'(E_NW, E_W, E_SW)	δ'(E_NW, E_W, E_SW, E_N, E_M, E_S)	δ'(E_N,E_M, E_S)	δ'(E_N,E_M, E_S,E_NE, E_E, E_SE)	δ'(E_NE, E_E, E_SE)
$Q_{iy}<F_y$	δ'(E_SW, E_N, E_SE)	E_SW	δ'(E_SW, E_S)	E_S	δ'(E_S, E_SE)	E_SE
$Q_{iy}=F_y<T_{1y}$	δ'(E_NW, E_N, _NE, E_W, E_M, E_E)	δ'(E_SW, E_W)	δ'(E_W, E_M, E_SW, E_S)	δ'(E_S, E_M)	δ'(E_M, E_E, E_S, E_SE)	δ'(E_E, E_SE)
$F_y<Q_{iy}<T_{1y}$	δ'(E_W, E_M, E_E)	E_W	δ'(E_W, E_M)	E_M	δ'(E_M, E_E)	E_E
$F_y<Q_{iy}=T_y$	δ'(E_W, E_M, E_E, E_SW, E_S, E_SE)	δ'(E_NW, E_W)	δ'(E_NW,E_N, E_W, E_M)	δ'(E_N, E_M)	δ'(E_N,E_NE, E_M, E_E)	δ'(E_NE, E_E)
$Q_{iy}>T_y$	δ'(E_NW, E_N, E_NE)	E_NW	δ'(E_NW, E_N)	E_N	δ'(E_N, E_NE)	E_NE

is scale-dependent (Clementini et al. 1997). When we say a place is "close" to another place in the urban scale, it might be much farther than the concept of "far" in the campus scale. Generally, we consider it is "far" when the metric distance between two objects is close to the analysis scale, while we believe it is "near" when the metric distance is rather small compared with the scale size.

Compared with qualitative distance, quantitative distance has the following three axioms:

1. $d(x,x) = 0$ (reflexivity)
2. $d(x,y) = d(y,x)$ (symmetry)
3. $d(x,y) + d(y,z) \geq d(x,z)$ (triangle inequality)

where $d(x,y)$ is the quantitative distance function from x to y. However, these three rules will not be well satisfied any more for qualitative distances. Usually qualitative distances are asymmetric (Egenhofer & Mark 1995) and do not follow the triangle inequality rule. In the context of ICD relations, the container determines the background scale for inferring qualitative distance. Using the distance measure method presented in Goyal and Egenhofer (2001), distance is defined in the ICD framework based on the shortest path when we assume that an object moves from one tile to another. In this paper, the qualitative distance is quantified into three distinctions: close (Cl), commensurate (Cm) and far (F). Their ordinal relation is $Cl \leq Cm \leq F$.

In an ICD relation based positioning framework, let b, c be two geometries inside a, they both have atomic ICD relations to a. If the shortest path from $\bar{b}$ to $\bar{c}$ passes one tile ($\bar{b}$ and $\bar{c}$ are equal), then the QD relations is Cl. If it includes two parts, then the relation may be Cm or Cl. Lastly, if more than two parts are involved, the possible relations are F and Cm. The compositions are shown in Table 3.

If $ICD(b,a)$ or $ICD(c,a)$ is a complex relation, it means that $\bar{b}$ or $\bar{c}$ includes more than one ICD part. Assume $\bar{b} = \cup_{i=1}^{l} R_{bi}(a)$ and $\bar{c} = \cup_{i=1}^{m} R_{ci}(a)$, then we have:

$$QD(b,c) = \min\{QD(R_{bi}(a), R_{cj}(a)) | 1 \leq i \leq l, 1 \leq j \leq m\} \qquad (5)$$

For instance, if $b\,I\_N{:}I\_NE{:}I\_M\,a$ and $b\,I\_S{:}I\_SW\,a$, then $QD(b,c)$ is Cl. A shortcoming of this approach is that the sizes of b and c are neglected.

211

Table 3. Composition table of ICD and ICD to get qualitative distance relations.

	I_N	I_NE	I_E	I_SE	I_S	I_SW	I_W	I_NW	I_M
I_N	Cl	Cm, Cl	Cm, Cl	F, Cm	F, Cm	F, Cm	Cm, Cl	Cm, Cl	Cm, Cl
I_NE	Cm, Cl	Cl	Cm, Cl	F, Cm	F, Cm	F, Cm	F, Cm	F, Cm	Cm, Cl
I_E	Cm, Cl	Cm, Cl	Cl	Cm, Cl	Cm, Cl	F, Cm	F, Cm	F, Cm	Cm, Cl
I_SE	F, Cm	F, Cm	Cm, Cl	Cl	Cm, Cl	F, Cm	F, Cm	F, Cm	Cm, Cl
I_S	F, Cm	F, Cm	Cm, Cl	Cm	Cl	Cm, Cl	Cm, Cl	F, Cm	Cm, Cl
I_SW	F, Cm	F, Cm	F, Cm	F, Cm	Cm, Cl	Cl	Cm, Cl	F, Cm	Cm, Cl
I_W	Cm, Cl,	F, Cm	F, Cm	F, Cm	Cm, Cl	Cm, Cl	Cl	Cm, Cl	Cm, Cl
I_NW	Cm, Cl	F, Cm	F, Cm	F, Cm	F, Cm	F, Cm	Cm, Cl	Cl	Cm, Cl
I_M	Cm, Cl	Cm, Cl	Cm, Cl	Cm, Cl	Cm, Cl	Cm, Cl	Cm, Cl	Cm, Cl	Cl

3.4 QSR about ICD and topological relations

Topological relations are related to the connection between spatial objects. One of their important characteristics is that they remain invariant under topological transformations, such as rotation, translation and scaling. There are two main approaches to describe topological relations formally, i.e. Region Connection Calculus (RCC) (Randell et al. 1992) and the point set based intersection model (Egenhofer & Franzosa 1991).

As mentioned earlier, if b and c are inside a, then $\bar{b}$ and $\bar{c}$ can be viewed as upper approximations of b and c. Consequently, the topological relation between $\bar{b}$ and $\bar{c}$ plays a filter role to filter some impossible relations. When discussing topological relations, the point, line and area geometries should all be considered. Since $\bar{b}$ and $\bar{c}$ are regions, we assume that b and c are both connected areal objects. Thus, RCC-8 is used to represent the topological relations in the following part.

In the RCC-8 model, there are eight jointly exhaustive and pairwise disjoint topological base relations between two areal geometries. They are DC (DisConnected), EC(Externally Connected), PO (Partial Overlap), EQ (EQual), TPP (Tangential Proper Part), $NTPP$ (Non-Tangential Proper Part), TPP^{-1} (converse TPP), and $NTPP^{-1}$ (converse $NTPP$). In terms of ICD model, determining the topological relation between $\bar{b}$ and $\bar{c}$ is easier than testing the relation between two actual regions. If $ICD(b,a)$ and $ICD(c,a)$ are atomic relations, then the relation between $\bar{b}$ and $\bar{c}$ is somewhat simple, since there are only three possible topological relations: DC, EC, and EQ. For example, $I\_N(a)\ DC\ I\_S(a)$, $I\_N(a)\ EC\ I\_NE(a)$, and $I\_N(a)\ EQ\ I\_N(a)$. Otherwise, if they are complex ICD relations, the relation can be determined according to their constituent parts. As shown in Figure 8, the ICD relations are $b\ I\_N{:}I\_NE{:}I\_M{:}I\_E\ a$ and $c\ I\_S{:}I\_SE{:}I\_M{:}I\_E\ a$ respectively. We thus have:

$$\bar{b} = I\_N(a)\cup I\_NE(a)\cup I\_M(a)\cup I\_E(a)$$
$$\bar{c} = I\_S(a)\cup I\_SE(a)\cup I\_M(a)\cup I\_E(a) \qquad (6)$$

Obviously, the topological relation is $\bar{b}\ PO\ \bar{c}$. That makes the possible topological relations between b and c include DC, EC and PO; meanwhile, the other five relations are filtered.

The topological relation filter can be extended to the other situations of upper approximations, such as MBRs, instead of ICD cells. Table 4 lists all possible cases of the topological relation between $\bar{b}$ and $\bar{c}$, where the function TR is designed to test the topological relation between two objects. Some conclusions can be drawn from Table 4:

1. Ordered by their capacities to exclude impossible topological relations, we have: $DC > EC > PO > NTPP = NTPP^{-1} = TPP = TPP^{-1} > EQ$. The DC relation gives the strongest constraint, while EQ has the weakest constraint, i.e. EQ cannot exclude any relation.
2. Considering the relations after being filtered, the DC relation can be realized most easily; in other words, it cannot be excluded by any relation. In terms of the times appearing in the filtered relations, a sequence

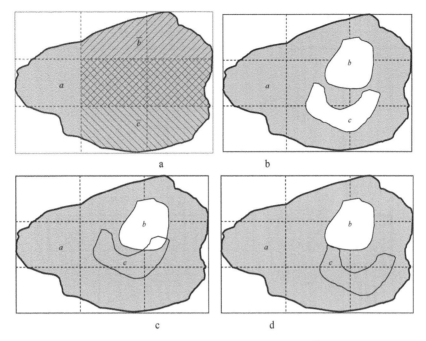

Figure 8. Filtering impossible topological relations based on ICD relations (a. $\bar{b}\ PO\ \bar{c}$, b. $b\ DC\ c$, c. $b\ PO\ c$, d. b $EC\ c$).

can also be created. It is $DC > EC > PO > NTPP = NTPP^{-1} = TPP = TPP^{-1} > EQ$, which is the same as the one above.

The study of these features is beyond the scope of this paper. Nevertheless, what should be pointed out is that all ECD relations are possible if $\bar{b}\ EQ\ \bar{c}$ (as shown in Table 1). This characteristic is similar to that of topological relations. Such a conclusion is reasonable. For example, we could not deny any possible spatial relation between two objects with the given knowledge that they are located in the same areal place.

4 CONCLUSIONS

The internal cardinal relation is a type of spatial relation that should be emphasized in both QSR and GIS. The container object determines the analysis scale and the containee objects have part-whole relations with the former. That makes ICD relations play an important role for representing and transferring spatial knowledge, especially survey knowledge.

Using the ICD-9 model, this paper summarized some composition cases for ICD relations involved in qualitative reasoning. We focus on the possible relations, including ECD relations, qualitative distance relations and topological relations between two containee objects based on their respective ICD relations to the container. Using some tables, the compositions for ICD relations are presented. Although the compositions are not in their normal forms, they still are valuable for representing and processing spatial knowledge due to the characteristics of ICD relations. In addition, the simplification rule for a complex ICD relation and the case of composing two nesting ICD relations in different scale levels are also discussed. As a conclusion, since the other types of spatial relations can be deduced based on ICD relations, we believe that a container, as well as the associated ICD relations, forms a positioning framework for spatial knowledge.

Table 4.　Topological relations inferred from ICD relations.

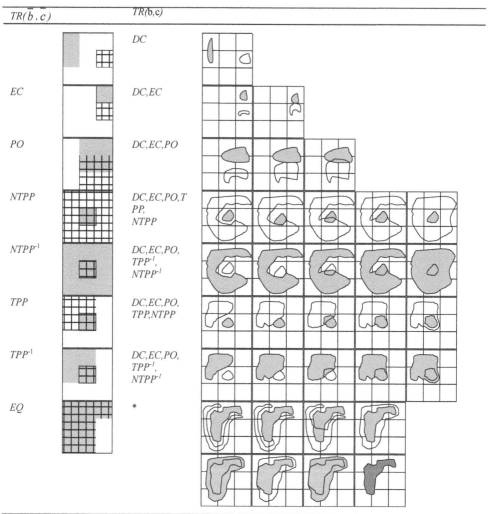

$TR(\bar{b},c)$		$TR(b,c)$
		DC
EC		DC,EC
PO		DC,EC,PO
NTPP		DC,EC,PO,TPP, NTPP
$NTPP^{-1}$		DC,EC,PO, TPP^{-1}, $NTPP^{-1}$
TPP		DC,EC,PO, TPP,NTPP
TPP^{-1}		DC,EC,PO, TPP^{-1}, $NTPP^{-1}$
EQ		*

Planned further work on ICD relations includes:

1. studying the ICD-5 and ICD-13 models and associated QSRs;
2. developing some quantitative approaches to ICD relation and exploring the related reasoning and computability.

Eventually, together with research on the other spatial relations, we hope the research will lead to a unified formal approach to represent survey knowledge consistently.

REFERENCES

Allen, J.F. 1983. Maintaining knowledge about temporal intervals. *Communications of the ACM*, 26(11): 832–843.
Behr, T. & Schneider, M. 2001. Topological relationships of complex points and complex regions. In H.S. Kunii, S. Jajodia & A.E. Solvberg (eds), *Lecture Notes in Computer Science* Vol. 2224: 56–69. Berlin: Springer-Verlag.

Clementini, E., Felice, P. & Hernandez, D. 1997. Qualitative representation of positional information. *Artificial Intelligence* 95: 317–356.

Duckham, M. & Worboys, M. 2001. Computational structure in three–valued nearness relations. In D.R. Montello (ed.), *COSIT 2001. Lecture Notes in Computer Science* Vol. 2205: 76–91. Berlin: Springer-Verlag.

Düntsch, I., Wang, H. & McCloskey, S. 2001. A relation–algebraic approach to the region connection calculus. *Theoretical Computer Science* 255: 63–83.

Egenhofer, M.J. & Al–Taha, K.K. 1992. Reasoning about gradual changes of topological relationships. In A.U. Frank, I. Campari & U. Formentini (eds), *Theory and methods of Spatio–temporal Reasoning in Geographic Space, Lecture Notes in Computer Science* Vol. 639: 1–24. Berlin: Springer-Verlag.

Egenhofer, M.J. & Franzosa, R. 1991. Point–set topological spatial relations. *International Journal of Geographical Information Systems* 5: 161–174.

Egenhofer, M.J. & Mark, D.M. 1995. Naive geography. In A.U. Frank & W. Kuhn (eds), *Spatial Information Theory A Theoretical Basis for GIS, International Conference COSIT'95. Lecture Notes in Computer Science*, Vol. 988: 1–15. Berlin: Springer-Verlag.

Frank, A.U. 1991. Qualitative spatial reasoning about cardinal directions. In D. Mark and D. White (eds), *Proceedings of the 7th Austrian Conference on Artificial Intelligence*: 157–167. Baltimore: Morgan Kaufmann.

Frank, A.U. 1992. Qualitative spatial reasoning about distances and directions in geographic space, *Journal of Visual Languages and Computing* 3: 343–371.

Frank, A.U. 1996. Qualitative spatial reasoning: cardinal directions as an example. *International Journal of Geographic Information Systems* 10(3): 269–290.

Freksa, C. 1992. Using orientation information for qualitative spatial reasoning. In A.U. Frank, I. Campari & U. Formentini (eds), *Proceedings of the International Conference GIS–From Space to Territory: Theories and Methods of Spatio–Temporal Reasoning on Theories and Methods of Spatio–Temporal Reasoning in Geographic Space. Lecture Notes in Computer Science*, Vol. 639: 162–178. Berlin: Springer-Verlag.

Giritli, M. 2003. Who can connect in RCC? In A. Gunter, R. Kruse & B. Neumann (eds), *KI 2003. Lecture Notes in Artificial Intelligence*, Vol. 2821: 565–579. Berlin: Springer-Verlag.

Goodchild, M.F. 2001. A geographer looks at spatial information theory. In D.R. Montello (ed.), *COSIT 2001. Lecture Notes in Computer Science*, Vol. 2205: 1–13. Berlin: Springer-Verlag.

Goyal, R.K. & Egenhofer, M.J. 2000. Consistent queries over cardinal directions across different levels of detail. In A.M. Tjoa, R. Wagner & A. Al–Zobaidie (eds), *Proceedings of IEEE 11th International Workshop on Database and Expert Systems Applications*: 876–880.

Goyal, R.K. & Egenhofer, M.J. 2001. Similarity of cardinal directions. In C.S. Jensen, M. Schneider, B. Seeger & V.J. Tsotras (eds), *Advances in Spatial and Temporal Databases, Lecture Notes in Computer Science*, Vol. 2121: 36–58. Berlin: Springer-Verlag.

Isli, A., Cabedo, L., Barkowsky, T. & Moratz, R. 2000. A topological calculus for cartographic entities. In C. Freksa, W. Brauer, C. Habel & K.F. Wender (eds), *Spatial Cognition II, Lecture Notes in Artificial Intelligence*, Vol. 1849: 225–238. Berlin: Springer-Verlag.

Isli, A. & Moratz, R. 1999. Qualitative spatial representation and reasoning: algebraic models for relative position. *Technical Report FBI–HH–M–284/99*. University at Hamburg, FB Informatik, Hamburg.

Ligozat, G. 1998. Reasoning about cardinal directions. *Journal of Visual Languages and Computing* 9: 23–44.

Ligozat, G. 1999. Simple models for simple calculi. In C. Freksa & D.M. Mark (eds), *COSIT'99. Lecture Notes in Computer Science*, Vol. 1661: 173–188. Berlin: Springer-Verlag.

Ligozat, G. & Renz, J. 2004. What is a qualitative calculus? A general framework. In C. Zhang, H.W. Guesgen & W.K. Yeap (eds), *PRICAI 2004: Trends in Artificial Intelligence, 8th Pacific Rim International Conference on Artificial Intelligence, Lecture Notes in Computer Science*, Vol. 3157: 53–61. Berlin: Springer-Verlag.

Liu, Y., Wang, X., Jin, X. & Wu, L. 2005. On internal cardinal direction relations. In A.G. Cohn & D.M. Mark (Eds), *Spatial Information Theory: Cognitive and Computational Foundations, COSIT'05. Lecture Notes in Computer Science*, Vol. 3693: 283–299. Berlin: Springer-Verlag.

Mennis, L.M., Peuquet, D.J. & Qian, L. 2000. A conceptual framework for incorporating cognitive principles into geographical database representation. *International Journal Geographical Information Science* 14: 501–520.

Montello, D.R. 2001. Scale in geography. In N.J. Smelser & P.B. Baltes (eds), *International Encyclopedia of the Social & Behavioral Sciences*: 13501–13504. Oxford: Pergamon Press.

Moratz, R., Nebel, B. & Freksa, C. 2003. Qualitative spatial reasoning about relative position: the tradeoff between strong formal properties and successful reasoning about route graphs. In C. Freksa, W. Brauer, C. Habel and K.F. Wender (Eds), *Spatial Cognition III, Lecture Notes in Artificial Intelligence* Vol. 2685: 385–400. Berlin: Springer-Verlag.

215

Randell, D.A., Cui, Z. & Cohn, A.G. 1992. A spatial logic based on regions and connection. In *The 3rd Int. Conf on Knowledge Representation and Reasoning*: 165–176. CA: San Mateo.

Renz, J. 2002. Qualitative spatial reasoning with topological information. *Lecture Notes in Artificial Intelligence*, Vol. 2293: 41–50. Berlin: Springer-Verlag Berlin.

Skiadopoulos, S., Giannoukos, C., Vassiliadis, P., Sellis, T. & Koubarakis, M. 2004. Computing and handling cardinal direction information. In E. Bertino, S. Christodoulakis, D. Plexousakis, V. Christophides, M. Koubarakis, K. Bohm and E. Ferrari (eds), *EDBT 2004. Lecture Notes in Computer Science*, Vol. 2992: 329–347. Berlin: Springer-Verlag.

Skiadopoulos, S. & Koubarakis, M. 2001. Composing cardinal direction relations. In C.S. Jensen, M. Schneider, B. Seeger & V.J. Tsotras (eds), *SSTD 2001. Lecture Notes in Computer Science*, Vol. 2121: 299–317. Berlin: Springer-Verlag.

Advances in Spatio-Temporal Analysis – Tang et al. (eds)
© 2008 Taylor & Francis Group, London, ISBN 978-0-415-40630-7

Data mining technology in predicting the demand for cultivated land

Zuohua Miao & Yaolin Liu
Resource and Environment Science Department of Wuhan University of China, Wuhan, China
Key Laboratory of Graphic Information Systems, Ministry of Education of China, Wuhan University, China

ABSTRACT: Although data mining is a relatively young technique, it has been used in a wide range of problem domains during the past few decades. In this paper, the authors present a new model that applies the data mining technique to forecast the demand for cultivated land. The new model is called the fuzzy Markov chain model with weights. It applies data mining techniques to extract useful information from the enormous quantities of historical data and then applies the fuzzy sequential cluster method to set up the dissimilitude fuzzy clustering sections. The new model regards the standardized self-correlative coefficients as weights based on the special characteristics of correlation among the historical stochastic variables. The transition probabilities matrix of the new model is obtained by using fuzzy logic theory and statistical analysis. The experimental results show that the ameliorative model, combined with the technique of data mining, is more scientific and practical than traditional predictive models.

1 INTRODUCTION

Data mining is the non-trivial extraction of implicit, previously unknown, and potentially useful information from the mass of incomplete, fuzzy, uncertain and stochastic data. Data mining is the important step in Knowledge Discovery in Databases (KDD) (Li 2001). It has three major components, namely, clustering or classification, association rules and sequence analysis. The main tasks of data mining are association analysis, clustering, classification, prediction, time-series pattern and deviation analysis. Data mining encompasses a number of different technical approaches such as statistical methods, artificial neural networks, decision trees, genetic algorithms, nearest neighbour methods, rough set theory and fuzzy logic theory.

At the present time, data mining is mainly applied in the commercial areas such as banking, telecom, insurance, transportation and retailing. It solves problems in database marketing, customer segmentation and classification, credit scoring and fraud detection, etc.

Geo-spatial data mining, a sub-field of data mining, is a process to discover interesting and potentially useful spatial patterns embedded in spatial databases. Efficient tools to extract information from massive geo-spatial datasets are crucial for organizations that own, generate and manage geo-spatial datasets. These organizations are spread across many domains including ecology and environmental management, public safety, transportation, public health and tourism (Hinzle 2004).

The techniques of data mining are a relatively young research field and new models and theories appear every year. Many problems and challenges are presented for data mining such as enhancing the efficiency of models, data mining from dynamic sets, data mining on the Web, fuzzy spatial association rules mining, etc.

As a new kind of data analysis technique, data mining developed fast. Many kinds of dataset can be the object of data mining. Because time series data are very common in datasets, Time Series Data Mining (TSDM) has been one focus of current data mining research (Murray 1998). The cultivated land demand data are time series data but research on how to apply the technique of data mining to predict cultivated land use have not been discovered at the present time. In this paper, the authors present a new model that adopts data mining techniques to predict the demand for cultivated land in land use planning.

2 DATA MINING AND PREDICTION OF DEMAND FOR CULTIVATED LAND

Prediction is one of the tasks for the data mining technique. It is a process to find the mutative rule from the mass of fuzzy and stochastic data.

Prediction of the demand for cultivated land is an important stage in land use planning. The results of prediction, whether accurate or not, relate directly to the quality of land use planning work. Sun (2000) researched the quantity and area prediction of cultivated land for Anhui province by applying a linear regression prediction model and a trend prediction model. These models were the only mathematical methods available to analyse time series data and did not considered the "fuzzy and uncertain" effect of natural and social factors. Jia and Wan (2002) applied the grey model to cultivated land demand prediction in Wuhu city. The GM(1,1) exhibited more accuracy and reliability than the linear regression prediction model. But the GM(1,1) had a lag error for larger time intervals and quickly changing data and lacked a scientific basis for smaller time interval data. Liu and Liu (2004) presented the Grey-Markov Model for the prediction of cultivated land demand in Hubei province. The Grey-Markov model was more rational and scientific than the foregoing models.

Fuzzy logic theory is one of the technical approaches of data mining. It is a non-classical mathematical theory for uncertain problems.

The Markov chain model can predict a good many problems, but it does have shortcomings. The traditional Time Homogeneous Finite Markov Chain model transacts time series data with methods of pure algebra and does not consider the effect of fuzzy, uncertain information embedded in the data. This paper presents a new Markov chain model, combined with fuzzy logic theory.

Cultivated land demand data are stochastic time series data, so the Markov chain model can be applied to forecast the future data making use of the historical data. Generally, time series data can be divided into a continuous real number zone. In order to use the Markov chain model, the continuous real number zone should be divided into a finite number of unambiguous state sets. But the states during the process of predicting cultivated land demand were not unambiguous but fuzzy. Therefore, the authors believe that using the fuzzy state sets will better describe the classified states Factors that affected cultivated land demand were diverse, complicated and uncertain, so the authors concluded that the new model should adopt the technique of data mining. In this paper, the authors combine fuzzy logic theory and the data mining technique with the traditional Markov chain model to improve the prediction precision. In order to effectively employ the historical data, the new model also calculates the weights according to the correlations among the series of historical data.

3 CONSTRUCTION OF THE NEW MODEL

The calculation of transition probabilities is the important step in the Markov chain model process. Transition probabilities were calculated based on the unambiguous state sets. The authors solved this problem by extending the transition probabilities matrix from unambiguous state sets to fuzzy state sets. The fuzzy sequential cluster model is a useful method to classify time series data. In this paper, it was applied to divide the historical data into fuzzy state zones and then construct the transition probabilities matrix of the Markov chain model using fuzzy logic theory and data mining techniques.

3.1 *Fuzzy sequential cluster model*

In order to make the divided zones reasonable, the new model applies the fuzzy sequential cluster model to divide the cultivated land demand data into several fuzzy mutative zones based on the analysis of the data structure. The Fisher algorithm was the traditional algorithm of the fuzzy sequential cluster model (Hu 1990). The fundamentals of the Fisher algorithm can be described by the following equation:

$$\overline{X_{ij}} = \frac{1}{j-i+1}\sum_{l=i}^{j} x_l \tag{1}$$

where $\overline{X_{ij}}$ is the average vector, $\{x_i \cdots x_j\}$, $j \geq i$ is one possible fuzzy sequential clustering zone of stochastic variables series $x_1 \cdots x_n$.

The correlation among data satisfies the following equation:

$$D(i,j) = \sum_{l=i}^{j} (x_l - \overline{x_{ij}})'(x_l - \overline{x_{ij}}) \tag{2}$$

where $D(i,j)$ is the diameter of $\{x_i \cdots x_j\}$, $j \geq i$. It denotes the discrepant degree among variables located at one and the same clustering zone. It shows that when $D(i,j)$ is smaller, the discrepancy among data series is smaller and the correlation stronger; otherwise the discrepancy is larger and the correlation is weaker. In order to compare the effect among different possible clusterings, the Fisher algorithm defines the error function in the following way:

$$e[P(n,K)] = \sum_{j=1}^{k} D(i_j, i_{j+1} - 1) \tag{3}$$

where $e[P(n,K)]$ is the error numerical value, K is the number of clustering zones, $D(i_j, i_{j+1} - 1)$ is obtained from Equation 2, and $P(n,K)$ is one possible clustering that can be expressed as follows:

$$P(n,K): \{i_i = 1, \cdots, i_2 - 1\}; \{i_2, \cdots, i_3 - 1\}; \cdots; \{i_k, \cdots, n\} \tag{4}$$

From Equation 3, it can be seen that when $e[P(n,K)]$ has a minimal value, the best fuzzy clustering division will be obtained. At the same time, the number K can be obtained at the inflexion of the correlation graph between $e[P(n,K)]$ and K.

3.2 Time homogeneous finite Markov chain model

A time homogeneous finite Markov chain model is a stochastic process with finite sets $S = \{1, 2, \cdots, n\}$ of state (Breikin 1997). How to calculate these transition probabilities is the important step of the Markov chain model. The transition probability only depends on the state of the previous step. It can be described as follows:

$$p\{x_n = j | x_{n-1} = i\} \tag{5}$$

where j is the series number of state among state sets, n and $n - 1$ are the time parameters. In Equation 5, $x_n = j$ expresses the Markov chain process located at j state at n time and $x_{n-1} = i$ express the process located at i state at $n - 1$ time. $p_{ij}(n)$ denotes the transition probability from j state to i state at n time. These transition probabilities form the transition probability matrix, which is an $n \times n$ non-negative matrix (such a matrix is called a stochastic matrix). The transition probabilities matrix can be described as follows:

$$P = (P_{ij}) = \begin{bmatrix} P_{11} & P_{12} & \cdots & P_{1n} \\ P_{21} & P_{22} & \cdots & P_{2n} \\ \cdots & \cdots & \cdots & \cdots \\ P_{n1} & P_{n2} & \cdots & P_{nn} \end{bmatrix} \tag{6}$$

4   FUZZY MARKOV CHAIN MODEL WITH WEIGHTS BASED ON DATA MINING

As the above sections analyse, the authors present a new model adopting the technique of data mining. The new model is summarized in the following steps:

Calculate the self-correlative coefficient of cultivated land demand data

The cultivated land demand data series depend on each other. In order to describe the correlation, the new model calculates the self-correlative coefficients among data. The equation can be expressed as:

$$r_k = \sum_{t=1}^{n-k} (x_t - \bar{x})(x_{t+k} - \bar{x}) \Big/ \sqrt{\sum_{t=1}^{n-k}(x_t - \bar{x})^2 \cdot \sum_{t=1}^{n-k}(x_{t+k} - \bar{x})^2} \tag{7}$$

where r_k is the self-correlative coefficients of k grade (step is k), x_t is the cultivated land demand data at time t, $\bar{x}$ is the average value of all cultivated land demand data series and n is the data number of cultivated land demand data series.

Standardizing self-correlative coefficients data as weights

Different fuzzy clustering zones have different effects on predictive values. In order to describe these differences, the new model takes the standardized self-correlative coefficient data as weight. The weight can be calculated by the following equation:

$$\varpi_k = |r_k| \Big/ \sum_{k=1}^{m} |r_k| \tag{8}$$

where ϖ_k = weight, r_k can be obtained with Equation 7.

Calculate the cultivated land using trend coefficients

The cultivated land use trend coefficient (LUTC) is an important parameter to express the cultivated land utilized during a unit of time. The coefficient can be calculated by the following equation:

$$S = ds/s_i \times 100 \times (1/t) \times 100/100 \tag{9}$$

where s_i is the area of cultivated land at the beginning of the time period, ds is the reduction in cultivated land during the time period, t is the length of the time period and S is the cultivated land use trend coefficient, which was scaled up to 100 for convenience.

Divide the fuzzy clustering zones

With step 3 the cultivated land use trend coefficients were obtained. The new model applies the fuzzy sequential cluster model to divide these coefficients into several fuzzy clustering zones. Thus, the fuzzy clustering zone state for each historical cultivated land demand data can be confirmed.

Calculate the transition probability

Calculation of transition probabilities is the main step of the new model. The transition probability for each different step is calculated using fuzzy logic theory in the following manner.

Assume $X_t : x_1 \cdots x_n$ is the stochastic time series data of cultivated land demand, $E_k : e_1 \cdots e_k$ is one fuzzy clustering set for X_t. The new model obtains the transition probability by the following equations:

$$M_i^{(m)} = \sum_{l=1}^{n-m} \mu_{e_i}(x_l), i = 1, 2 \cdots k \tag{10}$$

and

$$M_{ij}^{(m)} = \sum_{l=1}^{n-m} \mu_{e_i}(x_l) \cdot \mu_{e_j}(x_{l+m}) \quad (11)$$

where M_i is the number of data which locate at the fuzzy clustering state $e_i(e_i \subset E_k)$ among X_t, $M_{ij}^{(m)}$ is the data number transitioned from e_i fuzzy clustering state to e_j state through m steps, m is the step number of transition, $\mu_{e_i}(x_l)$ is the fuzzy membership function, which denotes the degree of x_l belonging to e_i fuzzy state.

From Equations 10 and 11, the transition probability from e_i fuzzy clustering state to e_j state is obtained by the following equation:

$$P_{ij}^{(m)} = \frac{M_{ij}^{(m)}}{M_i^{(m)}} \quad (12)$$

Construct the new transition probabilities matrix

In the transition probabilities matrix, each line vector is the different step transition probabilities of historical data, which was calculated with Equation 12.

Calculate the predictive value

The data of each row vector belonging to same state in the new transition probabilities matrix is summed by the following equation:

$$P_i = \sum_{k=1}^{m} \varpi_k P_i^{(k)} \quad (13)$$

where $P_i^{(k)}$ is the transition probability of k step, i is the serial number of the fuzzy clustering zone, ϖ_k is the weight obtained with Equation 8. In this paper, the fuzzy clustering state of the predictive value was obtained according to the maximum subordination principle, that is: $\max \{P_i, i \in I\}$. The predictive value can be calculated with the following equation:

$$\hat{Y}_{(k)} = \frac{Q_{1d} + Q_{2d}}{2} \quad (14)$$

where $\hat{Y}_{(k)}$ is the predictive value, Q_{1d} is the left value of the fuzzy clustering value zone and Q_{2d} is the right value.

Quality assessment of the new model

The relative error between the predictive value and actual value is taken as the quality assessment parameter of the new model. The relative error can be calculated by the following equation:

$$\sigma = (\hat{Y}_{(k)} - Y_{(k)})/Y_{(k)} \times 100/100 \quad (15)$$

where $\hat{Y}_{(k)}$ is the predictive value obtained with Equation 14 and $Y_{(k)}$ is the actual value.

5 EXPERIMENTAL RESULTS AND ANALYSES

The historical cultivated land demand data of one county in Hubei province of China (Table 1) during 1950–2003 were obtained from the Hubei Province Statistical Yearbooks.

- Calculate the self-correlative coefficient of cultivated land demand data of the research area, that is: $r_1 = 0.980$, $r_2 = 0.941$, $r_3 = 0.905$, $r_4 = 0.871$.

Table 1. The historical cultivated land demand data and states during 1950–2003. (Unit: hm².).

Year	1950	1951	1952	1953	1954	1955
Area /hm²	63646.57	65866.61	72073.33	73480.01	73493.30	75440.02
LUTC		3.488	9.423	1.951	0.018	2.649
States		1	1	1	2	1
Year	1956	1957	1958	1959	1960	1961
Area /hm²	74406.67	74673.43	73226.68	70020.02	66973.33	70626.61
LUTC	−1.370	0.359	−1.937	−4.379	−4.351	5.455
States	4	2	4	4	4	1
Year	1962	1963	1964	1965	1966	1967
Area /hm²	67006.59	67180.01	67179.80	67179.80	65846.67	65333.33
LUTC	−5.126	0.259	0.000	0.000	−1.984	−0.780
States	4	2	2	2	4	3
Year	1968	1969	1970	1971	1972	1973
Area /hm²	60073.38	58926.61	56866.69	56400.06	56720.08	55986.67
LUTC	−8.051	−1.909	−3.496	−0.821	0.567	−1.293
States	4	4	4	3	2	3
Year	1974	1975	1976	1977	1978	1979
Area /hm²	55566.39	55380.07	55333.33	55086.59	55073.33	55073.33
LUTC	−0.751	−0.335	−0.084	−0.446	−0.024	0.000
States	3	3	3	3	3	2
Year	1980	1981	1982	1983	1984	1985
Area /hm²	55046.68	54933.31	54866.55	54906.60	54806.58	54739.51
LUTC	−0.048	−0.206	−0.122	0.073	−0.182	−0.122
States	3	3	3	2	3	3
Year	1986	1987	1988	1989	1990	1991
Area /hm²	54553.33	54366.62	54086.57	53906.48	53146.37	53059.21
LUTC	−0.340	−0.342	−0.515	−0.333	−1.410	−0.164
States	3	3	3	3	4	3
Year	1992	1993	1994	1995	1996	1997
Area /hm²	52986.37	52419.11	51939.02	51233.13	50805.66	50453.31
LUTC	−0.137	−1.071	−0.916	−1.359	−0.834	−0.694
States	3	3	3	4	3	3
Year	1998	1999	2000	2001	2002	2003
Area /hm²	49500.02	48952.89	48259.91	47606.62	46959.82	46327.01
LUTC	−1.889	−1.105	−1.416	−1.354	−1.359	−1.347
States	4	3	4	4	4	3

- Standardizing the self-correlative coefficients data as weights, that is: $\varpi_1 = 0.266$, $\varpi_2 = 0.255$, $\varpi_3 = 0.245$, $\varpi_4 = 0.236$.
- Applying the fuzzy sequential cluster model to divide the cultivated land demand data into fuzzy clustering zones. The cultivated land demand data is arranged from low to high order. Figure 2 shows the correlation between error value and the number of category.
- Confirming the fuzzy state of each historical data. The result is described in Table 1.
- Calculate the transition probability matrix of each transition step, as follows:

$$P_1 = \begin{bmatrix} 2/5 & 1/5 & 0/5 & 2/5 \\ 1/8 & 2/8 & 3/8 & 2/8 \\ 0/24 & 3/24 & 16/24 & 5/24 \\ 1/14 & 2/14 & 5/14 & 5/14 \end{bmatrix} \quad P_2 = \begin{bmatrix} 2/5 & 3/5 & 0/5 & 0/5 \\ 0/8 & 1/8 & 4/8 & 3/8 \\ 0/24 & 2/24 & 17/24 & 5/24 \\ 1/13 & 2/13 & 3/13 & 5/13 \end{bmatrix}$$

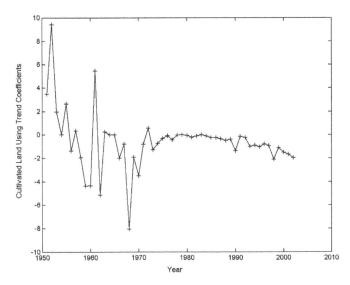

Figure 1. The graph of cultivated land used trend degree. In the graph, negative values indicate that the area of cultivated land was increasing, positive values indicate decreases. The change in the area of cultivated land was great before 1979 and slow after 1979.

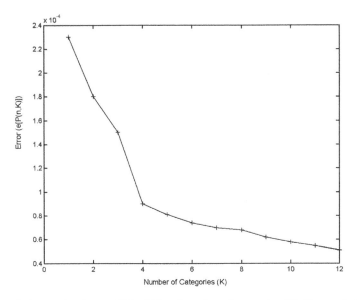

Figure 2. The correlation graph of error $(e[P(n, K)])$ and number of categories (K). The graph shows that the best division category is 4.

$$
P_3 = \begin{bmatrix} 1/5 & 2/5 & 0/5 & 2/5 \\ 0/8 & 1/8 & 4/8 & 3/8 \\ 0/24 & 2/24 & 17/24 & 3/24 \\ 1/12 & 3/12 & 3/12 & 3/12 \end{bmatrix} \quad P_4 = \begin{bmatrix} 1/5 & 2/5 & 0/5 & 2/5 \\ 1/8 & 1/8 & 3/8 & 3/8 \\ 0/23 & 1/23 & 17/23 & 3/23 \\ 0/12 & 3/12 & 4/12 & 4/12 \end{bmatrix}
$$

Table 2. The results of fuzzy clustering zones using the method of fuzzy sequential clustering.

States	Levels	Distinguishing standards
1	Increase Quickly	$X \geq 0.591$
2	Increase Slowly	$0 \leq x < 0.591$
3	Decrease Slowly	$-1.349 \leq x < 0$
4	Decrease Quickly	$x < -1.349$

Table 3. The predicted value of cultivated land demand data in 2003.

Year	States	Step	Weight	Transition probabilities			
				1	2	3	4
1990	3	4	0.236	0/23	1/23	17/23	3/23
2000	4	3	0.245	1/12	3/12	3/12	3/12
2001	4	2	0.255	1/13	2/13	3/13	5/13
2002	4	1	0.266	1/14	2/14	5/14	5/14
The sum of P_i with weight				0.062	0.154	0.395	0.295

Table 4. The predicted value of cultivated land demand data in 2004.

Year	States	Step	Weight	Transition probabilities			
				1	2	3	4
2000	4	4	0.236	0/12	3/12	4/12	4/12
2001	4	3	0.245	1/12	3/12	3/12	3/12
2002	4	2	0.255	1/13	2/13	3/13	5/13
2003	3	1	0.266	0/24	3/24	16/24	5/24
The sum of P_i with weight				0.04	0.21	0.42	0.31

- Predict the cultivated land demand data in 2003 based on the historical data and the corresponding transition probabilities matrix during the 1999–2002 period. Table 3 shows the result of prediction using the proposed new model.

As Table 3 shows, the max $\{p_i\} = 0.395$ and $i = 3$. This means that the predictive data locate at the third fuzzy sequential clustering zone (decrease slowly). The fuzzy zone of predictive value can be calculated, that is: $[46, 326.33, 46, 959.82]$. Thus, the predictive value $\widehat{Y}_{(k)}$ in 2003 is $46,642.84$ hm^2 while the actual value obtained from the statistical yearbook is $46,327.01$ hm^2. It is easy to calculate the relative error of only 0.68 per cent. It indicates that applying the reformed model (fuzzy Markov chain model with weights) to predict the cultivated land demand will effectively improve the precision of prediction.

At the same time, the new model can reduce the complexity of calculation. With the same process, the authors predicted the cultivated land demand in 2004 based on the historical data and the corresponding transition probabilities matrix during 2000–2003. Table 4 shows the result of prediction using the proposed new model.

As Table 4 shows, the max $\{p_i\} = 0.342$ and $i = 3$. This means that the predictive data locates at the third fuzzy sequential clustering zone (decrease slowly). The fuzzy zone of predictive value can be calculated, that is: [45702.06, 46327.01]. Thus, the predicted value for 2004 is 46,014.53 hm^2.

The precision of the new model was greater than the traditional model. The new model extracted useful information from the time series of historical data, and found the appropriate information to improve the prediction precision.

6 CONCLUSIONS AND DISCUSSION

The traditional models for the prediction of cultivated land demand did not effectively deal with fuzzy and uncertain information. A new kind of data analysis technique, data mining, can extract the useful information from the enormous store of historical data. Thus, the paper presents a new model to reform the general Time Homogeneous Finite Markov Chain model. The new model adopts data mining techniques such as fuzzy logic theory, weights, statistical analysis and the cultivated land use trend coefficients model.

As experimental results and analyses have shown, the reformed model can divide the time series data more reasonably and effectively describes the distribution rule existing in the data series. The new model takes a standardized, self-correlative coefficient of each transition step as a weight. The concept of the new model is clear and calculation is simple. The new model provides a grouping method for enhancing the prediction precision.

Although the fuzzy Markov chain model with weights has many advantages, it also has some disadvantages, including: the precision of the model depends on an enormous amount of historical data; the new model only describes the quantitative change from the temporal aspect; and it cannot describe spatial change.

At the same time, the authors show that the influence of other fuzzy state zones should be considered, in addition to the maximum subordination state zone. The corresponding weight for each fuzzy clustering state zone should be defined and then the predictive value calculated with the average value of all weights. Adopting another theory and model into the calculation of weights for fuzzy clustering state zones is the research topic for the next stage of our work.

REFERENCES

Breikin, T.V., Arkov, V.Y. & Kulikov, G.G. 1997. On Stochastic System Identification: Markov Models Approach. In *Proceedings of the 2nd Asian Control Conference, Seoul, Korea*: 775–778.

Cheng, B.X. & Yu, J.S. 2004. Feasibility Study of Applications of Data Mining to Weather Forecast. *Applied Science and Technology* 31(3): 48–50.

Feng, Y.L. & Han, W.X. 1999. The Application of Weighted Markov-Chain to the Prediction of River Run off State. *Systems Engineering–Theory and Practice* 10(10): 89–93.

Helma, C. 2004. Data mining and knowledge discovery in predictive toxicology. *11th International Workshop on Quantitative Structure–Activity Relationships in the Human Health and Environmental Sciences* 15(6): 367–383.

Hinzle, F. & Sester, M. 2004. Derivation of implicit information from spatial data sets with data mining. In *International Society for Photogrammetry and Remote Sensing, Istanbul, Turkey* Vol. 35: 335–341.

Hu, G.D. & Zhang, R.C. 1990. *Analyzing Multivariate Data–Transacting with Pure Algebra Method*. Tianjing: Nankan University Press.

Jia, H.J. & Wan, R.R. 2002. Application of Grey System on Forecast of Cultivated Land Resources. *Area Research and Development* 21(4): 55–60.

Kamber, M. & Jia, W.H. 2001. *Data Mining: Concept and Techniques*. Beijing: China Mechanical Industry Press.

Lan, R.Q., Lin, L.X. & Cheng, L.Y. 2004. Status and Progress of Spatial Data Mining and Knowledge Discovery. *Geographical Information* 2(3): 19–21.

Li, D.R. & Tao, C. 1994. Knowledge discovery from GIS. *The Canadian Conference on GIS* Vol. 1: 1001–1012.

Li, D.R., Wang, S.L. & Shi, W.Z. 2001. On Spatial Data Mining and Knowledge Discovery. *Geomatics and Information Science of Wuhan University* 21(6): 491–499.

Liang, W.Q. & Jiang, K.Q. 2004. Research on Fuzzy Cluster Analysis and Application in Data Mining. *Journal of Anqing Teachers College (Natural Science)* 10(2): 65–68.

Liu, Y.L., Liu, Y.F. & Zhang, Y.M. 2004. Prediction of Gross Arable Land Based on Grey-Markov Model *Geomatics and Information Science of Wuhan University* 29(7): 575–580.

Murray, A.T. & Estivill-Castro, V. 1998. Clustering Discovery Techniques for Exploratory Spatial Data Analysis. *International Journal of Geographical Information Science* 12(5): 431–443.

Nedeljkovic, I. 2004. Image classification based on fuzzy logic. In *International Society for Photogrammetry and Remote Sensing, Istanbul, Turkey* Vol. 34, Part XXX: 83–89.

Sun, H.Z. 2000. Forecast and countermeasure of the temporal change on AnHui province cultivated land resources. *Journal of AnHui Agricultural Sciences* 28(5): 643–645.

Advances in Spatio-Temporal Analysis – Tang et al. (eds)
© 2008 Taylor & Francis Group, London, ISBN 978-0-415-40630-7

Statistical data mining for NFGIS database content refinement

Liang Zhai & Xinming Tang
Key Laboratory of Geo-informatics of State Bureau of Surveying and Mapping,
Chinese Academy of Surveying and Mapping, Beijing, China

Lan Wu
State Bureau of Surveying and Mapping, Beijing, China

Lin Li
School of Resource and Environment Science, Wuhan University, Wuhan, China

ABSTRACT: China has established a series of National Fundamental Geographical Information System (NFGIS) databases. However, with the progress in national information dissemination, NFGIS has been confronted by many new requirements from GIS users, including requests for its refinement and updating. This paper is mainly concerned with the content refinement for the National Fundamental Geographical Information System (NFGIS) 1:50,000 DLG database using statistical data mining technology. We propose a methodology employing a clustering strategy in database content refinement. This approach is very suitable to explore the survey data and obtain useful information. A user survey was conducted to collect users' requirements of the NFGIS database, such as data content and attribute information, and the survey results analysed further by clustering analysis. Through this research, we can reach the following conclusions: (1) Clustering analysis is essential for survey results; (2) Spatial analysis is supplementary to clustering analysis; (3) Using different clustering methods ensures the correctness of clustering results.

1 INTRODUCTION

The term data mining has been used mostly by statisticians, data analysts and the management information systems (MIS) communities (Fayyad et al. 1997). Data mining is a step in the KDD (knowledge discovery in databases) process consisting of applying computational techniques that, under acceptable computational efficiency limitations, produce a particular enumeration of patterns (Fayyad et al. 1996). It is emerging as a new active area of research that combines methods and tools from the fields of statistics, machine leaning, database management and data visualization (Feelders et al. 2000). Data mining techniques have been applied to many real-life applications, and new applications continue to drive research in the area. Many statistical models exist for explaining relationships in a data set or for making predictions: cluster analysis, discriminant analysis and non-parametric regression can be used in many data mining problems (Hosking et al. 1997).

In this paper, we propose a methodology employing clustering strategy in database content refinement. This approach is very suitable for exploring survey data and getting useful information. The remainder of the paper is organized as follows. Section 2 gives an overview on data mining techniques and different methods and techniques are discussed. Section 3 presents the application: content refinement for the National Fundamental Geographical Information System (NFGIS) databases and shows the clustering algorithm used with the user survey results. Section 4 presents the refinement plan and discusses some further analysis. Section 5 concludes the presented work.

2 DATA MINING TECHNIQUES

Many different methods have been used to perform data mining tasks. These techniques not only require specific types of data structures, but also imply certain types of algorithmic approaches (Dunham 2003). In the following, we categorize and describe some typical data mining techniques.

2.1 *Statistical analysis*

Statistics is used as the most common approach for analysing categorical or quantitative data and statistical analysis is a well-studied area, where there exists a large number of algorithms including various optimization techniques. It handles numerical data well and usually comes up with realistic models of spatial phenomena. However, it is a kind of technique that can only be used by experts with a fair amount of domain knowledge and statistical expertise (Jiang et al. 2003).

Clustering analysis is a kind of statistical analysis and is used to identify clusters embedded in the data, where a cluster is a collection of data objects that are 'similar' to one another. It can be expressed by distance functions, specified by users or experts. A good clustering method produces high quality clusters to ensure that the inter-cluster similarity is low and the intra-cluster similarity is high. For example, one may cluster the parcels according to their land use, cover, soil type, ownership and geographical locations. Section 3.1.2 will further illustrate this method.

2.2 *Generalization-based mining*

Data and objects in databases often contain detailed information at primitive concept levels (Chen et al. 1996). It is often desirable to summarize a large set of data and present it at a high concept level. For example, one may wish to summarize the detailed traffic information at different times in a day and show the general traffic pattern. This requires generalization-based mining, which first abstracts a large set of relevant data from a low concept level to relatively high ones and then perform knowledge extraction on the generalized data.

2.3 *Fuzzy sets method*

Fuzzy sets have been used in many computer science and database areas. In the classification problem, all records in a database are assigned to one of the predefined classification areas. A common approach to solving classification problems is to assign a set membership function to each record for each class. The record is then assigned to the class that has the highest membership function value. Similarly, fuzzy sets may be used to describe other data mining functions. Association rules are generated given a confidence value that indicates the degree to which it holds in the entire database. This can be thought of as a membership function (Dunham 2003).

2.4 *Rough set method*

Rough set theory was first proposed by the Polish scientist Z. Pawlak in 1982. It is one approach to knowledge-based decision support and has been widely used in uncertain information classification and knowledge discovery and, of course, it can be applied in data mining. It provides a new way to attribute information analysis in GIS, such as attribute consistency, importance, dependence and classification.

2.5 *Cloud theory*

Cloud theory is a new theory for dealing with uncertainties and consists of a cloud model, reasoning under uncertainty and a cloud transform. This theory combines randomness and fuzziness; so it compensates for the inherent limitation of the membership function, which is the basis for rough set theory, and makes it possible to join quality and quantity in data mining.

Besides the above, there are many other methods used in data mining, such as raster based analysis, decision trees, genetic algorithms, visual data mining modelling and artificial neural networks, etc. These methods have always been used together in knowledge discovery.

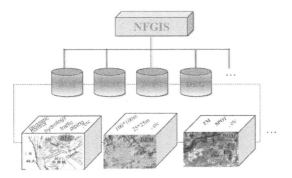

Figure 1. NFGIS databases.

3 CASE STUDY: DATABASE CONTENT REFINEMENT

Since the 1990s, China has been establishing a series of NFGIS databases that provide a united and authoritative space platform for GIS user communities and play an important role in social development, economy and state safety. These databases include the Digital Linear Graphics (DLG) database, Digital Elevation Model (DEM) database, Digital Orthography Model (DOM) database, Digital Raster Graphics (DRG) database, etc. (see Fig. 1). Together with the provincial databases, they have made up the National Fundamental Geo-Spatial Framework (Li 2003, Wang 2003).

However, with the rapid progress being made in the economy, many geo-spatial information users have promoted many new and higher demands on the National Fundamental Geo-Spatial Information and most of these require NFGIS database refinement. Questions such as which kind of dataset or database is most valuable to the users; what kinds of dataset or database are being used; and who are the users or potential users, should have been already answered (Steven 1995). Also, are the current databases satisfactory for the users? What features (attributes) should the database include? Among these questions, the last one is of great importance. We adopt the methodology of a user survey with questionnaires to fulfil these aims.

This survey aims at gaining a thorough understanding of the NFGIS applications in sectors such as hydrology, agriculture, forestry, transportation, economic statistics, land use, marine studies, environment, meteorology, cultural relics, civil engineering, scientific surveying, geology, mining, seismology, and surveying and mapping, etc., and identifying the features and attributes most needed by them. The questionnaire was designed according to the National 1:50,000 Relief Map Specification (China). Features having great significance, like boundaries, double lane railways, perennial rivers, important buildings, etc. are not involved in the questionnaire in order to decrease the number of problems. The respondents only need to check what they need.

The questionnaire is in two parts. The first part is background information, including some common questions, such as the respondent's name, organization, contact information, reason for using NFGIS information; data precision and updating. The second part concerns geodetic control, hydrology, residential areas and buildings, traffic, piping, relief, soil texture and vegetation. Here are two examples:

(1) Factory Buildings include these features:
   – transformer substation
   – sewage treatment works
   – industrial well (oil/gas/salt. . .)
   – liquid/gas storage equipment
   – tower building (distillation tower, chimney, water tower, watch tower. . .)
   – strip mine, excavated field
   – saltern
   please select what you need.

Table 1. Summary of the data content response (part).

Residential area and building	Feature	Agree	Total
House	Shed	122	205
	Cave	60	205
	Mongolian Tent	47	205
Population	Population	121	205

(2) Please select the attribute of swamp that you or your organizations have used. If the given choices are not relevant, please add.
 – not used
 – passable/not passable
 – water depth
 – depth of ooze layer
 – undefined boundary
 – others_____

3.1 *Survey results summary*

This survey employed two kinds of strategy: Focus Groups and Web-Survey (Ke 2001). The questionnaire has been issued on the Websites: http://lab.casm.ac.cn/ and http://www.csi.gov.cn/ investigate/gisurvey.asp. From 1 August to 1 October 2004, 295 replies were received, of which 274 were valid. The response rate was 93 per cent.

In this questionnaire, there are 108 features and attributes (each feature or attribute is regarded as a case or sample). Table 1 is the summary of data content responses.

By summing up the responses, it was discovered that users from different departments have different requirements and applications of the NFGIS data and qualitative conclusions can be drawn. For example, the geological sectors use relief maps in geology investigations and fieldwork, and they wish that the NFGIS 1:50,000 DLG database could include all relief features and attributes. They are in great need of independent features for river crossings when looking for water, residences to help with orientation; labels such as 'footbridge', 'ferry', 'spring' and 'well'. Departments of hydrology use relief maps in water resource planning and drainage area measurement, so they wish that the database could contain all kinds of rivers and water resource facilities, including banks and dams. At the same time, transformer substations and power lines are needed by the electricity supply industry; seismic departments use geo-spatial information in identifying seismic centres, fault locations and monitoring stations and need residential areas with population, hydrology, railways, highways, piping, power lines, springs (especially ascending springs), earth flows, earth slides and volcanos. City and country planning departments are more concerned about roads, contours, residential areas, rivers, power lines and land use, etc.

3.2 *Survey results analysed by statistical data mining*

3.2.1 *Selection of statistical data mining methods*
From the above, we know that the responses are huge and a methodology should be adopted to mine beneficial conclusions. In other words, in order to make the NFGIS 1:50,000 database content refinement plan, summing up the responses and getting some qualitative conclusions are far from enough. Since each department hopes that the database will include the features or attributes having close relation with their daily work and speciality, we should take the users' needs into account as a whole and select features or attributes according to the degree to which they are required. Statistical data mining seems to be a good choice. One possibility is the clustering method: features and attributes having the same degree may be clustered.

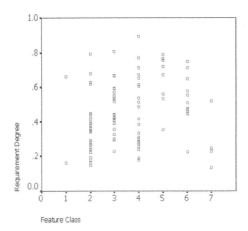

Figure 2.   Data distribution based on the feature class.

Clustering analysis is the searching for groups (clusters) in the data, in such a way that samples belonging to the same cluster resemble each other, whereas samples in different clusters are dissimilar. Generally speaking, there are two categories of methods for clustering (Wu 2002, Zhang 2002, Theodore 2002):

(1) Partitioning Algorithms. A partitioning algorithm describes a method that divides the data set into k clusters, where the integer k needs to be specified by the user. Typically, the user runs the algorithm for a range of k-values. Algorithms of this type include k-means, partition around medoids, fuzzy clustering, etc.
(2) Hierarchical Algorithms. A hierarchical algorithm describes a method of yielding an entire hierarchy of clusters for the given data set. Agglomerative methods start with the situation where each object in the data set forms its own little cluster, and then successively merges clusters until only one large cluster remains, which is the whole data set. Divisive methods start by considering the whole data set as one cluster, and then split up clusters until each object is separate.

The difference between these two algorithms is the following: Partitioning algorithms are based on specifying an initial number of groups, and iteratively reallocating samples between groups until some equilibrium is attained. In contrast, hierarchical algorithms proceed by combining or dividing existing groups, producing a hierarchical structure displaying the order in which groups are merged or divided.

Users from different sectors have different requirements, but we can hypothesize that there are three clusters representing different degrees of requirement, namely, high, medium and low. Thus, we take the partitioning algorithm, for it is not possible to specify the number of clusters as required by the hierarchical algorithm.

In the following, we will discuss the two methods and their clustering results.

3.2.2   Distribution of survey data
Figure 2 shows the distribution of the survey results after normalization. It is the distribution grouped by feature class. There are seven feature classes, each of which contains several "features and attributes". The vertical axis denotes the requirement degree: the percentage of the 'Agree/Total' of each feature, and the horizontal axis denotes 'geodetic control', 'hydrology', 'residential area and building', 'traffic', 'piping', 'relief and soil texture' and 'vegetation', respectively. For example, in geodetic control, there are 'monument points' and 'astronomical points', with the former having a higher degree of requirement.

3.3   Clustering analysis

3.3.1   Clustering of survey data: based on partitioning algorithm
Among all partitioning algorithms, the k-means algorithm is the most widely used. It is applied in distinguishing relatively homogeneous sample sets. Since the sample set has some similarity in this survey—the

Table 2. Final cluster centres.

	Cluster		
	1	2	3
Agree per cent	.6925	.2573	.4637

Table 3. Distances between final cluster centres.

Cluster	1	2	3
1		.616	.324
2	.616		.292
3	.324	.292	

Table 4. Number of cases in each cluster.

Cluster	1	26
	2	41
	3	41
Valid		108

samples are all geographical features and they have close characters—the k-means algorithm is applied here. In (1), it partitions N data points into K disjoint subsets (clusters) R_i containing N_j data points so as to minimize the sum-of-squares criterion.

$$J = \sum_{i=1}^{k} \sum_{\overline{X}_j \in R_i} \| \overline{X}_j - \overline{W}_i \|^2 \qquad (1)$$

where $\overline{X}_j$ is a vector representing the jth data point and $\overline{W}_i$ is the geometric centroid of the data points in R_i. This algorithm initially takes the number of components of the population equal to the final required number of clusters. In this step itself, the final required number of clusters is chosen such that the points are mutually farthest apart. Next, it examines each component in the population and assigns it to one of the clusters depending on the minimum distance. The geometric centroid's position is recalculated every time a component is added to the cluster and this continues until all the components are grouped into the final required number of clusters.

SPSS for windows software (version 10.0 SPSS Inc.) is employed here for the analysis. At the beginning, the SPSS software selects the centres (geometric centroids) randomly and after the iteration process the centres are changed. Table 2 shows the centres after the clustering process. The values of the centres represent the means of the samples in every class. According to the values of centres, we can easily find that clusters 1, 3, 2 correspond to high, medium, and low requirement degrees, respectively.

Table 3 illustrates the Euclidean distances between the final cluster centres. The distance between cluster 1 and cluster 2 is bigger than the one between cluster 1 and cluster 3, because cluster 1 and cluster 2 represent high and low requirement degrees, respectively. From the semantic view, 'high' and 'low' have a big difference.

Table 4 shows the number of cases (samples) in each final cluster.

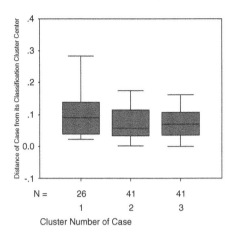

Figure 3.   Distance of each case from its classification cluster centre.

We need further analysis to see whether cluster number 3 is suitable. If not, we would have to cluster again. Through analysing distances of cases from its classification cluster centre, we can tell whether the cluster number 3 is suitable. If many cases were seriously out of centre, then the cluster number is not suitable. Figure 3 is a box diagram: the black bold line in the centre represents the mean; the rectangular box is the bound of interquartile ranges. We find that all cases are not seriously out of centre. To some degree, this indicates that cluster number 3 is suitable.

3.3.2   Clustering of survey data: based on hierarchical algorithm

The basic process of hierarchical clustering is:

Step 1:   Start by assigning each item to its own cluster, so that if you have N items, you now have N clusters, each containing just one item. Let the distances (similarities) between the clusters equal the distances (similarities) between the items they contain.

Step 2:   Find the closest (most similar) pair of clusters and merge them into a single cluster, so that you now have one less cluster.

Step 3:   Compute distances (similarities) between the new cluster and each of the old clusters.

Repeat steps 2 and 3 until all items are clustered into a single cluster.

Step 3 can be done in different ways, which is what distinguishes average-link clustering from single-link and complete-link.

The following gives the details of average-link, single-link and complete-link clustering.

- Average-link clustering:

$$d(R,Q) = \frac{1}{|R \parallel Q|} \sum_{i \in R, j \in Q} d(i,j) \qquad (2)$$

In (2), R and Q represent two different sets, d(i, j) is the distance between the two clusters. It satisfies d(i, i) = 0; d(i, j) ≥ 0; d(i, j) = d(j, i). The same terms are used in equations 3 and 4.

- Single-link clustering:

$$d(R,Q) = \min_{i \in R, j \in Q} d(j,j) \qquad (3)$$

- Complete-link clustering:

$$d(R,Q) = \max_{i \in R, j \in Q} d(j,j) \qquad (4)$$

233

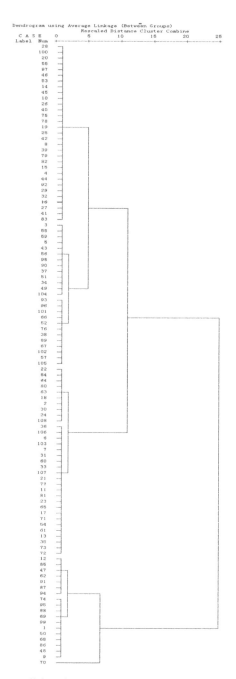

Figure 4. Dendrogram using average linkage between groups.

We adopt average-link clustering and obtain a dendrogram (see Fig. 4). From it, we can easily find that if the clustering number is 4, then there would be one single sample (alley) in a class. This is unreasonable. So the clustering number cannot be 4. If it were 2, then there would be 2 types of semantics: need and no need, which is also unreasonable. So the clustering number should be 3.

4 CONTENT REFINEMENT PLAN

4.1 *Plan 1: based on partitioning algorithm*

4.1.1 *Alternatives of different selections*
All of the cases (108) belong to different requirement degrees: high, medium and low.

Alternative 1: contains the 'high' (requirement degree) features only, 26 samples.
Alternative 2: contains the 'high' and 'medium' features, 67 samples.
Alternative 3: contains all the features, 108 samples.

Three alternatives are obtained through the above clustering. At the same time, we notice that many features or attributes have been omitted to simplify the questionnaire; so we should add such features or attributes.

4.1.2 *Comparison*
Alternative 1 is the intersection of the 3 alternatives, including the minimum features and attributes and they are the most important features and attributes that can be used as geographic references.

Alternative 3 is the most detailed, but it is not fit for the database refinement plan. As discussed above, each department hopes that the database would include the features or attributes having close relation with their daily work and speciality. It is impossible to satisfy the needs of each department for the sake of economy in China, because the cost of data collection, updating, distribution and maintenance would be enormous. It is recommended that alternative 3 be used as the reference for national relief data classification.

The content in alternative 2 can satisfy the needs of geology, hydrology, transportation and seismology departments by and large. For example, geology departments need 'footbridge', 'ferry', 'spring' and 'well'; seismic departments need residential areas with population, hydrology, railways, highways, piping, power lines, springs (especially ascending springs), earth flows, earth slides and volcanos; city and country planning departments are concerned with roads, contours, residential areas, rivers, power lines and land use, etc. Those requirements are all reflected in alternative 2.

4.1.3 *Further analysis*
By the above comparison, alternative 2 appears to be the basis for NFGIS 1:50,000 database content refinement. However, clustering analysis is conducted according to users' requirement degree on each feature or attribute and the integrity and relevancy among features or attributes are neglected. The rationality of alternative 2 should be further discussed in the context of spatial analysis.

Spatial analysis is a statistical description or explanation of either locational or attribute information, or both (Guo 2001, Wang 2001). Further analysis involves three facets: the integrity of features, logical consistency and attribute complexity consistency.

(1) Integrity of features analysis
    Some features and attributes are rejected in both alternative 1 and 2. But a few important features should be added, such as 'reef', which has ownership; fixed or seasonal Mongolian tents are significant to civil administration or planning departments. They should all be added.
(2) Logical consistency of features analysis
    'Altitude annotation' and 'water depth annotation' have the same importance, but in alternative 2 'water depth annotation' is missed. It should be added. If not, the database would include 'altitude annotation' without 'water depth annotation'. This could cause logical confusion.
(3) Attribute complexity consistency analysis
    The attributes of 'river', 'high watermark' and 'water line' are always used together; 'dam/bank', 'safety line of flood control' and 'warning line' are also used together; but in alternative 2 they were missed. Similarly, the attribute of 'highway'/'constructional materials' was missed. Since these missing attributes may break attribute complexity consistency and are not difficult to collect, they should be added.

4.2 *Plan 2: based on hierarchical algorithm*

Similarly, we can find that cluster 3, 1, 2 correspond to high, medium, and low requirement degrees, respectively, and there are 3 alternatives.

Alternative 1: The plan contains the 'high' (requirement degree) features only, 19 samples.
Alternative 2: The plan contains the 'high' and 'medium' features, 75 samples.
Alternative 3: The plan contains all the features, 108 samples.

4.3 *Discussion*

By comparing plan 1 and plan 2, we detect that there are 13 features or attributes (water level, high water level, river width/depth, length of dam or bank, shanty, population, etc) having different clustering results. After a discussion with experts and a careful comparison, we recommend alternative 2 in plan 1 with some modification for the final content refinement plan. The details of it can be found in the NFGIS Database Refinement Report.

5 CONCLUSIONS

In this paper, we have discussed data mining approaches in detail and applied a clustering algorithm in NFGIS database content refinement. From this research, we can reach the following conclusions:

(1) Clustering analysis is essential with survey results. How can we obtain the useful information from the survey results? It is not enough to only summarize it and get some qualitative conclusions. We need further analysis by adopting data mining methods: clustering analysis.
(2) Spatial analysis is supplementary to clustering analysis. Clustering analysis is conducted according to users' requirement degree on each feature or attribute, so the integrity and relevancy among features or attributes are neglected. The rationality of the alternatives should be further discussed using spatial analysis of the integrity of features, the logical consistency and attribute complexity consistency. After spatial analysis, more features and attributes are added.
(3) Using different clustering methods ensures the correctness of clustering results. In the above, we adopt both partitioning and hierarchical clustering algorithms, which to some degree helps find the correct and reasonable clustering results.

Furthermore, we can integrate the application model and establish the test database according to the proposed content refinement plan.

ACKNOWLEDGEMENTS

The authors thank Mr. Wang Zhongyuan (Wuhan University, China) and Ms. Gao Xiaoming (Chinese Academy of Surveying and Mapping, China) for their help in collecting questionnaire returns. Financial support for this research was provided by the open research subject of Key Laboratory of Geo-informatics of the State Bureau of Surveying and Mapping (No. 1469990424233) and the National Surveying and Mapping Fund (No. 1469990324221).

REFERENCES

Beauchaine, P.T. & Beauchaine III, J.R. 2002. A comparison of maximum covariance and k-means cluster analysis in classifying cases into known taxon groups. *Psychological Methods* 7: 245–261.
Chen, M.-S., Han, J.W. & Yu, S.P. 1996. Data Mining: An overview from a database perspective. *IEEE Transaction on Knowledge and Data Engineering* 8(6): 869–883.
Dunham, M.H. 2003. *Data Mining Introductory and Advanced Topics*: 23–24. Beijing: TsingHua University Press.

Fayyad, U., Piatetsky-Shapiro, G., Smyth, P. & Uthurusamy, R. 1996. *Advances in Knowledge Discovery and Data Mining*: 2–4. Massachusetts: MIT Press.

Fayyad, U. & Stolorz, P. 1997. Data mining and KDD: Promise and Challenges. *Future Generation Computer Systems* 13: 99–115.

Feelders, A., Daniels, H. & Holsheimer, M. 2000. Methodological and practical aspects of data mining. *Information & Management* 37: 271–281.

Frank, S.M., Goodchild, M.F., Onstrud, H.J. & Pinto, J.K. 1995. *Framework Data Sets for the NSDI*. http://www.ncgia.ucsb.edu/Publications/Tech_Reports/95/95–1.pdf (accessed 20 May 2003).

Guo, R. 2001. *Spatial Analysis*: 4–5. Beijing: Higher Education Press.

Hosking, J.R.M., Pednault, E.P.D. & Sudan, M. 1997. A statistical perspective on data mining. *Future Generation Computer Systems* 13: 117–134.

Jiang, L. & Cai, Z. 2003. Review and prospect of spatial data mining. *Computer Engineering* 29: 9–10.

Ke, H.X. 2001. Web–survey methodologies. *Modern Media* 4: 80–84.

Li, J.W. 2003. NFGIS 1:50.000 Database Establishment and Integration. In Chen Jun & Wu Lun (eds), *Digital China Geo-spatial Fundamental Framework*: 162–168. Beijing: Science Press.

Wang, D.H. 2003. NFGIS 1:50.000 DEM Database Establishment. In Chen Jun & Wu Lun (eds), *Digital China Geo-spatial Fundamental Framework*: 148–161. Beijing: Science Press.

Wang, J.Y. 2001. *Spatial Information System Theory*: 262–264. Beijing: Science Press.

Wu, L. 2002. *Methodology for selection of framework Data–Case study for NSDI in China*: 10–15. M.A. Thesis. ITC.

Zhang, G.J. 2002. *Applications of soft computing and data mining in electrical load forecasting*: 37–42. PhD. Dissertation. Zhejiang University.

237

Advances in Spatio-Temporal Analysis – Tang et al. (eds)
© 2008 Taylor & Francis Group, London, ISBN 978-0-415-40630-7

Author index

Bareth, Georg 123
Ben, Jin 41

Caccetta, Peter 155
Chen, Bin 59
Chen, Jianzhong 51
Cheng, Tao 135

Delavar, Mahmoud R. 85

Fang, Yu 59
Frank, Andrew U. 85
Furby, Suzanne 155

Gong, Jianhua 135
Gong, Yuchang 67

He, Jianhua 9
Huang, Fengru 59

Jin, Peiquan 67
Jin, Xin 203

Kainz, Wolfgang 1
Karimipour, Farid 85
Kraak, Menno J. 147

Lei, Bing 101
Li, Deren 171
Li, Lin 77, 193, 227
Li, Zhilin 135
Liu, Dayou 51, 111
Liu, Hanli 193
Liu, Jie 111
Liu, Yaolin 1, 9, 217
Liu, Yu 203

Meng, Li 41
Miao, Zuohua 217

Pan, Mao 21

Qu, Honggang 21

Rezayan, Hani 85

Shi, Wenzhong 29

Tang, Xinming 1, 101, 227
Tolpekin, Valentyn 147
Tong, Xiaochong 41
Turdukulov, Ulanbek D. 147

Wallace, Jeremy 155

Wang, Bin 21
Wang, Huibing 101
Wang, Shengsheng 111
Wang, Shuliang 171
Wang, Xiaoming 203
Wang, Xinying 111
Wang, Yong 21
Wang, Zhangang 21
Wu, Lan 227
Wu, Xiaofang 163
Wu, Xiaoliang 155

Xu, Zhiyong 163

Yan, Huiwu 163
Yin, Zhangcai 77
Yu, Qiangyuan 51
Yu, Zhenrong 123
Yue, Lihua 67

Zhai, Liang 101, 227
Zhang, Jixian 1
Zhang, Yi 203
Zhang, Yongsheng 41
Zhu, Guorui 163
Zhu, Haihong 77, 193
Zhu, Min 155

Printed and bound by CPI Group (UK) Ltd, Croydon, CR0 4YY

01/11/2024

01782599-0018